57830

# Hazardous Materials Chemistry *for* Emergency Responders

# Robert Burke

# Hazardous Materials Chemistry *for* Emergency Responders

**CRC** **LEWIS PUBLISHERS**

Boca Raton    New York    London    Tokyo

| | |
|---|---|
| Acquiring Editor: | Joel Stein |
| Project Editor: | Les Kaplan |
| Marketing Manager: | Greg Daurelle |
| Direct Marketing Manager: | Becky McEldowny |
| Typesetter: | Pamela Morrell |
| Cover design: | Denise Craig |
| Prepress: | Carlos Esser |
| Manufacturing: | Sheri Schwartz |

**Library of Congress Cataloging-in-Publication Data**

Burke, Robert.
    Hazardous materials chemistry for emergency responders : street
chemistry / by Robert Burke.
        p.    cm.
    Includes bibliographical references and index.
    ISBN 1-56670-174-0
    1. Hazardous substances.  I. Title.
T55.3.H3B87 1997
604.7—dc20
                96-31459
                CIP

© 1997 by CRC Press, Inc.
Lewis Publishers is an imprint of CRC Press

No claim to original U.S. Government works
International Standard Book Number 1-56670-174-0
Library of Congress Card Number 96-31459
Printed in the United States of America  1 2 3 4 5 6 7 8 9 0
Printed on acid-free paper

# Preface

There has been much discussion about the validity of emergency response personnel studying chemistry. Obviously, from the title of this book, the author's opinion is clear; however, the chemistry presented here is more "street chemistry" than college chemistry. I have tried to make the chemistry subject matter appropriate for response personnel and to express the information in understandable terms.

The book is organized into the U.S. Department of Transportation's (DOT) nine hazard classes, with which emergency response personnel should be intimately familiar. Each chapter reviews one of the nine hazard classes.

Almost every hazardous material presents more than one hazard; the DOT's placarding and labeling system only identifies the most severe hazard. Within the hazard classes, individual chemicals are discussed, along with their hazards and their physical and chemical characteristics, both as distinct chemicals and within chemical families. Additionally, the multiple hazards of hazardous materials, including "hidden" hazards, are studied throughout the book.

The Top 50 industrial chemicals, along with other common hazardous materials, are discussed throughout the book. Learning about these chemicals will provide responders with an overview of the varying dangers presented by hazardous materials, and will show the similarities and differences among chemical family members, as well as other hazardous materials. Reports of various incidents involving many of the chemicals are presented to emphasize the effects that chemical and physical characteristics can have on an incident outcome. Responders should become familiar with the chemical terminology that they will encounter when researching reference sources, including books, computer data bases, material safety data sheets (MSDS), and shipping papers. Basic chemistry provides emergency responders with a background that will help them not only to understand chemical terminology, but also to talk intelligently with CHEMTREC (Chemical Transportation Emergency Center), shippers, and industry representatives about chemicals involved in hazardous materials incidents.

**Robert Burke**
Hazardous Materials
Specialist/Instructor

# Dedication

This book is affectionately dedicated to my "Bunky" Clare, with whom I have spent the past 13 wonderful years. I love you, Clare. The old saying that "behind every successful man is a woman" was never truer. Thank you for all of your patience and support throughout the long hours, weeks, and months I spent at the keyboard. This book is for you!

# Acknowledgments

This book would not have been possible if not for the hundreds of emergency response personnel and fellow instructors from various organizations whom I have had the distinct pleasure to know, teach with, and learn from during classes at the National Fire Academy and across the country. It was their inspiration, friendship, and support that gave me the desire and dedication to put together a book of this type. Many of the ideas for material presented here have come from the input of these responders, students, and instructors; for them and the many lasting friendships that have developed over the years, I am grateful.

Specifically I would like to thank several people who played a major role in the assembling of information, photo opportunities, and photographs, as well as in hours of tireless editing of the manuscript.

I would like to thank Battalion Chief Bill Doty, a good friend; the firefighters of Battalion One in South Philadelphia; and the Philadelphia Fire Department for allowing me to ride along to take many of the pictures and to gather information for this book. The experiences of South Philadelphia, including many wonderful meals at various fire houses, the Ashland Chemical open house, the traditions of the area, and the many interesting incidents will long be remembered. Thanks, Bill! Captain Rudy Rinas, Jr., of the Chicago Fire Department, also a good friend, provided many photographic opportunities during several visits to Chicago and northern Indiana. Rudy found photo opportunities that I would never have been able to find on my own. I will always remember my most recent visit to Wrigley Field to watch my baseball team, the Cubs. Rudy and I sat for two hours eating hot dogs (the best in the world by the way), drinking some beer, talking war stories, and waiting through a driving rainstorm for a game that never got started. Rain checks don't do you much good if you live in Baltimore! Thanks, Rudy! Stan Totten, a good friend, with whom I have taught numerous Chemistry of Hazardous Materials courses, provided technical editing (and taught me many things), for which I will always be grateful. Actually Stan should be granted "sainthood," just for listening to my jokes over and over. Maybe he is still trying to figure out the punch lines! Thanks, Stan! Jan Kuczma of the National Fire Academy, a good friend and former instructor, has provided me with opportunities, inspiration, guidance, and, on many occasions, "fatherly" advice, for which I will never be able to thank him enough. He has helped me through several difficult periods of my life and career, and is one of the finest people I have ever had the pleasure of knowing. Thanks, Jan! Chris Waters, a retired firefighter from New York City and former Hazardous Materials Program Chair at the National Fire Academy, also provided me with guidance, understanding, and opportunity on many occasions. I have the greatest respect and admiration for Chris. I have gained a wealth of knowledge from him, and I am grateful for his friendship over the years. Thanks, Chris!

Last, but certainly not least, I would like to thank Walter Shipley (Uncle Ship) for his tireless hours of editing, and Bonnie Whisenant for trying to figure out where I learned grammar and punctuation! Special thanks to Irma Hill for the photographs and the book about the Cresent City, IL, train derailment, and Hank Graham, the railroad conductor who took the photographs of the Kingman, AZ propane explosion. To all of you and many others too numerous to mention, although you know who you are, thanks for the memories, friendships, and inspiration. This book is really about all of you!

# The Author

Robert A. Burke, born in Nebraska and raised in Illinois, earned an A.A. in Fire Protection Technology from Catonsville Community College and a B.S. in Fire Science from the University of Maryland. He has also completed graduate work at the University of Baltimore in Public Administration.

Mr. Burke has over 20 years experience in the emergency services as a career and volunteer firefighter, and has served as an Assistant Fire Chief for the Verdigris Fire Protection District in Claremore, OK, Deputy State Fire Marshal in the state of Nebraska; a private fire protection and hazardous materials consultant; and is currently a Exercise and Training Officer for the Chemical Stockpile Emergency Preparedness Program (CSEPP) for the Maryland Emergency Management Agency. He has served on several volunteer fire companies and is currently a member of the Earleigh Heights Volunteer Fire Company in Severna Park, MD, which is a part of the Anne Arundel County, MD Fire Department.

He is an adjunct instructor at the National Fire Academy in Emmitsburg, MD, and a hazardous materials instructor for the Maryland Fire and Rescue Institute and the Delaware County, PA Emergency Services Training Academy. Mr. Burke is also a contributing editor for *Firehouse Magazine*, and has had numerous articles published in *Firehouse, Fire Chief*, and *Fire Engineering* magazines. He has developed several training programs, including "Advanced Hazardous Materials for Emergency Responders," which is being produced by American Heat and The Fire and Emergency Television Network (FETN) as a video training program. Mr. Burke has also developed a 35-mm slide version of the same program. Additionally he has developed a unique modularized Hazardous Materials Awareness Program that can be used for fire, police, emergency medical services (EMS), and public-works personnel as separate training programs.

# Contents

# 1                          INTRODUCTION

Mention the word *chemistry* to the average firefighter or other emergency responder and it strikes terror in their hearts. Most emergency personnel do not want any part of it. Actually, chemistry is all around us. Everything on Earth, including the human body (Figure 1.1), is made up of one of the chemical elements or a combination of those elements from the Periodic Table. Firefighters learn about fire behavior as part of their basic training, in order to have a better understanding of fire. Fire behavior is part of the chemistry of fire, which is really a chemical chain reaction. Understanding fire and how it behaves helps firefighters extinguish fires more safely and effectively. Emergency medical personnel take courses to learn how to care for sick and injured patients. Sometimes drugs are used as part of the initial treatment. Drugs are chemicals and can be hazardous if they are not handled properly. Emergency medical personnel learn about drugs and treatment techniques in order to effectively treat their patients. Law enforcement personnel take courses in criminal justice to better prepare for their job. All response personnel receive some type of basic training to better prepare for their jobs as well. Responders are called upon daily to deal with incidents of all kinds that can and often do involve hazardous chemicals. In order for personnel to better understand the hazardous chemicals they face on a daily basis, an understanding of basic chemistry principles and terminology is as essential as their other training.

This book will present emergency response personnel with a view of chemistry as it applies to the hazardous materials encountered daily. Some of the concepts presented may bend the rules of chemistry a bit. However, the purpose of this book is not to educate chemists, but rather to teach response personnel about basic chemistry. This study of chemistry for the purpose of HazMat response may be considered "street chemistry." For this reason, some chemistry rules can be bent without being broken. The concepts will work in the street application of chemistry when dealing with hazardous materials. This book presents firefighters, police, and emergency medical services personnel (EMS) with some basic tools to better understand hazardous chemicals. This information may

**Chemical Makeup of the Human Body**

| | | | | |
|---|---|---|---|---|
| Oxygen | 65% | Sulfur | | |
| Carbon | 18% | Iron | | |
| Hydrogen | 10% | Sodium | | |
| Nitrogen | 0.3% | Zinc | | 1% |
| Calcium | 1.5% | Magnesium | | |
| Phosphorus | 0.1% | Silicon | | |
| | | Potassium | | |

**Figure 1.1**

help keep them from being injured or killed at the scene of a hazardous materials incident.

## THE BASICS OF CHEMISTRY

Chemistry is the study of matter. It can be divided into two sections: inorganic and organic. Inorganic chemistry involves acids, bases, salts, elements, and the physical state of matter in which they are found. Organic chemistry involves compounds that contain carbon. The definition of matter is "anything that occupies space and has mass." Matter can exist as a solid, liquid, or gas. Temperature and pressure can affect the physical state of a chemical, but not its properties. The hazards presented by a chemical may not be the same, depending on the physical state of the material. For example, only gases burn; solids and liquids do not burn, even though they may be listed as flammable. A solid or liquid must

Hazardous materials can have multiple hazards.

be heated until it produces enough vapor to burn. It is important to understand the states of matter in order to better understand the physical and chemical characteristics of hazardous materials. There can be some intermediate steps in the process of classifying solids, liquids, and gases. Some solids may have varying particle sizes from large blocks to filings, chips, and dusts. Particle sizes of vapors may vary, from vapors that are small enough to be invisible to mists that are readily visible.

A molecule is the smallest particle of a compound that can normally exist by itself. Molecules of compounds contain different types of atoms bonded together in fixed proportions. The molecules of solids are packed together very closely in an organized pattern. Because the molecules are packed tightly together, they can only vibrate very gently in a very small space. This is why solids have a definite size and shape. When particles are this close together, they attract each other, and it takes a lot of energy to pull them apart. When a solid is heated, the molecules start to vibrate faster and eventually pull apart. The particles in liquids are farther apart, but are still able to attract each other. They are not arranged in a regular pattern. Liquids do not have a shape of their own, so they conform to the shape of the container in which they are placed. Particles of a gas are moving very rapidly and are not attracted to each other. Gases have no shape of their own and conform to the space in which they exist.

Hazardous materials may undergo both chemical and physical changes. A chemical change involves a reaction that alters the composition of the substance and thereby alters its chemical identity. Chemical properties include reactivity, stability, corrosivity, toxicity, and oxidation potential. A new compound may be formed that may have different characteristics than the compounds or elements that make it up. Chlorine, for example, is a poison gas; sodium is a reactive metal. When they are combined, they form sodium chloride, which is neither a poison nor a reactive chemical. Physical changes involve changes in the physical state of the chemical, but do not produce a new substance, such as the physical transformation from a liquid to a gas or a liquid to a solid. Physical properties include specific gravity, vapor pressure, boiling point, vapor density, melting point, solubility, flash point, fire point, autoignition temperature, flammable range, heat content, pH, threshold limit value (TLV), and permissible exposure level (PEL).

## The Periodic Table

The basics of chemistry cannot be effectively discussed without studying the Periodic Table of the Elements. The properties of the elements repeat in a regular way when the elements are arranged by increasing atomic number. The Periodic Table is a method of organizing everything that is known about chemistry on one piece of paper. It shows the relationship between the elements by revealing the tendency of their properties to repeat at regular intervals. All chemicals are derived from the elements or combinations of elements from the Periodic Table. Symbols are used to represent the elements on the Periodic Table. It is composed of a series of blocks representing each element. Within each block is a symbol that corresponds to the name of the element, a type of shorthand for the name of the element. For example, the element gold is represented by the symbol Au, chlorine is represented by the symbol Cl, and potassium is represented by the symbol K. Each symbol represents one atom of that element. The symbols may be made up of a single letter or two letters together. A single letter is always capitalized; when there are two letters, the first is capitalized and the second is lower cased. When the second letter is also capitalized, this indicates that the material described is not an element but a compound. For example:

**CO** is the molecular formula for the compound carbon monoxide.
**Co** is the symbol for the element cobalt.

The symbols for the elements are derived from a number of sources. They may have been named for the person who discovered the element. For example, W, the symbol for tungsten, was named for Wolfram, the discoverer. Other elements are named for famous scientists, universities, cities, and states: Es is the symbol for einsteinium, named for Albert Einstein; Cm is the symbol for curium, named for Madame Curie; Bk is the symbol for berkelium, named for the city of Berkeley, CA; and Cf is the symbol for the element californium, named after the state of California. Other element names come from Latin, German, Greek, and English: in the case of sodium, Na comes from the Latin for *natrium*; Au, the symbol for gold, comes from *aurum*, meaning "shining down" in Latin; Cu (copper) is from the Latin *cuprum*, or *cyprium*, because the Roman source for copper was the island of Cyprus. Fe (iron) is from the Latin *ferrum*. Bromine means "stench" in Greek; rubidium means "red" (in color); mercury is sometimes referred to as quicksilver; sulfur is referred to as brimstone in the Bible.

### The Elements

There are 90 naturally occurring and 20 man-made elements. Not all of the elements on the Periodic Table are common or particularly hazardous to responders. There are, however, some 39 elements that we will call the HazMat elements. These elements are important to the study of the chemistry of hazardous materials. Most of the hazardous materials that will be encountered by response personnel include or are produced from these 39 elements (Figure 1.2). Hazardous materials personnel and students of HazMat chemistry should be familiar with these 39 elements by symbol and name. Elements with 83 or more protons are radioactive; many are rare and probably will not be encountered. The man-made elements are the result of nuclear reactions and research. These elements may have existed on earth at one time, but because they are radioactive and many half-lives have passed, they no longer exist naturally.

The elements on the Periodic Table can be divided into three groups: the primary elements, the transition elements, and the rare-earth elements. The primary elements have a definite number of electrons in the outer shell. This number is the number at the top of each column in the "towers" at each end of the Periodic

The HazMat Elements

| | | |
|---|---|---|
| H - Hydrogen | Cu - Copper | O - Oxygen |
| Li - Lithium | Ag - Silver | S - Sulfur |
| Na - Sodium | Au - Gold | F - Fluorine |
| K - Potassium | Zn - Zinc | Cl - Chlorine |
| Be - Beryllium | Hg - Mercury | Br - Bromine |
| Mg - Magnesium | B - Boron | I - Iodine |
| Ca - Calcium | Al - Aluminum | U - Uranium |
| Ba - Barium | C - Carbon | He - Helium |
| Ti - Titanium | Si - Silicon | Ne - Neon |
| Cr - Chromium | N - Nitrogen | Ar - Argon |
| Mn - Manganese | P - Phosphorus | Kr - Krypton |
| Fe - Iron | As - Arsenic | Xe - Xenon |
| Co - Cobalt | Pu - Plutonium | Pb - Lead |

**Figure 1.2**

Table. The transition metals may have different numbers of electrons in the outer shell. They are located in the "valley" between the towers. The rare-earth metals are relatively uncommon and all are radioactive. The horizontal rows are called periods and are numbered from 1 to 7. Atomic numbers increase by one as you go across the periods from left to right.

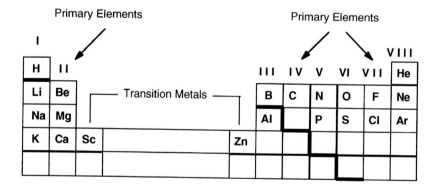

## Atomic Weight

The atomic weight of an element is listed on the Periodic Table. The atomic weight is the sum of the weight of the protons and neutrons in the nucleus of the atom. All of the weight of the element occurs in the nucleus of the atom. For the purpose of "street chemistry," the electrons do not have weight. The atomic weight is located on the Periodic Table above or below the symbol of the element. It is the number that is *not* a whole number; the location varies among periodic tables, so be sure you look for the number with the decimal point. When using atomic weight to determine molecular weight of a compound, always round it off to the nearest whole number. For example, oxygen has an atomic weight of 15.999; round the number to 16, so the atomic weight of oxygen becomes 16. Nitrogen has an atomic weight of 14.007, so you round the number to 14. The atomic weight is referred to in terms of atomic mass units (AMU).

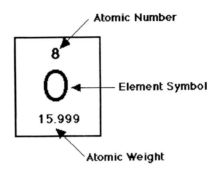

## Atomic Number

The other number on the Periodic Table is a whole number located either above or below the symbol, and is known as the atomic number of the element. The atomic number is equal to the number of protons in the nucleus. The atomic number also equals the total number of electrons in the orbits outside the nucleus of the atom. The protons have a positive (+) charge and the electrons have a negative (–) charge. There must be an equal number of protons and electrons in the atom of the element to maintain an electrical balance. It is the number of protons that identifies a specific element and does not change. If you change the number of protons, you change the element. The protons act as a kind of "social security number" to identify a specific element.

The Periodic Table is divided into two sections by a stairstepped line. The line starts under hydrogen, goes over to boron, and then stairsteps down one element at a time to astatine or radon, depending on which periodic table is being used.

| I | | | | | | | | | | | VIII |
|---|---|---|---|---|---|---|---|---|---|---|---|
| H | II | | | | | III | IV | V | VI | VII | He |
| Li | Be | | Transition Metals | | | B | C | N | O | F | Ne |
| Na | Mg | | | | | Al | | P | S | Cl | Ar |
| K | Ca | Sc | | | | Zn | | | | Br | Kr |
| | | | | | | | | | | I | Xe |
| | Ba | | | | | Hg | | Pb | | At | Ra |
| | | | | | | | | | | | |

Group I: all have 1 electron in the outer shell. Group II: have 2 electrons in the outer shell (transition metals will be discussed later). Group III: all have 3 electrons in the outer shell. Group IV: all elements have 4 electrons in the outer shell. Group V: all elements have 5 electrons in the outer shell. Group VI: all elements have 6 electrons in the outer shell. Group VII: all elements have 7 electrons in the outer shell. Group VIII: all elements have 8 electrons in the outer shell.

The 81 elements to the left and below the stairstepped line are metals. Metals make up about 75% of all the elements. Metals lose their outer-shell electrons easily to the nonmetals when forming compounds. Metals are malleable (they can be flattened), ductile (they can be drawn into a wire), and conduct heat and electricity very well. The farther to the left of the line you go, the more metallic the properties of the element; the closer to the line, the less metallic the properties of the element. Metallic properties increase as you go down a column on the Periodic Table. Metals are all solids, except gallium, mercury, francium, and cesium, which are liquids under normal conditions. The 17 elements to the right and above the line are nonmetals. Nonmetals have a strong tendency to gain electrons when forming a chemical bond. Nonmetals may be solids, liquids, or gases; all are poor conductors of heat and electricity. Solid nonmetals are either hard and brittle or soft and crumbly.

The numbers at the top of the vertical columns on the Periodic Table indicate the number of electrons in the outer shell of the elements in that column. The exception to this is the transition elements between the two towers on either end of the table. Unlike the primary elements, the numbers above the transition metal columns do not indicate the number of electrons in the outer shell of those elements. The transition elements can have differing numbers of electrons in their outer shells. The number of outer-shell electrons of the transition elements can be determined from the name or from a correct molecular formula. For example, copper may have one or two electrons in its outer shell. When copper loses one electron, it becomes $Cu^{+1}$, because it now has one more positive charge in the nucleus than negative charges around the outside. This extra positive charge creates a $+1$ charge on the element. If copper loses two electrons, it becomes $Cu^{+2}$, because it now has two more positive charges in the nucleus than negative charges around the outside. If chlorine atoms were to pick up these electrons from copper, copper I or II chloride will be formed. If copper loses only one electron, it becomes $Cu^{+1}$ with the molecular formula CuCl and would be called copper I chloride. When copper loses two electrons to chlorine, it becomes copper II chloride with the molecular formula $CuCl_2$. Note that the roman numeral in the name indicates the charge of the copper metal.

There is an older, alternate naming system for the transitional metals. There may be chemicals encountered still using this older naming system, therefore responders should be familiar with it. In this system, the suffixes "ic" and "ous" are used to indicate the higher and lower valence numbers (outer-shell electrons) of a transitional metal. For example, if copper I combines with oxygen, the name would be cuprous oxide. The lowest number of electrons in the outer shell of copper is one. When the metal with the lowest number of electrons is used, the suffix in the alternate naming system is "ous." When the metal with the highest number of electrons is used, the suffix is "ic." For example, copper II combined with phosphorus would create cupric phosphide. The importance of this concept of transitional metals will be discussed further in the section on salts.

The vertical columns of the Periodic Table contain elements that have similar chemical characteristics in their pure elemental form. These elements have the same number of electrons in the outer shell, which is why they have similar chemical behaviors. These similar elements are sometimes referred to as families. Some of the more important families include the *alkali metals* in column one, the *alkaline earth metals* in column two, the *halogens* in column seven, and the *noble or inert gases* in column eight. The transition elements in the center of the Periodic Table are also similar in the context that most of them have the possibility of differing numbers of electrons in their outer shells.

The alkali metals in column one begin with lithium and continue through sodium, potassium, rubidium, cesium, and francium. These elements are all solids, except for cesium and francium, which are liquids at normal temperatures. The alkali metals are water reactive: they react violently with water, producing flammable hydrogen gas and enough heat to ignite the hydrogen gas. These elements are so reactive that they do not exist in nature in pure form, but are found as compounds of the metal, such as potassium oxide and sodium chloride. Some isotopes of cesium and all isotopes of francium are radioactive. These elements are somewhat rare so you are not likely to see them on the street.

The alkaline earth metals in column two are less reactive than the alkali metals in column one. Beryllium does not react with water at all. The others have varying reactions with water. The alkaline earth metals are all solids. They have to be burning or in a smaller physical form before they become water reactive. Magnesium, for example, is violently water reactive when it is involved in fire. The application of water to a magnesium fire will cause violent explosions that can endanger responders. If it is necessary to fight fires involving magnesium, water should be applied from a safe distance with the use of unmanned appliances. The other elements in column two are also water reactive to varying degrees. These include calcium, strontium, barium, and radium, which is radioactive.

The halogens in column seven are nonmetals. They may be solids, liquids, or gases. Fluorine and chlorine are gases at normal temperatures and pressures. Bromine is a liquid to 58°C and produces vapor rapidly when above that temperature. Iodine is a solid. Astatine is also in column seven. It is radioactive; however, such a small amount has ever been found that you are not likely to encounter it. The halogens are all toxic and are also very strong oxidizers. Fluorine is a much stronger oxidizer than oxygen; in fact, fluorine is the strongest oxidizer known. In their pure elemental form, the halogens do not burn; however, they will accelerate combustion much like oxygen because they are oxidizers. Some halogen compounds are components of fire-extinguishing agents that are called halons, which are being phased out because of the damage they cause to the ozone layer of the atmosphere.

The elements in column eight are all gases. They are nonflammable, nontoxic, and nonreactive. Family eight elements are referred to as the inert or noble gases. Normally they do not react chemically with themselves or any other chemicals. Helium is used for weather balloons and airships. Neon is used for lighting and beacons. Argon is used in electric lightbulbs. Krypton and xenon are used in special lightbulbs for miners and in lighthouses. Radon is radioactive and is used in tracing gas leaks and treating some forms of cancer. The noble gases have a complete outer shell of electrons, two in helium and eight in the rest: an "octet" is eight electrons in the outer shell; a "duet" is two electrons in the outer shell.

# CONTENTS OF AIR

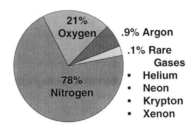

**Figure 1.3**

Both the octet and duet are stable configurations of electrons in an atom. It is because of the complete outer shell of electrons that the noble gases do not react chemically. All of the other elements on the Periodic Table try to reach the same stable electron arrangement as the nearest noble gas, which is usually in the same period. That is why there are chemical reactions: the elements are trying to reach stability. The column eight elements are all nonflammable and nontoxic gases; however, they can displace the oxygen in the air inside of buildings or in confined spaces and cause asphyxiation. Most of the noble gases are present in the air, like nitrogen and oxygen (Figure 1.3). Inert gases are commonly shipped and stored as compressed gases or cryogenic liquids.

## Compounds and Mixtures

Two or more elements that combine chemically and form chemical bonds are referred to as compounds. All compounds must be electrically neutral. The ions in an ionic compound always occur in a ratio such that the total positive charges from the nucleus equal the total negative charges from the electrons. Compounds can be identified by names, molecular formulas, and, in some cases, structures. Structures generally apply to nonmetal compounds, including hydrocarbons and hydrocarbon-derivative families. The molecular formula is made up of the symbols of the elements that are in the compound and the number of atoms of each. The number of atoms is represented by a subscript number behind the symbol. The molecular formula is a kind of recipe for the compound. It lists all of the ingredients and the proportions of each so that the compound can be reproduced if the formula is followed. For example, sodium chloride, table salt, has the molecular formula NaCl. The compound has one atom of sodium and one atom of chlorine. Aluminum oxide has a molecular formula of $Al_2O_3$ (two atoms of aluminum and three atoms of oxygen).

Chemical compounds may also exist in the form of mixtures. A mixture is two or more compounds blended together without any chemical bonding taking place. Each of the compounds retains its own characteristic properties. There are two types of mixtures, homogeneous and heterogeneous. Homogeneous means "the same kind" in Latin. In a homogeneous mixture, every part is exactly like every other part. For example, water has a molecular formula of $H_2O$. Pure water is homogeneous; it contains no substances other than hydrogen and oxygen. Loosely translated to include mixtures, homogeneous refers to two or more compounds or elements that are uniformly dispersed in each other. A solution is another example of a homogeneous mixture. Heterogeneous means "different kinds" in Latin. In a heterogeneous mixture, the different parts of the mixture have different properties. A heterogeneous mixture can be separated mechanically into its component parts. Some examples of heterogeneous mixtures are gasoline, the air we breathe, blood, and mayonnaise.

## Solubility

Solubility is another term associated with mixing two or more compounds together. The definition of solubility from the *Condensed Chemical Dictionary* is "the ability or tendency of one substance to blend uniformly with another, e.g., solid in liquid, liquid in liquid, gas in liquid, and gas in gas." Solubility may vary from one substance to another. When researching chemicals in reference sources, relative solubility terms may include very soluble, slightly soluble, moderately soluble, and insoluble. Generally speaking, nothing is absolutely insoluble. Insoluble actually means "very sparingly soluble," i.e., only trace amounts dissolve.

Compounds of the alkali metals of family one on the Periodic Table are all soluble. Salts containing $NH_4^+$, $NO_3^-$, $ClO_4^-$, $ClO_3^-$, and organic peroxides containing $C_2H_3O_2$ are also soluble. All chlorides ($Cl^-$), bromides ($Br^-$), and iodides ($I^-$) are soluble, except for those containing the metals $Ag^+$, $Pb_2^+$, and $Hg_2^{2+}$. All sulfates are soluble, except for those with the metals $Pb_2^+$, $Ca_2^+$, $Sr_2^+$, $Hg_2^{2+}$, and $Ba_2^+$. All hydroxides ($OH^-$) and metal oxides (containing $O^{2-}$) are insoluble, except those of family one on the Periodic Table and $Ca_2^+$, $Sr_2^+$, and $Ba_2^+$. When metal oxides do dissolve, they give hydroxides (their solutions do not contain $O^{2-}$ ions). The following illustration is an example of a metal oxide and water reaction. (This reaction is shown as an illustration and has not been balanced.)

$$Na_2O + H_2O \rightarrow NaOH$$

All compounds that contain $PO_4^{3-}$, $CO_3^{2-}$, $SO_3^{2-}$, and $S^{2-}$ are insoluble, except for those of family one on the Periodic Table and $NH_4$. Most hydrocarbon compounds are not soluble, such as gasoline, diesel fuel, pentane, octane, etc. Compounds that are polar, such as the alcohols, ketones, aldehydes, esters, and organic acids are soluble in water.

There are some factors that affect the solubility of a material. One is particle size: the smaller the particle, the more surface area that is exposed to the solvent; therefore, more dissolving takes place over a shorter period of time. Higher temperatures *usually* increase the rate of dissolving. The term miscibility is often used synonymously with the term solubility. Miscibility, solubility, and mixtures will be discussed as they pertain to specific chemicals and families of chemicals in other chapters of this book.

There are some elements that do not exist naturally as single atoms. They chemically bond with another atom of that same element to form "diatomic" molecules. The diatomic elements are hydrogen, oxygen, nitrogen, chlorine, bromine, iodine, and fluorine. One way to remember the diatomic elements involved is by using the acronym HONClBrIF, pronounced honk-le-brif, which includes the symbols for all of the diatomic elements. Oxygen is commonly referred to as $O_2$, primarily because oxygen is a diatomic element. Two oxygen atoms have covalently bonded together and act as one unit.

Much can be learned about a compound by looking at its elemental composition. Generally speaking, chemicals that contain chlorine in their formula may be toxic to some degree because chlorine is toxic. There are exceptions, such as sodium chloride with the formula NaCl. Sodium chloride is table salt. The toxicity is low; but even if you did not know sodium chloride was table salt and treated it as a toxic material because of the chlorine, your error would be on the side of safety. If you are going to make errors when dealing with hazardous materials, always attempt to err on the side of safety. You may have "egg on your face" afterward and take some ribbing, but no one has ever died from embarrassment. On the other hand, if you are not cautious and your error is not on the side of safety, it could be fatal!

## THE ATOM

An atom is the smallest particle of an element that retains all of its elemental characteristics. The word *atom* comes from the Greek, meaning "not cut." For example, take a sheet of paper and tear it in half. Keep tearing the paper in half until you cannot tear it any more. You could then take scissors and cut the paper into smaller pieces. Eventually you will not be able to cut the paper into a smaller piece. The atom is like that last piece of paper: you cannot have a smaller piece of an element than an atom. The symbols of the elements on the Periodic Table represent one atom of that element. A single atom cannot be altered chemically. To create a smaller part of an element would require that the atom be split in a nuclear reaction. Therefore, a single atom is the smallest particle of an element that would normally be encountered.

The atom is comprised of three major parts: electrons, protons, and neutrons (Figure 1.4). The atom is like a miniature solar system with the nucleus in the center and the electrons orbiting around the outside. These parts are referred to as subatomic particles. The atom is made up of positively (+) charged protons that are located in the nucleus along with neutrons. Neutrons do not have a charge, so they are neutral. Orbiting in shells or energy levels around the nucleus are varying numbers of electrons. Electrons are very important in discussing chem-

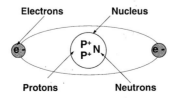

**Figure 1.4**

istry for hazardous materials responders. The shells or orbits about the nucleus of an atom have important characteristics:

1. Each energy level is capable of containing a specific number of electrons.
2. There is a limit on the number of electrons in the various shells.
3. The inner shell next to the nucleus **NEVER** holds more than two electrons.
4. The last shell **NEVER** holds more than eight electrons.
5. The electrons in the outer shell control chemical reactions.

The only elements that occur naturally with eight electrons in the outer shell are the inert gases. The actual number of outer-shell electrons that exist in any given element is represented by a roman numeral at the top of the vertical columns of the Periodic Table.

| I | | | | | | | VIII |
|---|---|---|---|---|---|---|---|
| **H** | **II** | | | | | | **He** |
| | | | | **III** | **IV** | **V** | **VI** | **VII** | |
| **Li** | **Be** | ⎯ Transition Metals ⎯ | **B** | **C** | **N** | **O** | **F** | **Ne** |
| **Na** | **Mg** | | **Al** | | **P** | **S** | **Cl** | **Ar** |
| **K** | **Ca** | **Sc** | | **Zn** | | | | **Br** | **Kr** |

The exception to the roman numeral designation for the number of outer-shell electrons is the transitional element metals. The transitional metals have varying numbers of outer-shell electrons and the process for identifying them will be discussed in the salts. Electrons have a negative (–) charge. Atoms are electrically neutral, so they must have an equal number of protons and electrons. The atom is held together by the strong attraction between the positive protons and the negative electrons. The inner-shell electrons are held tightly by the nucleus. The farther away the outer shells are from the nucleus, the less control the nucleus will have over those electrons. This makes large atoms tend to lose electrons and small atoms tend to gain electrons. Electrons of metals are generally farther away from the nucleus than are nonmetal electrons; therefore, metals give up electrons and nonmetals take on electrons.

The atomic weight of an element comes from the nucleus. The protons and neutrons in the nucleus each have a weight of 1 AMU. The total number of neutrons and protons equals the atomic weight. The atomic number of an element equals the number of protons in that element. The number of neutrons is determined by subtracting the number of protons from the atomic weight; what remains is the number of neutrons.

In chemistry, we are concerned more with the electrons orbiting the nucleus than the nucleus itself. We are particularly interested in the electrons in the outer-shell of the atom. Chemical activity takes place between the outer-shell electrons of elements, this chemical activity forms compounds. In radioactivity, the concern is with the nucleus where radiation is emitted. Radioactivity will be discussed further in Chapter 8.

## Formulas

A single element is the smallest portion of a chemical compound that can be encountered. Chemical compounds are made up of two or more elements in covalent or ionic bonds. Chemical compounds are represented by formulas much like elements are represented by symbols. According to the *Condensed Chemical Dictionary*, "a formula is a written representation using symbols of a chemical entity or relationship." There are three kinds of chemical formulas: empirical, molecular, and structural. Empirical formulas indicate composition, not structure, of the relative number and the kind of atoms in a molecule of one or more compounds, e.g., CH is the empirical formula for both acetylene and benzene. The molecular formula shows the actual number and kind of atoms in a chemical entity, e.g., the molecular formula for sulfuric acid is $H_2SO_4$. The structural formula indicates the location of the atoms in relation to each other in a molecule as well as the number and location of chemical bonds. The following illustration of the compound butyric acid is an example of a chemical structure:

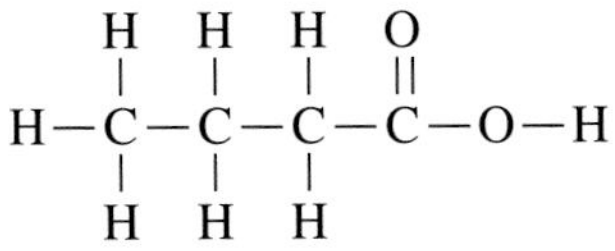

$$
\begin{array}{ccccc}
\text{H} & \text{H} & \text{H} & \text{O} & \\
| & | & | & \| & \\
\text{H}-\text{C}-\text{C}-\text{C}-\text{C}-\text{O}-\text{H} \\
| & | & | & & \\
\text{H} & \text{H} & \text{H} & &
\end{array}
$$

Ionic bonding

## IONIC BONDING

Electrons are shared or exchanged in the process of a chemical reaction. Once this exchange or sharing of electrons occurs, a chemical compound is formed. The opposite charges of metals and nonmetals make atoms want to come together to form salt compounds.

There are two basic groups of chemical compounds that are formed from elements. The first group is the salts. Salts are made up of a metal and a nonmetal. For example, when combined with the nonmetal chlorine (Cl), the metal sodium (Na) forms the salt compound sodium chloride, with the molecular formula NaCl. Metals generally do not bond together. Metals that are combined are melted and mixed together to form an alloy, e.g., copper and zinc are melted and mixed together to make brass. Brass is not an element. There is no chemical bond involved; rather it is a mixture of zinc and copper.

The second group of compounds is made up totally of nonmetal elements. For example, the nonmetal carbon combines with the nonmetal hydrogen to form a hydrocarbon. A typical hydrocarbon might be methane, with the molecular formula $CH_4$. Hydrocarbons will be discussed further in Chapters 3 and 4.

When the outer-shell electrons of a metal are given up to a nonmetal element, a salt compound is formed through a chemical bond. The outer shell of the metal is now empty, so the next shell becomes the outer shell. This shell will have two or eight electrons, which is a stable configuration. The metal is then stable and electrically satisfied. The nonmetal receives the electrons from the metal and now

Water transportation: intermodal containers being unloaded from a ship at a port facility, many containing hazardous materials.

has eight electrons in its outer shell. The nonmetal is now stable and electrically satisfied. The result is that a compound is formed.

The process of gaining or losing electrons is called ionization or ionic bonding. When the electrons are transferred, ions are formed. Like atoms, compounds must be electrically neutral. There must be an equal number of positive protons in the nucleus and negative electrons outside in each of the atoms. When a metal gives up electrons, the metal has a positive charge because there are now more positive protons in the nucleus than negative electrons outside the nucleus. There is an electrical imbalance. An element cannot exist with an electrical imbalance. The ion formed in the case of the metal is referred to as a positive cation, represented by a plus (+) sign and a superscript number to the right of the element symbol. The number represented is the number of electrons that were given up to the nonmetal element. When the nonmetal receives electrons from the metal, there are now more negative electrons outside the nucleus than there are positive protons in the nucleus. Again, there is an electrical imbalance. The ion of the nonmetal is referred to as negative anion, represented by a minus (−) sign and a superscript number to the right of the symbol. The number represents the electrons received from the metal. The superscript numbers must balance and cancel each other out for the compound to be electrically neutral. If the superscript numbers are not equal, additional atoms of the elements involved are needed to balance the formula. If that happens, the formula will have subscript numbers indicating how many of the atoms of the specific element are present in the compound. If there are no subscripts in the formula, it is understood that there is just one atom of each of the elements shown.

In the following example, each lithium atom has three electrons, two in the inner shell and one in the outer-shell. Lithium will give up that one outer-shell electron to oxygen. Lithium now has one less electron than the total number of protons. There is one more positive charge in the nucleus than negative charges around the outside of the nucleus. Because of this imbalance of charges, lithium has a +1 charge. Lithium has given up its one outer shell electron so the inner shell with two electrons now becomes the outer shell. Two electrons in the outer shell is a stable configuration, just like eight electrons in the outer-shell. Oxygen has six electrons in its outer-shell; it needs two electrons to complete its octet. Oxygen will take on two electrons in its outer shell and will have two more negative charges around the nucleus than it has positive charges in the nucleus. Because of this imbalance, oxygen has a –2 charge. Oxygen can get one of the electrons that it needs from lithium. Lithium has only one electron to give; however, oxygen needs another electron, so a second lithium atom is taken, which gives oxygen the two electrons it needs. The balanced formula for lithium, $Li_2O$, is two lithium atoms and one oxygen atom.

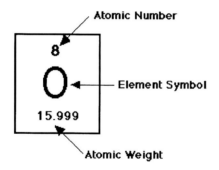

In the next example, calcium has two electrons in its outer shell. Calcium will give up those two electrons to sulfur and will have a +2 charge. Sulfur needs two electrons and will take the two electrons given up by calcium and have a –2 charge. Since calcium has a +2 charge and sulfur has a –2 charge, the charges cancel each other out and the balanced formula is CaS, calcium sulfide. Rules for naming will be covered in the Salts, Hydrocarbons, and Hydrocarbon Derivatives sections of this chapter.

$$Ca^{+2} + S^{-2} = CaS$$

It is important to understand how elements combine to form compounds. However, it is not the formula that will be seen most often in the real world of emergency response, but the name of the compound. When looking up hazardous materials in reference books, often the formula will be listed. The formula will look more familiar if you have a basic understanding of chemistry. This will help responders to better understand the characteristics of the hazardous material.

## COVALENT BONDING

The other family of compounds formed when elements combine is the nonmetal compounds. The nonmetals are comprised of two or more nonmetal elements combining to form a compound. Nonmetals may be solids, liquids, or gases. The most frequently encountered groups of hazardous materials are made up of just a few nonmetal materials. They are carbon, hydrogen, oxygen, sulfur, nitrogen, phosphorus, fluorine, chlorine, bromine, and iodine. In elemental form and in compounds, these elements make up the bulk of hazardous materials found by emergency responders in most incidents.

In the case of nonmetals, electrons are shared between the nonmetal elements. Approximately 90% of covalently bonded hazardous materials are made up of carbon, hydrogen, and oxygen; the remaining 10% are composed of chlorine, nitrogen, fluorine, bromine, iodine, sulfur, and phosphorus. It is still necessary that each atom of each element have two or eight electrons in the outer shell. However, there is no exchange of those electrons. When the bonding takes place, each atom of each element brings along its electrons and shares them with the other elements. This process of sharing electrons is called covalent bonding. In the following example, carbon has four electrons in its outer shell; carbon needs to have four more to become stable. Hydrogen has one electron in its outer shell. Hydrogen can share its one electron with carbon and hydrogen will think it has a complete outer shell. Hydrogen gets the second electron it needs by sharing with carbon. This covalent bond fulfills the duet rule of bonding. Carbon still needs three more electrons. One way carbon can get three more electrons is by sharing with three more hydrogens. When this happens, the carbon is complete. Carbon thinks it has eight electrons in the outer shell and each hydrogen thinks it has two. The octet and duet rules of bonding are satisfied. The compound formed between carbon and hydrogen is complete, the molecular formula is $CH_4$. In the following structure, the shared pairs of electrons are represented by dots between the hydrogens and the carbon. This is known as the Lewis Dot Structure.

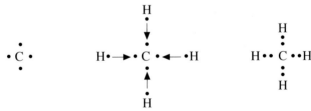

Carbon has four electrons, but needs four more.

Hydrogen has just one electron to share, so carbon needs four hydrogens.

Now each hydrogen has two and the carbon has eight; they are satisfied.

Electrons always share in pairs. Another way of representing pairs of shared electrons is by using a dash. The same one-carbon, four-hydrogen compound is shown with dashes representing the pairs of shared electrons.

$$
\begin{array}{c}
\text{H} \\
| \\
\text{H}-\text{C}-\text{H} \\
| \\
\text{H}
\end{array}
$$

A chemical compound becomes electrically stable through the process of sharing and exchanging electrons. The fact that compounds have become electrically stable does not mean they are no longer hazardous. Quite simply, elements combine and chemical reactions occur so that compounds can become electrically stable. These combinations of elements that form compounds create many new hazardous and nonhazardous chemicals.

Emergency responders may encounter elemental chemicals that are hazardous when released in an accident. However, most of the hazardous chemicals encountered by emergency responders will be in the form of compounds or mixtures. Compounds and mixtures present a broad range of hazards from explosive to corrosive. It is important for emergency responders at all levels to recognize these dangers. This book will use the Department of Transportation (DOT)/United Nations (UN) system of classifying hazardous materials. The remaining chapters of this book will examine the nine DOT/UN hazard classes and the types of hazardous materials they include. This hazard class system only identifies the most severe hazard presented by a material; almost every hazardous material has more than one hazard. This book will also focus on the hidden hazards of materials beyond the hazard class and the corresponding placard. The physical and chemical characteristics of hazardous materials found within each of the hazard classes will be discussed in detail in the remaining chapters.

## SALTS

Just as the Periodic Table of the elements has families of elements, compounds can also be divided into families. A family of materials has particular hazards associated with it. If you are able to recognize in which family a material belongs from the name or the formula, you should be able to determine the hazard even if you do not know anything else about the specific chemical. If you know the hazard, you know how to handle the material properly. The first family of materials we will look at is the salts (Figure 1.5). If a metal reacts with a nonmetal, the resulting compound is called a salt. In the following example, potassium, a metal, reacts with chlorine, a nonmetal. Potassium is located in the alkali metal family in column one of the Periodic Table. This means that potassium has one electron in the outer shell; therefore, potassium will give up that one electron and there will be eight electrons in the next shell. That shell will then become the

## SALT FAMILIES

| Binary | Binary Oxide | Peroxide | Hydroxide | Oxysalts |
|---|---|---|---|---|
| **M+NM** | M + Oxygen | M +$(O_2)^{-2}$ | M + $(OH)^{-1}$ | M + Oxy Rad |
| **Not Oxy** | Ends in Oxide | Ends in | Ends in | |
| **Ends in** | WR = CL,RH | Peroxide | Hydroxide | **Cyanide** |
| **-ide** | | WR = CL, | WR = CL, | M + CN |
| **General** | | RH, $RO_2$ | RH | Ends in |
| **Hazard** | | | | Cyanide |
| | | | | Poison, CL |

**Figure 1.5**

outer shell and will be stable with eight electrons. Chlorine has seven electrons in its outer shell. Chlorine will take the electron given by potassium to give it eight electrons in its outer shell. Chlorine will become stable (Figure 1.6). The name of the resulting compound is potassium chloride. Potassium chloride, much like sodium chloride (table salt), does not pose a serious risk to emergency responders.

$$K^{+1} + Cl^{-1} = KCl, \text{ potassium chloride, a binary salt}$$

Transition metals can also be combined with nonmetals to produce salts. Transition metals may have varying numbers of electrons in their outer shells (Figure 1.7). The Periodic Table does not list the number of electrons in the outer shells of the transitional elements as it does for the primary elements. The number of electrons has to be given in the name, such as copper II chloride, or from a balanced formula. The roman numeral II indicates two electrons in the outer shell of the copper. The number of outer shell electrons can also be determined from the molecular formula. First, look at the nonmetal element in a compound and determine what the charge is on that element. The charge can be found on the Periodic Table at the top of the appropriate column. The transition metal charge

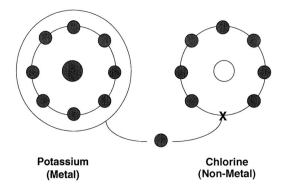

Potassium
(Metal)

Chlorine
(Non-Metal)

**Figure 1.6**

Transition Elements for HazMat Chemistry

| Element | Possible Charges | Example Compound | Naming of + Ion New | Old |
|---|---|---|---|---|
| Copper | +1 | CuCl | copper I | Cuprous |
| | +2 | CuCl$_2$ | copper II | Cupric |
| Iron | +2 | FeCl$_2$ | iron II | Ferrous |
| | +3 | FeCl$_3$ | iron III | Ferric |
| Mercury | +1 | HgCl | mercury I | Mercurous |
| | +2 | HgCl$_2$ | mercury II | Mercuric |
| Tin | +2 | SnCl$_2$ | tin II | Stannous |
| | +4 | SnCl$_4$ | tin IV | Stannic |

**Figure 1.7**

can then be determined by reversing the subscript numbers in the formula to the top of the elements. For example:

$$Fe_1 \quad Cl_3$$

The number that ends up at the top of the metal represents the number of electrons in the outer shell of that metal. Another way of viewing the same formula, Fe$_1$Cl$_3$, would be to look at the charge of the chlorine from family seven, which is −1. This −1 is placed above the Cl$^{-1}$ in the formula. There are three atoms of chlorine in the compound. Three times one equals three. There is one atom of iron in the compound. What number times one equals three? The answer is simple, three. So the charge of the iron in the compound is three. This is represented by a +3 above the Fe$^{+3}$ in the formula. The number of electrons in the outer shell of the iron in this compound is three. See the following illustration:

$$
\begin{array}{cc}
+3 & -1 \\
\times & \times \\
\hline
Fe_1 & Cl_3 \\
+3 & -3
\end{array}
$$

There are three positive charges and three negative charges. The compound is in balance by using one atom of iron and three atoms of chlorine. The number above the iron indicates the number of electrons in the outer shell of that atom of iron. If a compound containing a transitional metal does not have any subscript numbers, this indicates that the charges at the top were balanced. The compound did not require any additional atoms of those elements. In the following example, copper is combined with chlorine. From the Periodic Table, the charge on chlorine is –1. Therefore, the charge on the copper must be +1 in order to be electrically balanced, which means that the copper in this compound is copper I.

$$+1 \quad -1$$
$$Cu \quad Cl$$

There is an older naming system for the transitional metals that uses suffixes to indicate the higher or lower number of outer-shell electrons. The metal that contains the lower number of outer-shell electrons ends with the suffix "ous." The metal that contains the higher number of outer-shell electrons ends with the suffix "ic." For example, iron has the possibility of two or three electrons in the outer shell. When iron II is used in a compound, the name for the iron is *ferrous*; therefore, a salt compound containing iron II and chlorine would be called *ferrous* chloride. In this case, the "ous" indicates two electrons in the outer shell of the iron. In order to use this system, you must first know the possible numbers of outer shell electrons to determine which is the higher and which is the lower number of electrons in the outer shell. For example, mercury can have one or two, copper one or two, iron two or three, etc. If iron III were used in a compound, the name would end in "ic"; therefore, the iron III name would be *ferric*. If iron III was combined with chlorine, the name of the salt compound would be *ferric* chloride.

Salts have particular hazards, depending on which salt family they belong to. Salts generally do not burn, but can be oxidizers and support combustion.

Water transportation: barges transport many types of hazardous materials on the nation's waterways.

Some salts are toxic and some may be water reactive. Salt families can be divided into groups.

## Binary Salts

Binary (meaning two) salts are made up of two elements, a metal and a nonmetal, except oxygen. They end in "ide," such as potassium *chloride*. Binary salts as a group have varying hazards. They may be water reactive, toxic, and, in contact with water, may form a corrosive liquid and release heat. Chemical reactions often release heat, which is referred to as an exothermic reaction. The hazard of an individual binary salt cannot be determined by the family. To determine the hazards of the binary salts, they have to be researched in reference materials. This varying hazard applies to all binary salts, except for nitrides, carbides, hydrides, and phosphides. One helpful way to remember these four binary salts is by using the first letters of the element to form NCHP, which can be represented by "North Carolina Highway Patrol." These are compounds in which the metal has been bonded with one of the nonmetals (nitrogen, carbon, phosphorus, or hydrogen). These compounds have particular hazards associated with them when they are in contact with water: nitrides give off ammonia, carbides produce acetylene, phosphides give off phosphine gas, and hydrides form hydrogen gas. In addition, a corrosive base is formed from contact with water. The corrosive base will be the hydroxide of the metal that is attached to the nonmetal. For example, calcium carbide in contact with water will produce acetylene gas and the corrosive liquid calcium hydroxide. You have to look up the remaining binary salts to determine the hazard. In the first example, lithium metal is combined with chlorine. The resulting compound has a metal and a nonmetal other than oxygen and the name ends in "ide"; therefore, it fits the definition of a binary salt. If lithium chloride is researched in reference sources, it is found to be soluble in water. It is not water reactive; in fact, lithium chloride does not present any significant hazard in a spill. The DOT does not list lithium chloride on its hazardous materials tables.

$$Li^{+1} + Cl^{-1} = LiCl, \text{ lithium chloride, binary salt, varying hazard}$$

The second example combines the metal calcium and the nonmetal phosphorus. The compound name ends in "ide"; therefore, this is also a binary salt. Binary salts have varying hazards. Calcium phosphide, however, is one of the exceptions. Phosphides are one of the salts that have a known hazard: it gives off phosphine gas and calcium hydroxide liquid, which is a corrosive base. Even so, this is just a preliminary estimate. The material should still be researched for additional hazards. Phosphine is a dangerous fire risk and highly toxic by inhalation.

$$Ca^{+2} + P^{-3} = Ca_3P_2, \text{ calcium phosphide, binary salt}$$

Binary salts have varying hazards; they have to be researched to determine the exact hazards of a particular compound. Notice the varying hazards of the following examples: lithium fluoride is a strong irritant to the eyes and skin; potassium bromide is toxic by ingestion and inhalation; sodium chloride is table salt, a medical concern when ingested in excess, but certainly of no significant hazard to emergency responders. However, if sodium chloride is washed into a farmer's field as the result of an incident, the farmer may not be able to grow crops in that field for many years.

## Binary Oxides (Metal Oxides)

The next group of salts is known as the binary or metal oxides. They are also made up of two elements, a metal and a nonmetal, but in this case, the nonmetal can only be oxygen. They end in "oxide," such as aluminum oxide. As a group, they are water reactive and, when in contact with water, almost always produce heat and form a corrosive liquid. However, they do not give off oxygen, because there is not an excess of oxygen. There is only one oxygen atom and the oxygen atom is held tightly by the oxide salt and is not released as free-oxygen gas. In the following example, potassium metal is combined with one oxygen atom. The name of the salt compound is potassium oxide. This compound meets the definition of a binary oxide salt.

$K + O = K_2O$, potassium oxide, binary oxide salt, when in contact with water releases heat and potassium hydroxide, a corrosive liquid

## Peroxide Salts

Peroxides are composed of a metal and a nonmetal peroxide radical, $O_2^{-2}$. The prefix "per" in front of a compound or element name means that the material is "loaded" with atoms of a particular element. In the case of the peroxides, they are loaded with oxygen. A radical in the salt families is two or more nonmetals covalently bonded together acting as a single unit with a particular negative (–) charge on the radical. The charge will not be found on the Periodic Table. The charges must be found on a listing of radical charges that are in Figures 1.5 and 1.8. In the case of the peroxide salt, two oxygen atoms have bonded together and are acting as one unit with a –2 charge. When a peroxide comes in contact with water, heat is produced, a corrosive liquid is formed, and oxygen is released. This makes peroxides particularly dangerous in the presence of fire. Peroxides release oxygen because, unlike the oxide salts, there is an excess of oxygen present. In the following example, sodium metal combines with the peroxide radical to form the compound sodium peroxide. The name ends in "peroxide," so this is a peroxide salt. When in contact with water, sodium peroxide is a dangerous fire and explosion risk and a strong oxidizing agent.

# OXYRADICALS

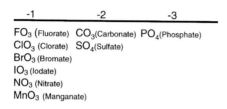

**Figure 1.8**

$Na^{+1} + O_2^2 = Na_2O_2$, sodium peroxide, a peroxide salt, gives off heat and a corrosive liquid, sodium hydroxide, and releases oxygen when in contact with water

## Hydroxide Salts

Hydroxide salts are made up of a metal and the nonmetal hydroxide radical $-OH^{-1}$. The name always ends with the word "hydroxide." They are water reactive and, when in contact with water, release heat and form a corrosive liquid. In the following example, calcium metal is combined with the hydroxide radical; the resulting compound is calcium hydroxide, a hydroxide salt.

$Ca^{+2} + OH^{-1} = Ca(OH)_2$, calcium hydroxide, a hydroxide salt, releases heat and forms a corrosive liquid, calcium hydroxide, in contact with water

## Oxysalts

The oxysalts are made up of a metal and an oxyradical. The names end in "ate" or "ite," and may have the prefixes "per" or "hypo." Generally, as a group, they do not react with water; they dissolve in water. Some of the "hypo–ites" technically do react with water to release chlorine, but the reaction is mild. Oxysalts are oxidizers as a family, they will release oxygen, which accelerates combustion if fire is present. Another hazard occurs when oxysalts dissolve in water and the water is soaked into another material, such as packaging or firefighter turnouts. The water will evaporate and the oxysalt will be left in the material. If the material is then exposed to heat or fire, the material will burn very rapidly because the oxysalt in the material accelerates the combustion. Nine oxysalt radicals will be presented with this group. There are other oxyradicals, but the ones chosen are considered most important to emergency response personnel.

The first six oxyradicals all have $-1$ charges: $FO_3$ (fluorate), $ClO_3$ (chlorate), $BrO_3$ (bromate), $IO_3$ (iodate), $NO_3$ (nitrate), and $MnO_3$ (manganate). The next two have −2 charges: $CO_3$ (carbonate) and $SO_4$ (sulfate). The last oxyradical is $PO_4$ (phosphate), which has a −3 charge (Figure 1.8). All of the radicals listed above are considered to be in their base state. The base state has to do with the number of oxygen atoms that are present in the radical. The base state is the normal number of oxygen atoms present in that oxyradical. When a metal is added to any oxyradical in the base state, the compound suffix is "ate," such as sodium *phosphate*. In the following example, the metal potassium is combined with the oxyradical carbonate; the resulting compound is potassium carbonate. The compound does not have a prefix on the oxyradical and the suffix is "ate"; therefore, it is the base state of the compound.

$$K^{+1} + CO_3^{-2} = K_2CO_3, \text{ potassium carbonate, an oxysalt}$$
$$\text{in the base state, an oxidizer}$$

Oxyradicals may be found with varying numbers of oxygen atoms. There may be more or less oxygen atoms than the base state. Regardless of the number of oxygen atoms on the oxyradical, the charge of the radical **does not** change. When naming the compounds with an additional oxygen atom, the prefix *per* is used to indicate excess oxygen over the base state (Figure 1.9); the suffix is still "ate." An example is sodium *persulfate*. In the following example, the metal potassium is combined with the oxyradical perchlorate: the resulting compound is potassium perchlorate. The level of oxygen is one above the base state. Notice that the charge on the oxyradical is still −1, even though the number of oxygen atoms has changed. Potassium perchlorate is a fire risk in contact with organic materials, a strong oxidizer, and a strong irritant.

$$K^{+1} + ClO_4^{-1} = KClO_4, \text{ potassium perchlorate, an oxidizer}$$

When the number of oxygen atoms is one less than the base state of an oxyradical, the suffix of the oxyradical name will be "ite"; an example is magnesium *sulfite*. In the following example, the metal sodium is combined with the oxyradical phosphite. There is now one less oxygen than the base state. The charge on the oxyradical has not changed. In addition to being an oxidizer, sodium phosphite is also used as an antidote in mercuric chloride poisoning.

$$Na^{+1} + PO_3^{-3} = Na_3PO_3, \text{ sodium phosphite, an oxidizer}$$

## NAMING OXYSALTS

| | | |
|---|---|---|
| +1 Oxygen Prefix | Per- _____ | ate |
| Base State Ending | _____ | ate |
| -1 Oxygen Ending | _____ | ite |
| -2 Oxygen Prefix | Hypo _____ | ite |

**Figure 1.9**

Finally, an oxyradical can have two less oxygen atoms than the base state. The oxyradical name will now have a *hypo* prefix and the suffix will be "ite." An example would be aluminum *hypo*phosph*ite*. In the following example, calcium is combined with the oxyradical hypochlorite; the resulting compound is calcium hypochlorite, a common swimming pool chlorinator. Calcium hypochlorite is an oxidizer and a fire risk when in contact with organic materials.

$$Ca^{+2} + ClO^{-1} = Ca(ClO)_2, \text{ calcium hypoclorite, an oxidizer}$$

## Cyanide Salts

The last salt family that will be discussed are the cyanide salts. They are made up of a metal and the cyanide radical, CN. The name of the resulting compound ends in the word "cyanide"; an example is potassium cyanide. Cyanide salts are toxic materials that dissolve in water to form a hydrogen cyanide solution. Hydroxide ions are produced that will make the solution basic. Cyanide salts react with acids to produce hydrogen cyanide gas, which has an almond-like odor. Hydrogen cyanide gas is used in gas chambers. The cyanides are deadly poisons and can be found as salts or in solution and may produce a toxic gas when heated. When the cyanide ion enters the body, it forms a complex ion with the copper ion located in the cells. These copper ions are essential to the enzyme that allows the cell to use oxygen from the blood. This enzyme is deactivated by the binding of the cyanide and copper ions. In the following example, sodium metal is combined with the cyanide radical. The resulting compound is sodium cyanide. Sodium cyanide is toxic by ingestion and inhalation.

$$Na^{+1} + CN^{-1} = NaCN, \text{ sodium cyanide, a poison}$$

## NONMETAL COMPOUNDS

Nonmetal compounds are combinations of nonmetallic elements that combine in a covalent bonding process. When elements bond covalently, the bonding electrons are shared between the elements; the resulting compound is carbon disulfide. Carbon may bond with itself to satisfy its need for electrons. This happens frequently in the hydrocarbon families (to be discussed in detail in Chapters 3 and 4). The naming of the hydrocarbon and hydrocarbon-derivative families in this book will focus primarily on the trivial naming system, which uses prefixes indicating the number of carbons in the compound (Figure 1.10). The suffix of the hydrocarbon name reflects the family and type of bond between the carbons in the compound (Figure 1.11). Another naming system, called the IUPAC system, was developed by the International Union of Pure and Applied Chemistry. (An outline of the IUPAC system for naming organic compounds is located in the Appendix of this book.) In the following example, the carbon is sharing four electrons with the two sulfur atoms. The result is that the carbon thinks it has eight electrons in its outer shell and each sulfur thinks it has eight

## Hydrocarbon Prefixes

| | |
|---|---|
| Meth/Form | 1 Carbon |
| Eth/Acet | 2 Carbons |
| Prop | 3 Carbons |
| But | 4 Carbons |
| Pent | 5 Carbons |
| Hex | 6 Carbons |
| Hept | 7 Carbons |
| Oct | 8 Carbons |
| Non | 9 Carbons |
| Dec | 10 Carbons |

**Figure 1.10**

electrons in its outer shell. The carbon and the sulfur atoms are satisfied. The compound formed is carbon disulfide, which is a poison by absorption; it is a highly flammable, dangerous fire and explosion risk, has a wide flammable range from 1 to 50%, and can be ignited by friction. Carbon disulfide also has a low ignition temperature and can be ignited by a steampipe or a lightbulb.

$$S=C=S$$

Nonmetals and their compounds may be solids, liquids, or gases. Some may burn; some may be toxic; and they may also be reactive, corrosive, and oxidizers. The largest quantities of hazardous materials encountered are made up of non-metal (nonsalt) materials. These materials can also be divided into families. The most commonly encountered hazardous material is hydrocarbon fuel. Products such as gasoline, diesel fuel, and fuel oil are all used to power our vehicles or heat our homes and businesses. Propane, butane, and natural gas are also fuels but are not mixtures; they are pure compounds. They are transported frequently and stored in large quantities, and can present frequent problems to responders. They form a family called the hydrocarbons, because they are made up primarily of carbon and hydrogen. These materials are flammable, and may be toxic or cause asphyxiation by displacing the oxygen in the air. There are four subfamilies

## Hydrocarbon Families

| Alkanes | Alkenes | Alkynes | Aromatics |
|---|---|---|---|
| Single Bond | Double Bond | Triple Bond | Resonant Bond |
| Saturated | Unsaturated | Unsaturated | Acts Saturated |
| Ends in -ane | Ends in -ene | Ends in -yne | **BTX** Benzene |
| $CnH2n+2$ | $CnH2n$ | $CnH2n-2$ | Toluene Xylene |

**Figure 1.11**

## Hydrocarbon Derivative Families

| FAMILY | GENERAL FORMULA | HAZARD |
|---|---|---|
| Alkyl Halides | R - X     X = (F,Cl,Br,I) | Toxic/Flammable |
| Nitro | R - NO$_2$ | Explosive |
| Amine | R - NH$_2$<br>R$_2$ - NH<br>R$_3$ - N | Toxic/Flammable |
| Ether | R - O - R | WFR/Anesthetic |
| Peroxide | R - O - O - R | Explosive/Oxidizer |
| Alcohol | R - O - H | WFR/Toxic |
| Ketone | R - C - O - R | Flammable/Narcotic |
| Aldehyde | R - C - H - O | WFR/Toxic |
| Ester | R - C - O - O - R | Polymerize |
| Organic Acid | R - C - O - O - H | WFR/Toxic |

**Figure 1.12**

of hydrocarbons, known as alkanes, alkenes, alkynes, and aromatics. (These families will be discussed in detail in Chapters 3 and 4.) The alkane and aromatic families of hydrocarbons occur naturally; the alkenes and alkynes are manmade. Both types of hydrocarbons are used to make other families of chemicals, known as hydrocarbon derivatives. Radicals of the hydrocarbon families are made by removing at least one hydrogen from the hydrocarbon and replacing it with a nonmetal other than carbon or hydrogen. Ten of these hydrocarbon derivatives will be discussed in detail in the appropriate chapters associated with their major hazards: alkyl halides, nitros, amines, ethers, peroxides, alcohols, ketones, aldehydes, esters, and organic acids (Figure 1.12). Just a few elements in addition to carbon and hydrogen are combined to make the 10 hydrocarbon derivative families discussed in this book. These materials can be toxic, flammable, corrosive, explosive, reactive; some may be oxidizers and some may polymerize.

## Physical and Chemical Terms

Physical and chemical terms, such as boiling point, flash point, ignition temperature, pH, sublimation, water reactivity, spontaneous combustion, and others will be discussed in the remaining chapters of the book where appropriate. Many of the reference books used by emergency responders contain physical and chemical terms, and it is important that responders understand the significance of each of them. There are numerous hazards that chemicals can present to emergency responders; very few hazardous materials have only one hazard. In

## Chemical Incompatibilities

| Chemical | Incompatible With |
|---|---|
| Acetylene | Bromine, chlorine, fluorine, copper, silver, mercury |
| Ammonia | Mercury, hydrogen fluoride, calcium hypochlorite, chlorine |
| Flammable Liquids | Ammonium nitrate, chromic acid, hydrogen peroxide, sodium peroxide, nitric acid, and the halogens |
| Oxygen | Oils, grease, hydrogen, flammable liquids, solids, gases |
| Nitric Acid | Acetic acid, hydrogen sulfide, flammable liquids and gases |
| Phosphorus | Air, oxygen |
| Sulfuric Acid | Potassium chlorate, perchlorate, permanganate |
| Alkali Metals | Carbon tetrachloride, carbon dioxide, water, halogens |
| Alkaline Earth Metals | Carbon tetrachloride, halogens, hydrocarbons |

**Figure 1.13**

addition to the nine DOT/UN hazard classes, some materials have hidden hazards. Not all placards will tell that a material is water reactive, air reactive, may be explosive, or may polymerize. These hidden hazards will be discussed throughout the remaining chapters of this book.

Chemicals may also have incompatibilities with each other and produce hazardous chemical reactions when combined (Figure 1.13). For example, when acids come in contact with cyanides, toxic gases are produced. This can occur in a facility that may store and use both types of chemicals, such as a plating company. In contact with ammonia, chlorine produces nitrogen trichloride, which explodes when heated to 200°F or exposed to sunlight. NFPA Standard 491M lists hazardous chemical reactions for selected chemicals. The remaining chapters of this book will discuss the hazards, hidden hazards, and physical and chemical characteristics of some of the more common hazardous materials.

## TOP 50 INDUSTRIAL CHEMICALS

Each year, *Chemical and Engineering News* compiles its list of the Top 50 industrial chemicals manufactured in the United States (Figure 1.14). Through case studies, specific members of the Top 50 and other common hazardous chemicals will be studied in the remaining chapters of this book. The Top 10 consists of three corrosives (sulfuric acid, phosphoric acid, and sodium hydroxide); two poisons (anhydrous ammonia and chlorine); nitrogen; two flammable gases (ethylene and propylene); oxygen; and lime. Naturally these chemicals, along with the rest of the Top 50, will frequently be transported and stored in a wide variety of industrial and other types of facilities. They will frequently be encountered in incidents, along with the fuel gases and liquids.

# Organics outpaced inorganics as Top 50 chemicals production rose overall

| Rank 1995 | Rank 1994[d] | | Billions of lb 1995 | Billions of lb 1994 | Common units[b] 1995 | Common units[b] 1994 | Average annual change 1994-95 | Average annual change 1993-94 | Average annual change 1990-95 | Average annual change 1985-95 |
|---|---|---|---|---|---|---|---|---|---|---|
| 1 | 1 | Sulfuric acid | 95.36 | 89.63 | 47,681 tt | 44,813 tt | 6.4% | 12.5% | 1.5% | 1.8% |
| 2 | 2 | Nitrogen | 68.04 | 63.91 | 939 bcf | 882 bcf | 6.5 | 10.8 | 4.6 | 3.7 |
| 3 | 3 | Oxygen | 53.48 | 50.08 | 646 bcf | 605 bcf | 6.8 | 10.6 | 6.9 | 5.1 |
| 4 | 4 | Ethylene | 46.97 | 44.60 | 46,966 mp | 44,602 mp | 5.3 | 11.5 | 5.2 | 4.6 |
| 5 | 5 | Lime[c] | 41.23 | 38.37 | 20,617 tt | 19,184 tt | 7.5 | 3.0 | 3.8 | 2.7 |
| 6 | 6 | Ammonia | 35.60 | 34.51 | 17,801 tt | 17,256 tt | 3.2 | 0.4 | 0.9 | 0.3 |
| 7 | 7 | Phosphoric acid | 26.19 | 25.58 | 13,096 tt | 12,792 tt | 2.4 | 11.1 | 1.7 | 2.2 |
| 8 | 8 | Sodium hydroxide | 26.19 | 25.11 | 13,094 tt | 12,555 tt | 4.3 | 0.5 | 1.7 | 1.9 |
| 9 | 10 | Propylene | 25.69 | 23.94 | 25,691 mp | 23,943 mp | 7.3 | 11.5 | 2.9 | 5.6 |
| 10 | 9 | Chlorine | 25.09 | 24.37 | 12,544 tt | 12,187 tt | 2.9 | 0.9 | 1.2 | 1.9 |
| 11 | 11 | Sodium carbonate[d] | 22.28 | 20.56 | 11,138 tt | 10,278 tt | 8.4 | 3.8 | 2.0 | 2.7 |
| 12 | 18 | Methyl tert-butyl ether | 17.62 | 13.61 | 17,620 mp | 13,610 mp | 29.5 | 5.5 | 14.7 | 25.0 |
| 13 | 14 | Ethylene dichloride | 17.26 | 16.76 | 17,263 mp | 16,762 mp | 3.0 | -6.6 | 4.5 | 3.6 |
| 14 | 12 | Nitric acid | 17.24 | 17.22 | 8,621 tt | 8,611 tt | 0.1 | 4.3 | 1.7 | 1.6 |
| 15 | 13 | Ammonium nitrate[e] | 15.99 | 17.03 | 7,993 tt | 8,517 tt | -6.2 | 2.8 | 0.7 | 1.7 |
| 16 | 16 | Benzene | 15.97 | 15.27 | 2,168 mg | 2,074 mg | 4.5 | 24.0 | 5.1 | 5.5 |
| 17 | 15 | Urea[f] | 15.59 | 15.90 | 7,796 tt | 7,952 tt | -2.0 | -4.5 | -1.0 | 1.6 |
| 18 | 17 | Vinyl chloride | 14.98 | 13.85 | 14,976 mp | 13,850 mp | 8.1 | -2.6 | 7.1 | 4.7 |
| 19 | 22 | Ethylbenzene | 13.66 | 10.76 | 13,656 mp | 10,758 mp | 26.9 | 15.2 | 10.3 | 6.3 |
| 20 | 21 | Styrene | 11.39 | 11.29 | 11,386 mp | 11,294 mp | 0.8 | 17.7 | 9.9 | 4.1 |
| 21 | 19 | Methanol | 11.29 | 12.18 | 11,292 mp | 12,176 mp | -7.3 | 15.9 | 6.2 | 8.5 |
| 22 | 20 | Carbon dioxide[g] | 10.89 | 11.80 | 5,446 tt | 5,899 tt | -7.7 | -13.5 | 2.0 | 1.8 |
| 23 | 23 | Xylene | 9.37 | 9.06 | 1,301 mg | 1,258 mg | 3.4 | 3.7 | 8.6 | 5.8 |
| 24 | 24 | Formaldehyde[h] | 8.11 | 8.17 | 8,110 mp | 8,165 mp | -0.7 | -0.3 | 3.8 | 3.8 |
| 25 | 25 | Terephthalic acid[i] | 7.95 | 7.58 | 7,950 mp | 7,575 mp | 5.0 | 6.6 | 3.7 | 2.0 |

Figure 1.14

| | | | | | | | | | | |
|---|---|---|---|---|---|---|---|---|---|---|
| 26 | 27 | Ethylene oxide | 7.62 | 7.24 | 7,621 mp | 7,238 mp | 5.3 | 35.8 | 6.8 | 3.4 |
| 27 | 26 | Hydrochloric acid | 7.33 | 7.47 | 3,663 tt | 3,734 tt | -1.9 | 6.9 | 3.1 | 2.7 |
| 28 | 28 | Toluene^l | 6.73 | 6.75 | 927 mg | 931 mg | -0.4 | 5.8 | 1.6 | 2.9 |
| 29 | 29 | p-Xylene | 6.34 | 6.26 | 6,342 mp | 6,255 mp | 1.4 | 8.0 | 4.0 | 2.9 |
| 30 | 31 | Cumene | 5.63 | 5.22 | 5,625 mp | 5,217 mp | 7.8 | 18.7 | 5.9 | 5.3 |
| 31 | 32 | Ammonium sulfate | 5.24 | 5.18 | 2,619 tt | 2,588 tt | 1.2 | 6.4 | 0.8 | 2.3 |
| 32 | 30 | Ethylene glycol | 5.23 | 6.09 | 5,230 mp | 6,090 mp | -14.1 | 17.1 | 0.6 | 2.3 |
| 33 | 33 | Acetic acid | 4.68 | 3.98 | 4,683 mp | 3,984 mp | 17.5 | 19.7 | 4.5 | 4.9 |
| 34 | 34 | Phenol^k | 4.16 | 3.92 | 4,163 mp | 3,920 mp | 6.2 | 22.9 | 3.3 | 4.1 |
| 35 | 35 | Propylene oxide | 4.00 | 3.70 | 4,000 mp | 3,700 mp | 8.1 | 12.8 | 9.0 | 5.2 |
| 36 | 36 | Butadiene^l | 3.68 | 3.38 | 3,682 mp | 3,376 mp | 9.1 | 8.3 | 3.6 | 4.6 |
| 37 | 37 | Carbon black | 3.32 | 3.25 | 3,315 mp | 3,250 mp | 2.0 | 0.8 | 2.9 | 2.6 |
| 38 | 39 | Isobutylene | 3.23 | 3.08 | 3,235 mp | 3,078 mp | 5.1 | 186.7 | 21.7 | 12.4 |
| 39 | 38 | Potash^m | 3.22 | 3.08 | 1,460 tmt | 1,399 tmt | 4.4 | -7.1 | -3.1 | 1.2 |
| 40 | 41 | Acrylonitrile | 3.21 | 3.03 | 3,207 mp | 3,026 mp | 6.0 | 21.6 | 3.7 | 3.2 |
| 41 | 40 | Vinyl acetate | 2.89 | 3.04 | 2,893 mp | 3,036 mp | -4.7 | 9.5 | 1.7 | 3.2 |
| 42 | 42 | Titanium dioxide | 2.77 | 2.76 | 1,383 tt | 1,380 tt | 0.2 | 7.8 | 5.1 | 4.9 |
| 43 | 43 | Acetone | 2.76 | 2.66 | 2,761 mp | 2,664 mp | 3.6 | 9.6 | 3.4 | 4.4 |
| 44 | 45 | Butyraldehyde | 2.68 | 2.19 | 2,679 mp | 2,194 mp | 22.1 | 11.5 | 7.2 | 7.6 |
| 45 | 44 | Aluminum sulfate | 2.41 | 2.21 | 1,204 tt | 1,107 tt | 8.8 | 3.3 | -0.4 | -0.5 |
| 46 | 46 | Sodium silicate | 2.25 | 2.01 | 1,125 tt | 1,007 tt | 11.7 | 0.1 | 6.6 | 5.0 |
| 47 | 47 | Cyclohexane | 2.13 | 1.96 | 2,134 mp | 1,962 mp | 8.8 | -1.9 | -2.8 | 2.6 |
| 48 | 48 | Adipic acid | 1.80 | 1.80 | 1,797 mp | 1,797 mp | 0.0 | 6.5 | 2.1 | 2.1 |
| 49 | — | Nitrobenzene | 1.65 | 1.44 | 1,650 mp | 1,435 mp | 15.0 | 34.5 | 7.0 | 6.1 |
| 50 | 49 | Bisphenol A | 1.62 | 1.70 | 1,623 mp | 1,702 mp | -4.7 | 32.4 | 7.1 | 5.5 |
| | | TOTAL ORGANICS | 285.89 | 270.40 | | | 5.7% | 10.0% | 5.3% | 5.0% |
| | | TOTAL INORGANICS | 464.10 | 444.15 | | | 4.5% | 6.2% | 2.7% | 2.4% |
| | | GRAND TOTAL | 749.99 | 714.55 | | | 5.0% | 7.6% | 3.6% | 3.3% |

a Revised  b tt = thousands of tons, bcf = billions of cubic feet, mp = millions of pounds, mg = millions of gallons, tmt = thousands of metric tons.  c Except refractory  d Natural and synthetic.  e Original solution.  f 100% basis.  g Liquid and sold only.  h 37%; by weight.  i Includes both acid and ester without double counting.  j All grades.  k Synthetic only.  l Rubber grade  m $K_2O$ basis.  Note: Average annual change calculation based or common units.

**Figure 1.14** (continued)

In addition to the Top 50 industrial chemicals, other chemicals frequently transported and released from highway or rail incidents will be studied, as well as chemicals frequently released from fixed facilities. While there may be some variations by region of the country, statistics generally hold true for most of the United States.

## CHEMICAL RELEASE STATISTICS

The Chemical Abstract Service lists over 63,000 chemicals used outside the laboratory environment. The DOT regulates over 3800 hazardous materials in transportation. The Occupational Safety and Health Administration (OSHA) regulates the occupational exposure of over 600 hazardous substances. There are many other lists of chemicals that are regulated and compiled by various governmental regulatory agencies, such as the U.S. Environmental Protection Agency (EPA). The EPA has several listings of hazardous materials, depending on whether they are in storage or are hazardous wastes. The Resource Conservation and Recovery Act (RCRA) deals with hazardous wastes exhibiting the characteristics of ignitibility, corrosivity, and reactivity. Regulations are found in 40 CFR 261.33. The Comprehensive Environmental Response, Compensation, and Liability Act of 1980 (CERCLA) lists hazardous substances in 40 CFR Part 302, Table 302.4. The Emergency Planning and Community Right-to-Know Act of 1986 (EPCRA) deals with a list of extremely hazardous substances, most of which are poisons. The Clean Air Act (CAA) Section 112r also has a list of toxic chemicals for accidental-release prevention. The EPA has a listing of the Top 15 chemicals released from chemical accidents in the United States (Figure 1.15). These 15 chemicals account for two thirds of all chemical releases. Eleven of the Top 15 released chemicals are also in the Top 50 industrial chemicals produced each year. During the statistical period from 1988 to 1992, over·34,500 accidents involving toxic chemicals were reported in the United States. Those accidents resulted in more than 450 deaths and 100,000 persons being evacuated from their homes. Almost 15% of those releases were to surface waters, 38% into the air, and 47% to land. These accidents occurred on an average of 575 times per month or over 19 times per day, releasing toxic substances into the environment and potentially exposing workers, communities, and wildlife. Over half of all acci-

**EPA TOP 15 CHEMICAL ACCIDENTS**

| | |
|---|---|
| 1) Polychlorinated Biphenyls | 9) Benzene |
| 2) Anhydrous Ammonia | 10) Hydrogen Sulfide |
| 3) Sulfuric Acid | 11) Sodium Hydroxide |
| 4) Chlorine | 12) Vinyl Chloride |
| 5) Hydrochloric Acid | 13) Toluene |
| 6) Ethylene Glycol | 14) Mercury |
| 7) Sulfur Dioxide | 15) Ethylene Oxide |
| 8) Radioactive Material | |

**Figure 1.15**

## TOP 23 HAZARDOUS MATERIALS SHIPPED BY RAIL

1) Liquified Petroleum Gas
2) Sodium Hydroxide
3) Sulfuric Acid
4) Molten Sulfur
5) Anhydrous Ammonia
6) Elevated Temperature Material
7) Chlorine
8) Methyl Alcohol
9) Fuel Oil (Flammable)
10) Vinyl Chloride
11) Phosphoric Acid
12) Stryene Monomer, Inhibited
13) Carbon Dioxide Cryogenic
14) Hydrochloric Acid
15) Fuel Oil (Combustible Liquid)
16) Ammonium Nitrate
17) Sodium Chlorate
18) Gasoline
19) Butadiene
20) Phenol Molten
21) Methyl Tert Butyl Ether
22) Adipic Acid
23) Crude Oil, Petroleum

**Figure 1.16**

dents occurred in just 10 states including California, Texas, Louisiana, Pennsylvania, Ohio, Illinois, Kentucky, Florida, Michigan, and New Jersey. Nearly 11 accidents occur per day in these 10 states. These chemicals are listed in detail in the appropriate chapters of this book.

### Transportation Releases

According to the Association of American Railroads (AAR), over 70% of hazardous materials are transported in tank cars. During 1994, over 1,761,636 tank cars of hazardous materials were transported in the U.S. In 1994, there were 1196 nonaccident releases of hazardous materials from rail cars. Figure 1.16 shows the Top 23 accident and nonaccident releases. Tank cars were responsible for 88% of those releases. In 1994, there were 537 train accidents involving hazardous materials cars. In 266 of those cases, the hazardous materials cars were damaged or derailed. Thirty-six of those incidents resulted in a release, and 20 of those required an evacuation. The railcar shipments by DOT hazard class, with number of leaks by hazard class, are listed in Figure 1.17.

The DOT reports that there were 6 deaths, 399 injuries, and damage costs of over $28 million from 14,688 incidents involving hazardous materials in 1995. This is a decrease of over 1400 incidents from 1994; the number of deaths and injuries also decreased during the same period. The incidents are broken down by mode of transportation: air 812, highway 12,710, railway 1154, and water 12. Deaths, injuries, and damage occurred most often in incidents involving flammable liquids. The top 10 commodities released during the incidents in 1995 were all flammable or corrosive liquids (Figure 1.18).

### DOT/UN Hazard Classes of Hazardous Materials

The DOT lists materials that it regulates in the hazardous materials tables in 49 CFR Parts 100–199. OSHA also lists chemicals that are dangerous in the workplace, as part of the Worker Right-To-Know regulations. This book will use

Railroad transportation: railyard with many different types of railcars, many containing hazardous materials.

the nine DOT/UN hazard classes to group hazardous materials (Figure 1.19). It is important to note that the placard on a transport vehicle depicts only the most severe hazard of a material. When a material has more than one hazard, the DOT prioritizes the hazard that will be placarded. These hazards are listed by the DOT in 49 CFR Part 173.2a (Figure 1.20) to determine which hazard will be assigned to a particular material when the material has multiple hazards. Almost every

## TANK CAR LEAKS BY HAZARD CLASS

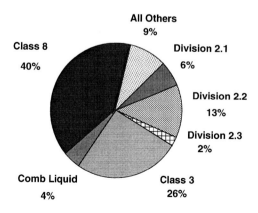

**Figure 1.17**

Highway transportation: major interstate highway system with many types of hazardous materials being transported.

## TOP 10 DOT  TRANSPORTATION INCIDENTS

| | |
|---|---|
| 1 | Corrosive Liquids N.O.S. |
| 2 | Flammable Liquids N.O.S. |
| 3 | Resin Solution |
| 4 | Sodium Hydroxide |
| 5 | Hydrochloric Acid |
| 6 | Sulfuric Acid |
| 7 | Adhesives |
| 8 | Gasoline |
| 9 | Isopropanol |
| 10 | Paint or Paint Related |

**Figure 1.18**

## DOT /UN  HAZARD CLASSES

| | |
|---|---|
| Class 1 | Explosives |
| Class 2 | Compressed Gases |
| Class 3 | Flammable Liquids |
| Class 4 | Flammable Solids |
| Class 5 | Oxidizers |
| Class 6 | Poisons |
| Class 7 | Radioactives |
| Class 8 | Corrosives |
| Class 9 | Miscellaneous HazMat |

**Figure 1.19**

**DOT CLASSIFICATION OF MATERIALS WITH
MORE THAN ONE HAZARD**

1-Radioactive
2-Division 2.3 Poison Gas
3-Division 2.1 Flammable Gas
4-Division 2.2 Non-Flammable Gas
5-Division 6.1 Poisonous Liquid, Inhalation Hazard

The above chart does not apply to the following:
Class 1 Explosives, 5.2 Organic Peroxides,
Division 6.2 Infectious Substances, and Wetted
Explosives such as Picric Acid.

**Figure 1.20**

hazardous material has more than one hazard. As an emergency responder, you must be familiar with the other potential, many times hidden hazards that materials may present. Figure 1.21 is a chart that shows the potential hidden hazards of the nine hazard classes. For example, some corrosive materials are classified as oxidizers, such as perchloric acid above 50% concentration. The placard on perchloric acid above 50% will be an oxidizer placard. Do not forget that perchloric acid is also a strong corrosive material. Do not focus only on the hazard depicted by the placard. Thoroughly research the chemical and find out about all of the physical and chemical characteristics and hazards associated with the material.

The Federal agencies that regulate hazardous materials have begun to stress accident prevention in decreasing the frequency of hazardous materials releases across the country. Emergency response organizations can also work to reduce accidents, particularly in fixed facilities within their jurisdictions. This can be accomplished through local legislation and the adoption of national codes to address hazardous materials storage and use issues. Through OSHA the Department of Labor, has compiled a listing of the Top 10 regulations violations that may result in accidental releases of chemicals, injuries, or deaths, if not corrected (Figure 1.22). The DOT has also issued a Top 10 listing of violations of transportation regulations that may lead to accidents resulting in releases of hazardous materials (Figure 1.23).

There are some general hints that can be used to identify when hazardous materials are present when looking at the names of chemicals. The suffixes of chemical names can indicate particular chemical families and potential hazardous materials. In the case of chemical families, if you can recognize the family, you will be able to recognize the hazards. Chemical families, their naming, and their hazards will be discussed further in the remaining chapters of this book. Figure 1.24 lists some hints from chemical names and suffixes that indicate hazardous materials.

## NFPA 704 Marking System

The National Fire Protection Association (NFPA) has developed a fixed-facility marking system to designate general hazards of chemicals (Figure 1.25), referred to as the NFPA 704 marking system. The system is designed to warn emergency responders of the presence of hazardous materials in fixed facilities

**BURKE PLACARD HAZARD CHART**

X - Primary Hazard
\* - Secondary Hazard

| PLACARD COLOR | EXPLOSIVE | FIRE HAZARD | WATER REACT | AIR REACTIVE | TOXIC | CORROSIVE | COMP GAS | THERMAL | POLYMERIZE | SPONT COMBUST | ASPHYXIANT | OXIDIZER | HEAT-SHOCK SEN | RADIOACTIVE |
|---|---|---|---|---|---|---|---|---|---|---|---|---|---|---|
| ORANGE | X | * | | | * | * | | | | | | | * | |
| RED | | X | | | * | * | * | * Cold | * | | * | | | |
| GREEN | | * | | | * | X | | * Cold | | | * | * | | |
| RED & WHITE STRIPES | * | X | * | * | * | | | | | | | | | |
| WHITE OVER RED | | * | | | * | * | | | | X | * | | | |
| BLUE | | * | X | * | * | * | | | | | | | | |
| YELLOW | * | | * | * | * | * | * | * Cold | | | | X | | |
| WHITE | X | X | | X | * | X | * | * Cold | X | | | X | | |
| YELLOW OVER WHITE | | | | | * | | | | | * | * | | | X |
| WHITE OVER BLACK | * | * | | | * | X | | | | * | * | | | |
| BLACK & WHITE STRIPES ** | | * | * | * | * | * | | * Hot | * | * | | * | | * |
| RED WHITE RED | * | * | * | * | * | * | * | * Cold | * | * | * | * | * | * |

Figure 1.21

and to give them some general information about the hazards of the materials. It does not identify any specific chemicals. The system utilizes a diamond-shaped placard with four colored sections. The blue section indicates health hazard; the red section flammability; the yellow section reactivity; the white section contains special information, such as, "oxy" for oxidizer, "cor" for

## Top Ten OSHA Violations

1) Improper storage of gas cylinders
2) Lack of employee training
3) Failure to ground electrical systems
4) No OSHA posters in work place
5) Hard hats not worn
6) Improper storage of combustible liquids
7) No accident-prevention program
8) No chemical inventory
9) No shoring system in excavations
10) Lack of safety inspections by competent person

**Figure 1.22**


## Top Ten DOT Violations

1) Failure to conduct design qualifications testing on packages
2) Performing hydrostat tests with improper equipment
3) Failure to conduct a second retest at 100 psi or higher
4) Failure to keep accurate records of retest of cylinders
5) Unauthorized packaging of hazardous materials
6) Unapproved explosives offered for transportation
7) Failure to properly mark packages
8) Incorrect shipping paper information
9) Failure to train employees or keep records of Haz Mat training
10) Operating under expired exemptions

**Figure 1.23**


## Hints to Hazardous Material Names

### Names Ending In:

| - al  | - ate | - ane | - azo | - ene | - ine | - ite | - ol |
|-------|-------|-------|-------|-------|-------|-------|------|
| - one | - oyl | - yde | - ane | - yl  | - yne |       |      |

### Names That Include:

| acet  | acid    | alkali | amyl | azide  | bis     | caustic | cis |
|-------|---------|--------|------|--------|---------|---------|-----|
| hepyl | hydride | iso    | mono | naptha | oxy     | penta   |     |
| per   | tetra   | trans  | tris | vinyl  | nitrile | cyan    |     |

**Figure 1.24**

## NFPA 704 CLASSIFICATION SYSTEM

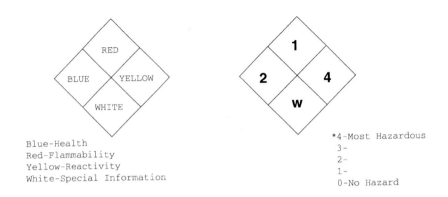

Blue-Health
Red-Flammability
Yellow-Reactivity
White-Special Information

\*4-Most Hazardous
3-
2-
1-
0-No Hazard

\*Complete information on the definitions are located in the appendix

Copyright © 1990, National Fire Protection Association, Quincy, MA 02269. This warning system is intended to be interpreted and applied only by properly trained individuals to identify fire, health and reactivity hazards of chemicals. The user is referred to certain limited number of chemicals with recommended classifications in NFPA 49 and NFPA 325M which would be used as a guideline only. Whether the chemicals are classified by NFPA or not, anyone using the 704 system to classify chemicals does so at their own risk.

**Figure 1.25**

NFPA 704 diamond placards on the outer fence of a compressed gas facility.

## 1995 HAZMAT INCIDENTS BY HAZARD CLASS

| Hazard Class | Incidents | Deaths | Injuries |
|---|---|---|---|
| Explosives | 16 | 0 | 0 |
| Flammable Gas | 240 | 2 | 13 |
| Non-Flammable Gas | 276 | 0 | 28 |
| Poison Gas | 51 | 0 | 0 |
| Flammable Liquids | 6360 | 4 | 122 |
| Flammable Solids | 87 | 0 | 2 |
| Spontaneously Combustible | 23 | 0 | 1 |
| Dangerous When Wet | 38 | 0 | 0 |
| Oxidizer | 361 | 0 | 3 |
| Organic Peroxide | 102 | 0 | 6 |
| Poison Solid/Liquid | 1209 | 0 | 52 |
| Radioactive | 9 | 0 | 0 |
| Corrosive | 5280 | 0 | 134 |
| Miscellaneous HazMat | 670 | 0 | 11 |
| Consumer Commodity-ORM-D | 54 | 0 | 0 |

**Figure 1.26**

corrosive, a "W" with a slash through it for water reactivity, and a radioactive propeller indicating radioactivity. Numbers from 0–4 are placed in the blue, red, and yellow sections, indicating the degree of hazard: 0 indicates no hazard, 4 indicates the most severe hazard. These diamond placards are placed on the outside of buildings, storage tanks, storage sheds, and on doors leading to areas where hazardous materials are present. Responders still have to obtain more information when the 704 diamonds are present. The diamonds act only as stop signs, indicating hazardous materials are present. The chemicals must be identified for specific hazards before responders enter. The NFPA 704 designation is not available for all hazardous materials. Those chemicals that have been assigned NFPA 704 designations are listed with the chemicals in each of the chapters.

## Chemical Listings and Incidents

Each of the remaining chapters will list the nine DOT hazard classes and the types of chemicals in each. Specific chemicals will be listed along with their physical and chemical characteristics. The chemicals are provided to illustrate the similarities of some chemicals in the same families and the differences of chemicals in other families. When reading through the specific chemicals, note the multiple hazards, many of which are not placarded for the hazard class and others that are not a hazard class at all. Along with the listings of chemicals, there will be excerpts from incidents that have occurred involving many of the chemicals listed. These are provided to show the dangers of the hazardous materials and how the physical and chemical characteristics contributed to the incident outcome. Figure 1.26 lists by hazard class the hazardous materials incidents, number of deaths, and injuries that occurred in 1995.

## REVIEW QUESTIONS

(Answers located in the appendix)

1. Chemistry is the study of _____.
   A. Isomers
   B. Matter
   C. Isotopes
   D. Reactions

2. Elements on the Periodic Table are represented by _____.
   A. The atomic number
   B. The atomic weight
   C. Symbols
   D. Molecules

3. Match the following family names with the column on the Periodic Table.
   A. Noble gases                     _____ Family I
   B. Bromine family                  _____ Family II
   C. Alkaline earth metals           _____ Family VII
   D. Alkali metals                   _____ Family VIII
   E. Halogens                        _____ Not a family

4. When metals and nonmetals combine, the bond that is formed is called
   _____.

5. When nonmetals combine, the bond that is formed is called _____.

6. Atoms are made up of subatomic particles called _____, _____,
   and _____.

7. An atom of an element is said to be electrically stable when it has
   _____ or _____ electrons in the outer shell.

8. The numbers in question 7 represent the _____ and the _____ rules of
   bonding.

9. Balance the formulas if needed, give the name, the salt family, and the hazards
   for the following salts. (Salts with transitional metals are already balanced.)

   $NaCl$      $CaPO_4$      $AlO_2$      $CuBr_2$      $KOH$
   $LiO$       $MgClO$       $HgO_2$      $NaF$         $FeCO_3$

10. Provide a balanced formula, family name, and hazard(s) for the following salts.

    Calcium hypochlorite      Aluminum chloride         Lithium hydroxide
    Copper II peroxide        Sodium oxide              Potassium iodide
    Magnesium phosphide       Mercury I perchlorate     Iron III fluorate

# 2

## EXPLOSIVES

The first Department of Transportation/United Nations (DOT/UN) hazard class deals with explosives. The DOT defines an **explosive** in 49 CFR 173.50 as, "any substance or article, including a device, which is designed to function by explosion or which, by chemical reaction within itself, is able to function in a similar manner even if not designed to function by explosion." This definition only applies to chemicals designed to create explosions. Another definition, taken from the National Fire Academy Tactical Considerations Student Manual, is, "a substance or a mixture of substances which, when subjected to heat, impact, friction, or other suitable initial impulse, undergoes a very rapid chemical transformation, forming other more stable products entirely or largely gaseous whose combined volume is much greater than the original substance."

There are other chemicals that have explosive potential under certain conditions but may not be placarded or recognized as explosives. These materials will be found in fixed facilities as well as in transportation, and may not have any markings or warnings that they have explosive properties. Some examples are ethers, potassium metal, formaldehyde, organic peroxides, and perchloric acid. Ethers are Class 3 flammable liquids that have a single oxygen atom in their composition; as ether ages, oxygen from the air can combine with this oxygen atom to form a peroxide molecule. An oxygen-to-oxygen single bond that is formed is a very reactive and unstable bond.

$$-O-O-$$

This same type of bond is present in nitro compounds that are explosive and will be presented later in this chapter. Peroxides formed in compounds in this manner are shock and heat sensitive; they may explode simply by being moved or shaken. Formaldehyde solutions are Class 3 flammable liquids and potassium metal is a Class 4.3 flammable solid, dangerous when wet. These materials form explosive peroxides just like ether and are equally as dangerous as they age. Organic peroxides are Class 5.2 oxidizers. Many organic peroxides are temperature sensitive; they are usually stored under refrigeration. If the temperature is elevated, organic peroxides may decompose explosively. Perchloric acid is a corrosive by hazard class. However, perchloric acid is also a strong oxidizer and will explode when shocked or heated. Chemical oxidizers are one of the necessary components for a chemical explosive to function. These chemical oxidizers should be treated with the same respect as the explosives in Class 1.

## DEFINITION OF EXPLOSION

An **explosion** is defined in the National Fire Protection Association (NFPA) *Fire Protection Handbook* as "a rapid release of high-pressure gas into the environment." This release of high-pressure gas occurs regardless of the type of explosion that has produced it. The high-pressure energy is dissipated by a shock wave that radiates from the blast center. This shock wave creates an overpressure in the surrounding area that can affect personnel, equipment, and structures (Figure 2.1). An overpressure of just 0.5 to 1 psi can break windows and knock down personnel. At 5 psi, eardrums can rupture and wooden utility poles can be snapped in two. Ninety-nine percent of people exposed to overpressures of 65 psi or more would die.

## PHASES OF EXPLOSIONS

There are two phases of an explosion, the positive and the negative. The positive phase occurs first as the blast wave travels outward, releasing its energy to objects it comes in contact with. This is also known as the blast pressure, generally the most destructive element of an explosion. If the explosion is a detonation, the waves travel equally in all directions away from the center of the explosion.

The negative phase occurs right after the positive phase stops. A partial vacuum is created near the center of the explosion by all of the outward movement of air from the blast pressure. During the negative phase, the debris, smoke, and gases produced by the blast are drawn back toward the center of the explosion origin, then rise in a thermal column vertically into the air, and are eventually carried downwind by the air currents. The negative phase may last up to three times as long as the positive phase.

If the explosion is a result of an exothermic (heat producing) chemical reaction, the shock wave may be preceded by a high-temperature thermal wave that can ignite combustible materials. Some explosives are designed to disperse projectiles when the explosion occurs, e.g., anti-personnel munitions and hand grenades. Other explosives may be in metal or plastic containers to provide the necessary confinement for an explosion to occur. These containers may become projectiles when the explosion occurs. The projectiles from an explosion travel out equally in all directions from the blast center just as the blast pressure does. In order for the waves or projectiles coming off the explosion to travel equally in all directions, the velocity (speed) of the release of the high-pressure gases

## Examples of Overpressure Damage to Property

| | |
|---|---|
| 0.5– 1 psi | Window glass breakage |
| 1-2 psi | Buckling of corrugated steel and aluminum Wood siding and framing blown in |
| 2-3 psi | Shattering of concrete or cinder block walls |
| 3-4 psi | Steel panel building collapse |
| | Oil storage tank rupture |
| 5 psi | Wooden utility poles snapping failure |
| 7 psi | Overturning loaded rail cars |
| 7-8 psi | Brick walls shearing and flexure failures |

## Examples of Injuries and Death to Personnel

| | |
|---|---|
| 1 psi | Knockdown of personnel |
| 5 psi | Eardrum rupture |
| 15 psi | Lung damage |
| 35 psi | Threshold for fatalities |
| 50 psi | 50% fatalities |
| 65 psi | 99% fatalities |

**Figure 2.1**

must be supersonic, or faster than the speed of sound. This occurs only in a detonation, not in a deflagration.


## CATEGORIES OF EXPLOSIONS

According to the NFPA *Fire Protection Handbook*, there are two general categories of explosions, **physical** and **chemical**. In a physical explosion, the high-pressure gas is produced by mechanical means, i.e., even if chemicals are present in the container, they are not affected chemically by the explosion. In a chemical explosion, the high-pressure gas is generated by the chemical reaction that takes place.

### Mechanical Overpressure Explosion

The two major explosion types can be divided further into four sub-types that result in the release of high-pressure gas. The first occurs as the result of the physical overpressurization of a container, causing the container to burst, as in the case of a child's balloon bursting when too much air is placed in it. The container fails because it can no longer hold the pressure being built up inside. This can occur in containers that may not have pressure relief valves, or if the pressure-relief valve fails to operate. This overpressure does not have to occur as a result of filling the container. As heat is applied to a container from ambient temperature increases, or from radiant heat, the pressure increases inside the container. If this increase in pressure is not relieved, the container may fail.

## Mechanical/Chemical Explosion

The second type of explosion occurs via physical/chemical means, as in the case of a hot-water heater or boiler type of explosion. The water inside the heater turns to steam when overheated, which results in a pressure increase inside the container, and the container fails at its weakest point. This is by far the most common type of accidental explosion. This type of container failure can also occur in containers that hold liquefied compressed gases. As the temperature of the liquid in the container increases, so does the vapor content in the container; as the vapor content increases, so does the vapor pressure. This increase in temperature may be the result of increases in ambient temperature, radiant heat from a fire or other heat source, such as direct flame impingement on the pressure container. If the container has been damaged in an accident, it may fail at the point of damage, before the relief valve can function to relieve the pressure build-up. The tank can also be weakened by direct flame impingement on the vapor space of the container. This flame impingement weakens the metal, the tank can no longer hold the pressure, and tank failure occurs. When the heat from flame impingement is on the liquid level, it is absorbed by the liquid so the tank is not weakened. However, as the liquid absorbs the heat, more liquid is turned to vapor. This increase in vapor increases the pressure in the container. Pressure-relief valves are designed to relieve increased pressure caused by increases in ambient temperature or from radiant heat sources. The increase in pressure caused by direct flame impingement may overpower the relief valve and again the container may fail.

## Chemical Explosion

The third type of explosion involves a chemical reaction, the combustion of a gas mixture. Many of the explosive materials that are regulated and allowed to be transported by the DOT can cause this type of explosion. Therefore, most explosions and explosive materials in this hazard class involve materials that explode by chemical means, with the high pressure being created by a chemical reaction. This type of explosion is really nothing more than a rapidly burning fire. The components that allow this fire to burn rapidly enough to produce an explosion are the presence of a chemical oxidizer and confinement of the material. There are other chemicals offered for transportation and found in fixed facilities that are not classified as explosives, but can explode through chemical reaction.

**Explosive Properties of Common Grain Dusts**

| Type of Dust | Ign Temp | LEL |
|---|---|---|
| Wheat/Corn/Oats | 806°F | 55 |
| Wheat Flour | 380°F | 50 |
| Cornstarch | 734°F | 40 |
| Rice | 824°F | 50 |
| Soy Flour | 1004°F | 60 |

**Figure 2.2**

Responders must be aware of the fact that most chemicals have multiple hazards that may not be depicted by placards or information on shipping papers.

## Dust Explosion

There is a subgroup of potentially explosive materials that involve a chemical-reaction type of explosion (Figure 2.2), sometimes referred to as combustible dusts (Figure 2.3). These are, in many cases, ordinary combustible materials or other chemicals that, because of their physical size, have an increased surface area (Figure 2.4). This increased surface area exposes more of the particles to oxygen when they are suspended in air. When these materials are suspended in air, they can become explosive if an ignition source is present. One of the major facilities where dust explosions occur is grain elevators; explosions occur when grain dust is suspended in air in the presence of an ignition source. The primary danger area where the explosion is likely to occur within most elevators is the "leg," or the inclined conveyor, the mechanism within the elevator that moves the grain from the entry point to the storage point.

For a dust explosion to occur, five factors must be present: an ignition source, a fuel (the dust), oxygen, a mixing of the dust and the oxygen, and confinement. The explosion will not occur unless the dust is suspended in air within an enclosure at a concentration that is above its lower explosive limit (see Figure 2.3).

## EXPLOSIVE DUSTS

| | | |
|---|---|---|
| Coals | Crude Rubber | Peanut hulls |
| Soy Protein | Sugar | Aluminum |
| Cork | Cornstarch | Flour |
| Magnesium | Pea Flour | Titanium |
| Zirconium | Walnut Shell | Silicon |

**Figure 2.3**

**SURFACE AREA COMPARISON**

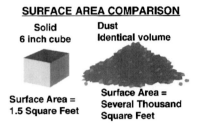

Figure 2.4

There are three phases in a dust explosion (Figure 2.5): initiation, primary explosion, and secondary explosion. Initiation occurs when an ignition source contacts a combustible dust that has been suspended in air. This causes the primary explosion, which shakes more dust loose from the confined area and suspends it in air. The secondary explosion then occurs, which is usually the larger of the two explosions because there is more fuel present.

Combustible dusts may be present in many different types of facilities. Common places for combustible dusts to be found are in grain elevators, flour mills, woodworking shops, and dry-bulk transport trucks. Dusts in facilities have caused many explosions over the years that have killed and injured employees. An explosion occurred in a facility on the East Coast that had many hazardous materials on site. At first, it was thought that one of the chemicals had exploded. The fire department and the HazMat team were called to the scene. Investigation revealed that the explosion occurred in a dust-collection system; it was a combustible-dust explosion. Dust explosions can be prevented by proper housekeeping and maintenance practices at facilities where these types of dusts are present.

## Nuclear Explosion

The fourth type of explosion is thermonuclear, which is the result of a tactical decision or a weapon that malfunctions. An *air burst* is designed to knock out all electronic equipment, disrupting communications and computer usage. This type of explosion does not create fallout, because it does not reach the ground and does not suck up debris in the negative phase of the explosion. A *ground burst* is designed for mass destruction of everything it contacts. Initially, during

**THREE PHASES OF A DUST EXPLOSION**

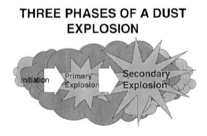

Figure 2.5

Dry-bulk truck transporting Class 1.5 blasting agent on the highway.

the positive phase of the blast, there is a thermal wave that is released first, followed by a shock wave. During the negative phase of the explosion, the debris from the explosion is drawn into the cloud, travels downwind, and then falls back to the ground. The debris, while in the cloud, is contaminated with radioactive particles, and radioactive fallout is created.

## COMPONENTS OF AN EXPLOSION

In simple terms, an explosive that functions via chemical reaction creates a rapidly burning fire that is made possible by the presence of a chemical oxidizer. Atmospheric oxygen does not provide enough oxygen for a chemical explosion to take place. Four components must be present for a chemical explosion to occur: fuel, heat (initiator, source of ignition), a chemical oxidizer, and confinement of the materials (Figure 2.6). The materials themselves can provide the confinement. Note the similarity between the requirements for an explosion and the fire triangle. There are other types of explosions that produce high-pressure gases, some of which will be discussed later.

### Types of Explosives

Explosives can be divided into two primary groups, high explosives and low explosives, based upon the speed with which the chemical transformation takes place, usually expressed in feet per second. Low explosives change physical state from a solid to a gas rather slowly. The low explosive burns gradually over a somewhat sustained period of time. This action is typically used as a pushing

**REQUIREMENTS FOR A
CHEMICAL EXPLOSION**

FUEL

HEAT

CHEMICAL
OXIDIZER    CONFINEMENT

**Figure 2.6**

and shoving action on the object against which it is placed. The primary uses of low explosives are as propelling charges and for powder trains, such as in time fuses. Examples of low explosives include black powder and smokeless powder. High explosives change from a solid to a gas almost immediately, an action referred to as detonation. A high explosive is detonated by heat or shock, which sets up a detonating wave. This wave passes through the entire mass of explosive material instantly. This sudden creation of gases and the extremely rapid extension produces a shattering effect that can overcome great obstructions. Examples of high explosives are trinitrotoluene (TNT) and dynamite.

The DOT explosives hazard class is divided into six subclasses 1.1 to 1.6 (49 CFR 173.20) (Figure 2.7). Because of their potential danger, subclasses 1.1 to 1.3 require placarding of the highway transportation vehicle regardless of the quantity of explosives being carried (49 CFR 172.504, Table 1) (Figure 2.8). Railroad shipments must always be placarded in all hazard classes, regardless of the quantity shipped.

Subclasses 1.4 to 1.6 (49 CFR 172.504, Table 2) (Figure 2.9) fall under the 1001-pound rule (49 CFR 172.504 (c)(1)), which requires 1001 lbs. or more of explosives on the vehicle before a placard is required. This means there could be 1000 pounds or less of a 1.4 to 1.6 explosive and no placard would be required at all!

## Explosive Subclasses

| | |
|---|---|
| Explosive 1.1 | Mass Explosion Hazard |
| Explosive 1.2 | Projection Hazard |
| Explosive 1.3 | Fire, Minor Blast, Minor Projection Hazards |
| Explosive 1.4 | Device With Minor Explosion Hazard |
| Explosive 1.5 | Very Insensitive Explosives |
| Explosive 1.6 | Extremely Insensitive Explosive |

**Figure 2.7**

# TABLE 1 MATERIALS

**EXPLOSIVES 1.1-1.3**

**POISON GAS 2.3**

**POISON 6.1**

**DANGEROUS WHEN WET 4.3**

**RADIOACTIVE  YELLOW III**

**Figure 2.8**

In the explosives hazard class, next to the hazard subclass number on the placard there will be a letter, known as the compatibility group letter. 49 CFR 173.52, Tables 1 and 2, identify the procedures for assigning compatibility group numbers to shipments. The compatibility groups are used to prevent the increase in explosive hazard should certain explosives be stored or transported together. This information is designed for the shippers of the explosive materials and has little if any emergency response value. All explosives should be treated as if they were Class 1.1, because all explosives will explode under certain conditions. Responders are not likely to recognize when an explosive material is in a condition to cause an explosion. Therefore, responders should treat all explosives as if they were the worst type of explosive they might encounter.

## Forbidden Explosives for Transportation

In addition to the explosives that are approved for transportation by the DOT, there are forbidden explosives that are too unstable or dangerous to be transported. The following are some examples of explosives that are forbidden to be transported. **Nitrogen triiodide** (black unstable crystals) explodes at the slightest touch when dry. When handled, it is kept wet with ether. It is too sensitive to be used

### TABLE 2 MATERIALS

Explosives 1.4- 1.6

Compressed Gases 2.1 - 2.2

Flammable Liquids 3

Flammable Solids 4.1 - 4.2

Oxidizers  5.1 - 5.2

Poison 6.1 (Non-Inhalation Hazard)

Corrosive 8

**Figure 2.9**

as an explosive, because it cannot be stored, handled, or transported. Azides, such as lead azide and hydrazoic azide, are very unstable. **Lead azide,** is a severe explosion risk, and should be handled under water; it is also a primary detonating compound. **Hydrazoic acid** or **hydrogen azide** is a dangerous explosion risk when shocked or heated. Metal fulminates, such as **mercury fulminate,** explode readily when dry. They are used in the manufacture of caps and detonators for producing explosions. Explosives that contain a chlorate along with an ammonium salt, or an acidic substance, including a salt of a weak base and a strong acid, are forbidden in transportation, e.g., **ammonium chlorate,** which is shock sensitive, can detonate when exposed to heat or vibration. It is used in the production of explosives. **Ammonium perchlorate,** is also shock sensitive, and may explode when exposed to heat or by spontaneous chemical reaction. This is the material that was involved in the explosion at the Pepcon plant in Henderson, NV. It is used in the production of explosives, pyrotechnics, etching and engraving, and jet and rocket propellants. Packages of explosives that are leaking, damaged, unstable, condemned, or contain deteriorated propellants are forbidden in transportation. Nitroglycerine, diethylene glycol dinitrate, or other liquid explosives are not authorized. **Diethylene glycol dinitrate** is a severe explosion hazard when shocked or heated. It is used as a plasticizer in solid rocket propellants. Other forbidden explosives include fireworks that combine an explosive and a detonator, fireworks that contain yellow or white phosphorus. Toy torpedoes exceeding 0.906-inch outside dimension, or containing a mixture of potassium chlorate, black antimony (antimony sulfide), and sulfur are prohibited if the weight of the explosive material in the device exceeds 0.01 oz. The Hazardous Materials Table in CFR 49 Part 172.101 lists all specific restricted explosives in various modes of transportation and those forbidden from shipment. Even though these materials are not transported, they may be found in fixed facilities of various types, such as research facilities, defense contractors, solid rocket fuel plants, explosives suppliers, and others.

## TYPES OF CHEMICAL EXPLOSIONS

**Detonation** is an instantaneous decomposition of the explosive material in which all of the solid material changes to a gas instantaneously with the release of high heat and pressure shock waves. Detonation is the only type of chemical explosion that will produce a true shock wave. A material that detonates is considered a high-yield explosive (Figure 2.10). Blast pressures can be as much as 700 tons/sq. in. Pressure and heat waves travel away from the center of the blast equally in all directions. The reaction occurs at supersonic speed or, in other words, faster than the speed of sound, which is 1250 ft/s. Many of the reactions occur between 3300 and 29,900 ft/s or over 20,300 miles/hr! Often the terms "explosion" and "detonation" are used interchangeably, which is not accurate. An explosion may very well be a detonation; however, an explosion can occur that is not a detonation. In either case, the explosions occur very rapidly, and the difference cannot be distinguished very well by the human senses. The only way

## CHARACTERISTICS OF HIGH AND LOW EXPLOSIVES

| | HIGH | LOW |
|---|---|---|
| Initiation Method | Primary by ignition<br>Secondary by detonation | By ignition |
| Conversion to Gas | Microseconds | Milliseconds |
| Consumption Velocity | 1 to 6 Miles/Second | Few inches to Feet<br>Per Second |
| Velocity of Flame Front | 1 to 6 Miles/Second | 1/3 to 1 Mile/Second |
| Pressure of Explosion | 50,000 to 4,000,000 psi | Up to 50,000 psi |

**Figure 2.10**

you can distinguish a detonation from a deflagration is by hearing the sound of the explosion. In a detonation, the explosion will be visualized, and the shock wave sent off before the explosion is actually heard. In a deflagration, the explosion will be heard almost immediately. The two terms only apply to the speeds of the explosions; they do not infer that one is any less dangerous than the other.

**Deflagration** is a very rapid autocombustion that occurs at a subsonic speed, less than 1250 ft/s. The solid material changes to a gas relatively slowly. A material that deflagrates is considered a low-yield explosive. A material that is designed to deflagrate may, however, under the right conditions produce a detonation. The explosion in Kansas City that killed six firefighters involved ammonium nitrate mixed with fuel oil. Ammonium nitrate is listed as a Class 1.5 insensitive explosive, designed to produce a deflagration. The ammonium nitrate–fuel oil mixture involved in Kansas City was on fire and, as a result, produced a detonation that may have been caused by the application of water by the firefighters.

## Explosive Effects

There are primary and secondary effects of explosions. The primary effects are blast pressure, thermal wave, and fragmentation. The blast pressure has two phases, the positive and negative. In the **positive phase,** the high-pressure gas, heat wave, and any projectiles travel outward. During the **negative phase,** a partial vacuum is produced, sucking materials back toward the area of origin. The next effect is fragmentation. Fragmentation may come from the container that held the explosive material and from materials in close proximity. Objects in the path of the explosion are broken into small parts by the force of the blast pressure, creating fragments. These fragments may have jagged or sharp edges. The fragments will travel away from the blast center at high speeds, greater than 2700 ft/s, faster than a speeding bullet! The final effect of an explosion is thermal, the generation of heat by the explosion. The amount of heat produced will depend on the type of material that is involved. There is a flash and a fireball associated

Explosives can be followed by fireballs, shock waves, and flying shrapnel.

with almost any explosion. The more rapid the explosion, the greater the effects that will be produced by the heat. Although the total heat produced may be similar in each explosion, a detonation will produce the most heat over a larger area, because of the speed of the explosion.

Secondary effects of an explosion are shock-wave modification, and fire and shock-wave transfer. There are three ways that a shock wave can be modified: it may be reflected, focused, or shielded. Reflection refers to the shock wave striking a solid surface and bouncing off. When a shock wave strikes a concave (curved) surface, the force of the shock wave is focused or concentrated on an object or small area once it bounces off the concave surface. This effect is similar to the principle behind satellite dishes. When a signal reaches a satellite dish from the satellite in space, the signal is focused on the electronic sensor protruding out of the front of the satellite dish. Shielding simply means that the shock wave encounters an object too substantial to be damaged by the wave so the shock wave goes around the object or is absorbed by it. The area immediately behind the object provides a place of shelter from the shock wave. Fire and shock-wave transfer involves the transfer of the shock wave energy and fire to other objects, causing fires and destruction.

## YIELD VS. ORDER

The yield of an explosive is associated with the rate or speed in which the explosion occurs. This is an indication whether an explosive will detonate or deflagrate. A high-yield explosive detonates, and the blast pressure shatters materials that it contacts. Examples of high-yield explosives are dynamite, TNT,

## COMMERCIAL EXPLOSIVES

Primary                                          Secondary

Mercury Fulminate                    Nitroglycerine
Lead Azide                               Ammonium Nitrate
Diazodinitrophenol                    Trinitrotoluene
Lead Styphnate                        Dinitrotoluene
Nitromannite                            Nitrostarch

**Figure 2.11**

nitroglycerine, detchord, C-3, C-4, and explosive bombs (Figure 2.11). A low-yield explosive deflagrates and is used to push and shove materials. Examples of low-yield explosives are black powder and commercial ammonium nitrate. Deflagrating materials are often used to move rocks in road construction, in quarries, and in mining.

Order has to do with the extent and rate of a detonation. A high-order detonation is one in which all of the explosive material is consumed in the explosion, and the explosion occurs at the proper rate. The proper rate in this case would be supersonic. A high-order detonation is one in which all of the explosive material is consumed at the proper rate. So a low-order explosion would occur as an incomplete detonation or at less than the desired rate. Yield involves the specific explosive material that is used, and order indicates the way in which the explosive detonated. The hazards to emergency responders are obvious. If an explosion is low order, not all of the explosive material has been consumed, and therefore the remaining material presents a hazard. Whether high or low yield, high or low order, all explosives should be treated as high-yield, high-order Class 1.1 explosives.

## EXPLOSIVES SUBCLASSES 1.1 TO 1.3

Division 1.1 (former Class A) explosives present a mass explosion hazard. They are sensitive to heat and shock and they may either detonate or deflagrate when they explode.

Division 1.2 (former Class A) explosives have a projection hazard, but not a mass explosion hazard.

Division 1.3 (former Class B) explosives have a fire hazard and either a minor blast hazard or a minor projection hazard or both, but not a mass explosion hazard.

### Nitro Compounds

There is one hydrocarbon derivative family that is classified as explosive, its primary hazard. However, there are some nitro compounds that have other primary hazards, such as nitrobenzene, which is a poison. This is an exception to the general hazard, and for safety purposes consider nitros explosive as a group. The nitro group is represented by a nitrogen covalently bonded to two oxygens. Nitrogen must have three connections to complete the octet rule of bonding. The oxygens have a single bond between themselves. This oxygen-to-oxygen single bond is very unstable and can come apart very explosively.

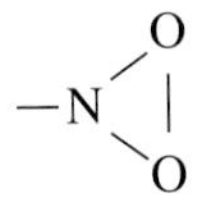

$$-N\begin{array}{c} O \\ | \\ O \end{array}$$

The other bonding spot on the nitrogen is attached to a hydrocarbon or hydrocarbon derivative backbone of some type. These backbones may include methane, benzene, toluene, or phenol, which is an alcohol, and others. When naming compounds from the nitro group, the word "nitro" is used first and the end is the hydrocarbon to which the nitro is attached. In the following example, one nitro functional group is attached to the hydrocarbon radical methane. The methane has had one hydrogen removed to create a place to attach the nitrogen on the nitro functional group. The compound is referred to as nitromethane.

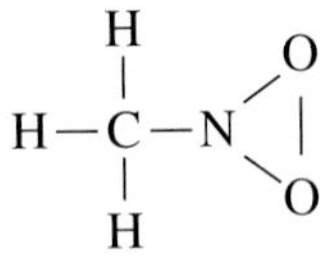

$$H-\underset{\underset{H}{|}}{\overset{\overset{H}{|}}{C}}-N\begin{array}{c} O \\ | \\ O \end{array}$$

Nitromethane
$CH_3NO_2$

**Nitromethane, $CH_3NO_2$,** is a colorless liquid that is soluble in water. The specific gravity is 1.13, which is heavier than water. Nitromethane is a dangerous fire and explosion risk, and is shock and heat sensitive. It may detonate from nearby explosions. The boiling point is 213°F, and the flash point is 95°F. The flammable range only lists a lower explosive limit, which is 7.3% in air; an upper limit has not been established. The ignition temperature is 785°F. Nitromethane may decompose explosively above 599°F if confined, and is a dangerous fire and explosion risk, as well as toxic by ingestion and inhalation. The threshold limit value (TLV) is 100 ppm in air. The four-digit UN identification number is 1261; the NFPA 704 designation is health 1, flammability 3, and reactivity 4. Nitromethane is used in drag racing to give the fuel in the engine an extra kick to increase speed. It is also used in polymers and rocket fuel.

Another example of a Division 1.1 to 1.3 explosive that is a nitro compound is **dynamite,** which is moderately sensitive to shock and heat and may also ignite when in contact with powerful oxidizing agents. It is primarily made up of nitroglycerin (straight dynamite) combined with a porous filler, such as sawdust as a desensitizer, or with special formulations made up of sensitized ammonium

Dynamite.

nitrate (ammonium gelatin dynamite) dispersed in carbonaceous materials. Dynamite is packaged in cylindrical cartridges, approximately 1 inch in diameter and 8 inch in length, enclosed in heavy, water-repellent paper. An indication that dynamite is old is the appearance of an oily substance on the casing of the cartridges or stains that appear on the wooden packing case. This can be attributed to the separation of the nitroglycerine from the porous base. Dynamite in this state is extremely sensitive. It should be destroyed immediately by explosives experts.

**Nitroglycerine, $CH_2NO_3CHNO_3CH_2NO_3$,** is a pale yellow, viscous liquid. It is slightly soluble in water with a specific gravity of 1.6, which is heavier than water. It is a severe explosion risk, and will explode spontaneously at 424°F. It is much less sensitive to shock when it is frozen. Nitroglycerine freezes at about 55°F. It is highly sensitive to shock and heat, and is toxic by ingestion, inhalation, and skin absorption. The TLV is 0.05 ppm in air. Nitroglycerine is forbidden in transportation unless sensitized. When in solution with alcohol at not more than 1% nitroglycerine, the four-digit UN identification number is 1204. When in solution with alcohol and more than 1%, but not more than 5% nitroglycerine, the four-digit UN identification number is 3064. The NFPA 704 designation for nitroglycerine is health 2, flammability 2, and reactivity 4. The primary uses are in explosives and dynamite manufacture, in medicine as a vasodilator, in combating oil-well fires, and as a rocket propellant. The structure for nitroglycerine follows. Notice that there are three nitro functional groups. These nitro groups were attached to the alcohol glycerol after three hydrogens were removed.

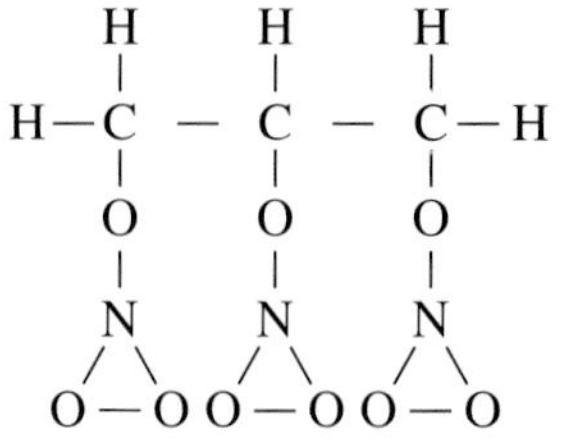

$$
\begin{array}{ccccccc}
 & H & & H & & H & \\
 & | & & | & & | & \\
H{-}C & {-} & C & {-} & C{-}H & \\
 & | & & | & & | & \\
 & O & & O & & O & \\
 & | & & | & & | & \\
 & N & & N & & N & \\
 & /\backslash & & /\backslash & & /\backslash & \\
O{-}O & & O{-}O & & O{-}O & 
\end{array}
$$

Nitroglycerine

$CH_2NO_3CHNO_3CH_2NO_3$

**Trinitrotoluene (TNT), $CH_3C_6H_2(NO_2)_3$,** is flammable, a dangerous fire risk, and a moderate explosion risk. It is light cream to rust in color, and is usually found in 0.5- or 1-lb. blocks. It is fairly stable in storage. TNT will detonate only if vigorously shocked or heated to 450°F; it is toxic by inhalation, ingestion, and skin absorption. The TLV is 0.5 mg/m³ of air. The four-digit UN identification number is 1356, when wetted with not less than 30% water. Other mixtures are listed in the hazardous materials tables with several ID numbers. TNT is one of the common ingredients used in military explosives and is used as a blast-effect measurement for other explosives. Trinitrotoluene is a member of the nitro hydro-carbon derivative family. In the following structure, toluene is the backbone for TNT. Three hydrogens are removed from the toluene ring and three nitro functional groups are attached.

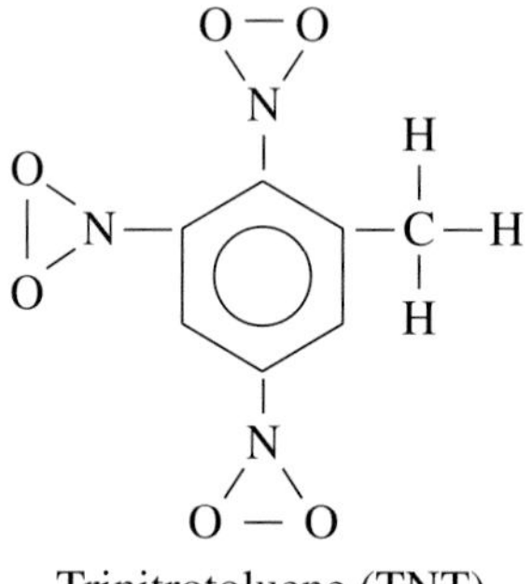

Trinitrotoluene (TNT)

$CH_3C_6H_2(NO_2)_3$

**Trinitrophenol (picric acid), $(NO_2)_3C_6H_2OH$,** composed of yellow crystals that are soluble in water. It is a high explosive, is shock and heat sensitive, and will explode spontaneously at 572°F. Trinitrophenol is reactive with metals or metallic salts and is toxic by skin absorption. The TLV is 0.1 mg/m³ of air. When shipped in 10 to 30% water, it is stable unless the water content drops below 10% or it dries out completely. The four-digit UN identification number is 1344 when shipped with not less than 10% water. The NFPA 704 designation is health 3, flammability 4, and reactivity 4. The primary uses are in explosives, matches,

Blasting caps used to detonate explosive materials.

electric batteries, etching copper, and textile dyeing. Picric acid is often found in chemical labs in high schools and colleges, and can be a severe explosion hazard if the moisture content of the container is gone. Picric acid was used by the Japanese during World War II as a main charge explosive filler. When in contact with metal, picric acid will form other picrates, which are extremely sensitive to

Dry, shock-sensitive picric acid on the shelf of a high school chemistry classroom.

heat, shock, and friction. Great care should be taken when handling World War II souvenirs, because of the possible presence of these picrates.

When the following structure of picric acid is compared with the structure of TNT, the only difference is the fuel that the nitro functional groups were placed on. The number of nitro groups is exactly the same. The explosive power of picric acid is very similar to that of TNT.

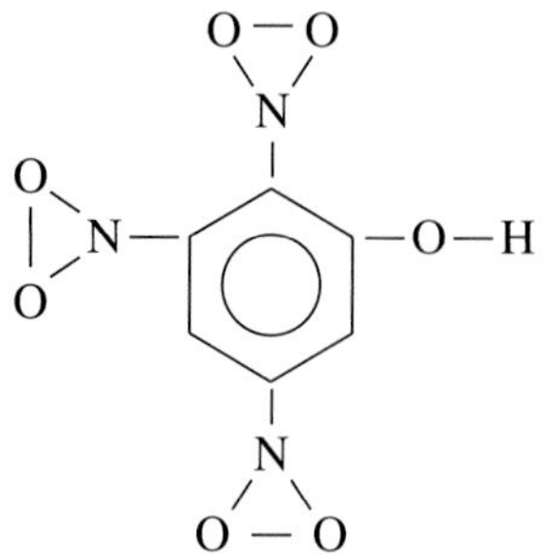

Trintrophenol (picric acid)
$(NO_2)_3C_6H_2OH$

There are other Class 1.1 to 1.3 materials that are nitro compounds and some that are made up of other chemicals. **Black powder** is a low-order explosive made up of a mixture of potassium or sodium nitrate, charcoal, and sulfur in 75, 15, and 10% proportions. It has an appearance of a fine powder to dense pellets, which may be black or have a grayish-black color. It is a dangerous fire and explosion risk, is sensitive to heat, and will deflagrate rapidly.

**Ammonium picrate,** $C_6H_2(NO_2)_3ONH_4$, is a high explosive when dry, and flammable when wet. It is composed of yellow crystals that are slightly soluble in water. The four-digit UN identification number is 1310, for ammonium picrate wetted with not less than 10% water. It is used in pyrotechnics and other explosive compounds. The structure and molecular formula for ammonium picrate are shown below. Notice the similarity to picric acid and TNT.

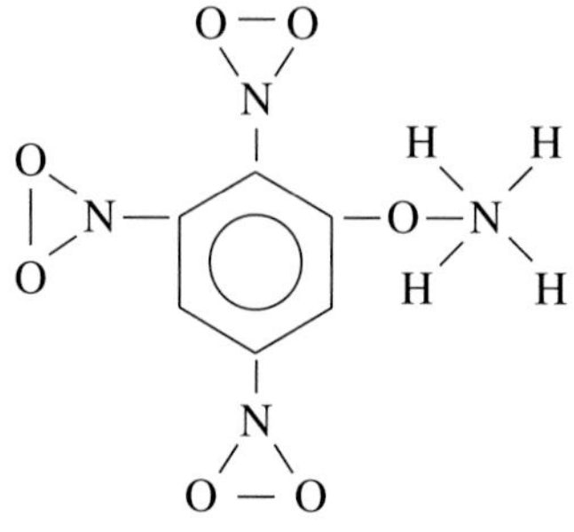

Ammonium picrate
$C_6H_2(NO_2)_3ONH_4$

## INCIDENTS

A number of accidents have occurred over the years involving 1.1 to 1.3 (former Class A and B explosive materials). In Waco, GA, an automobile collided with a truck carrying 25,414 lbs. of explosives, resulting in a fire and the explosion of the dynamite cargo. The fire started as a result of gasoline and diesel fuels spilling from the vehicles and then igniting. Heat transfer from the fires caused the nitroglycerine-based dynamite to detonate, killing two firefighters, a tow-truck driver, and two bystanders, injuring another 33 persons, and causing over $1 million in property damage.

The National Transportation Safety Board (NTSB) investigated the incident and, as part of their report, listed the following contributing factors that led to the injuries and deaths: (1) the lack of a workable system to warn everyone within the danger zone of an explosion, (2) the failure to notify emergency service personnel promptly and accurately of the hazards, (3) the decision of the fire-fighters to try and contain the hazardous fire, and (4) bystanders' disregard or lack of understanding of the truck driver's warnings.

Dynamite explosions also injured 13 people at Keystone, WV, and 9 at Lancaster, NY.

An explosion of up to a ton of smokeless powder, along with some black powder that was stored in the basement of a sporting goods store in Richmond, IN killed 41 persons, injured another 100, and caused over $2 million in property damage. Eleven buildings were destroyed by the explosion and four more from the resulting fires.

## EXPLOSIVES SUBCLASSES 1.4 TO 1.6

Division 1.4 (former Class C) is made up of explosives that present a minor explosion hazard. The explosive effects are largely confined to the package, and no projection of fragments of appreciable size or range is to be expected. An external fire must not cause virtually instantaneous explosion of almost the entire contents of the package.

Division 1.5 (former Blasting Agents) are considered very insensitive explosives by the DOT. This division comprises substances that have a mass explosion hazard, but are so insensitive that there is very little probability of initiation or of transition from burning to detonation under normal conditions of transport. (The probability of transition from burning to detonation is greater when large quantities are transported in a vessel or in storage facilities.)

Division 1.6 explosives are considered extremely insensitive by the DOT and do not have a mass explosion hazard. This division comprises articles that contain only extremely insensitive detonating substances and that demonstrate a negligible probability of accidental initiation or propagation. (The risk from articles of Division 1.6 is limited to the explosion of a single article.)

Most of the listings in 49 CFR 172.101 Hazardous Materials Tables for 1.4 to 1.6 explosives are for small premanufactured explosive devices: small arms ammunition, signal cartridges, tear gas cartridges, detonating cord, detonators for

Mixer truck used to transport ammonium nitrate and fuel oil and to mix into a blasting agent.

blasting, explosive pest control devices, fireworks, aerial flares, practice grenades, signals, railway track explosive smoke signals.

Recent changes in DOT rules now allow certain Class 1.1 and 1.2 explosives to be shipped in small quantities in special packages as Class 1.4. Normally Class 1.1 through 1.3 explosives must be placarded regardless of the quantity. The problem was that bomb squads and other explosives experts ordering small amounts of C-4, sheet explosives, DEXS caulk and slip boosters had to pay high shipping costs, as much as $1000 in some cases, to common carriers. Special packaging was designed and underwent thorough testing by the U.S. Department of Mines. The new DOT rules allow small amounts of explosives to be shipped by UPS, Federal Express, and other small-package shipping companies. Class 1.4 explosives do not require placarding until 1001 pounds or more are shipped.

One of the main chemical explosives listed from groups 1.4 to 1.6 is 1.5 ammonium nitrate. When mixed with fuel oil, it becomes a blasting agent. Subjected to confinement or high heat, it may explode, but does not readily detonate. Fertilizer-grade ammonium nitrate, which is a strong oxidizer above 33.5%, may also explode if it becomes contaminated. Fertilizer-grade ammonium nitrate was used in the bombings of the World Trade Center in New York City and the Federal Building in Oklahoma City.

## TOP 50 INDUSTRIAL CHEMICALS

**Ammonium nitrate** is the only explosive listed in the *Chemical Engineering News* Top 50 industrial chemicals (number 15, with 15.99 billion lbs. produced

Bomb trailer used to transport explosive devices and explosive materials found in populated areas.

in 1995), and it is classified as an oxidizer. It is a colorless or white to gray crystal that is soluble in water. It decomposes at 210°C releasing nitrous oxide gas. The four-digit UN identification number is 1942 with an organic coating, and 2067 as the fertilizer grade. There are a number of other mixtures of ammonium nitrate that have four-digit numbers; they can be found in the Hazardous Materials Tables, and in the DOT's *Emergency Response Guide*.

There are a number of other chemicals in the Top 50 that are not explosives, but have explosive potential under certain conditions. Oxygen (number 3, with 49.67 billion lbs. produced) is an oxidizer that causes organic materials to burn explosively. Chlorine, number 10, is also an oxidizer and may be explosive in contact with organic materials. There are three ethers in the Top 50: methyl *tert*-butyl ether, number 18, with 13.67 billion lbs.; ethylene oxide, number 26, with 6.78 billion lbs.; and propylene oxide, number 35, with 3.70 billion lbs. produced. Ethers may form explosive peroxides as they age. Butadiene, number 36, with 3.40 billion lbs. produced, may also form explosive peroxides when exposed to air.

## INCIDENTS

Several major incidents have occurred with ammonium nitrate: in Texas City, TX, 2280 tons of prilled ammonium nitrate (a commercial fertilizer) on board a ship in Galveston Harbor detonated after it caught fire. It is estimated that the blast caused 468 deaths, including 27 firefighters (the entire Texas City Fire Department), over 3000 injuries, and more than $50 million in property damage. The ammonium nitrate had been confined below deck on the ship and steam was injected into the sealed hold in an attempt to extinguish the fire. It is thought that the injection of steam into the confined ammonium nitrate caused the detonation to occur. The shock wave was felt as far away as Colorado. Before the incident was over, two more ships also exploded with results similar to the first explosion, compounding the emergency relief efforts.

In Kansas City, MO, commercial grade ammonium nitrate was inside a highway box trailer used for storage on a construction site. Someone set fire to a pickup truck on the property and the outside of the storage trailer. Firefighters responded to fight the fire. During the firefighting operations, the ammonium nitrate detonated, killing six firefighters and destroying their apparatus. There were no markings on the storage container indicating explosives were present. Because of this incident and others, the Occupational Safety and Health Administration (OSHA) now requires the use of the DOT placarding and labeling system for fixed-facility storage of hazardous materials. When they are transported and until the materials are used up or the containers purged, the placards and labels must remain on the containers.

Fire, thought to be caused by spontaneous heating following the blow-out of a tire, quickly spread to the cargo compartment of a 28-ft tractor-trailer truck

Heavily regulated storage bunkers for explosive materials.

## MILITARY EXPLOSIVES

### Primary

Mercury Fulminate
Lead Azide
Diazodinitrophenol
Lead Styphnate
Nitromannite

### Secondary

Nitroglycerine
Ammonium Picrate
Trinitrotoluene
RDX
Picric Acid

**Figure 2.12**

in Marshalls Creek, PA. The driver had disconnected the trailer off the roadway and had driven the tractor several miles to a service station. While the driver was gone, the tires ignited. The truck was carrying 4000 lbs. of 60% standard gelatin dynamite in boxes and 26,000 lbs. of nitrocarbonitrate blasting agent in 50-lb. bags. The compound is a mixture of 850 lbs. of ammonium nitrate fertilizer to 7 gal of No. 2 diesel fuel. Another passing tractor-trailer driver reported the fire to the Marshalls Creek Fire Department. The driver reported that there were no markings on the trailer. Three fire engines responded and an attack line was pulled to fight the blaze. As the firefighters approached the trailer, a detonation occurred. The fire had reached the explosive cargo and the resulting explosion killed six people, including three firefighters, the truck driver who reported the fire, and two bystanders. Property damage was over $600,000, including all three of the fire engines.

## MILITARY EXPLOSIVES

Military explosives are noted for their high shattering power accompanied by rapid detonation velocities (Figure 2.12). They must be very stable, because they are often kept in storage for long periods of time. Because of their intended use, they must detonate dependably after being stored and do so under a variety of conditions. The military explosives used most commonly are TNT, C-3, and C-4, and RDX cyclonite. These explosives release large quantities of toxic gases when they explode.

## SUMMARY

According to the NFPA *Fire Protection Handbook*, over 90% of explosives in the industrial world are used in mining operations, with the rest used in construction. When responding to fixed or transportation incidents in or around these types of operations, be on the look-out for explosives. When responding to transportation incidents, always consider the possibility of explosives being present.

Fire is the principal cause of accidents involving explosive materials. Look for explosive signs, such as placards and labels. Evacuate the area according to the distances listed in the *Emergency Response Guidebook* green section if the materials are not on fire, and in the orange section if they are. If no other evacuation information is available, a 2000-ft minimum distance should be observed, according to the NFPA *Fire Protection Handbook*. There is one rule of thumb in responding to incidents where explosives are involved: **DO NOT FIGHT FIRES IF THE FIRE HAS REACHED THE EXPLOSIVE CARGO.**

## REVIEW QUESTIONS

1. The two phases of an explosion are the _____ and _____.

2. There are five types of explosions: _____, _____, _____, _____, and _____.

3. Name the two types of chemical explosions: _____ and _____.

4. Name and draw structures and formulas as appropriate for the following nitro compounds:

   $CH_3NO_2$     Nitro propane     $C_6H_2(NO_2)_3CH_3$     Trinitrophenol

5. Name the four components necessary for a chemical explosion to occur: _____, _____, _____, and _____.

6. A chemical explosion is really nothing more than a rapidly burning fire. Which of the following make the rapid burning possible?
   A. High-yield explosives
   B. A chemical oxidizer
   C. The additive rapid burn
   D. Gasoline

7. A detonation occurs _____ than the speed of sound, and a deflagration occurs _____ than the speed of sound.

# 3        COMPRESSED GASES

Hazard Class 2 is composed of gases under pressure (Figure 3.1). The gases may also have other hazards, such as flammability, toxicity, and reactivity. Gases may be heavier or lighter than air; most gases are heavier. Seven lighter-than-air gases along with some common compressed gases are listed in Figure 3.1. The gases in this category may also be liquefied in order to ship larger quantities more economically.

Three terms are important in understanding the liquefaction of gases by pressure: critical point, critical temperature, and critical pressure. Critical point has to do with whether a gas will exist as a gas or as a liquid. Critical temperature and critical pressure come into play at the critical point to make this determination (Figure 3.2). Critical temperature is the maximum temperature that a liquid (in this case a liquefied gas) can be heated and still remain a liquid. For example, for butane the critical temperature is 305°F. As more heat is added, more of the liquid vaporizes. At the critical temperature, no amount of pressure can keep the liquid from turning into a gas. Critical pressure is the maximum pressure required to liquefy a gas that has been cooled to a temperature below its critical temperature. The critical pressure for butane is 525 psi. In order to liquefy any gas, it must be cooled to or below its critical temperature. For example, the critical temperature of butane is 305°F; at 305°F, it must be pressurized to 525 psi to become a liquefied compressed gas.

Liquefied gases have large liquid-to-gas expansion ratios (Figure 3.3), i.e., a very small amount of a liquid leaking from a container can form a very large gas cloud. Larger leaks will produce large vapor clouds. This increases the danger of flammability if an ignition source is present, and of asphyxiation or toxicity when vapor clouds form. Some liquefied gases, such as propane and butane, are ambient temperature liquids. They are shipped and stored in uninsulated tanks, and the temperature of the liquid inside is close to the ambient air temperature.

Cryogenic liquefied gases, such as hydrogen, are very cold liquids. The definition of a cryogenic liquid is any liquid with a boiling point below −130°F. Cryogenics will be discussed further in the Non-flammable gases section. When ambient temperatures are cold, the liquefied compressed gases will also be cold.

## COMMON CLASS 2 GASES

| Lighter Than Air | | Heavier Than Air | |
|---|---|---|---|
| Helium | He | Argon | Ar |
| *Acetylene | $C_2H_2$ | *Propane | $C_3H_8$ |
| *Hydrogen | $H_2$ | *Butane | $C_4H_{10}$ |
| *Ammonia | $NH_3$ | *Chlorine | $Cl_2$ |
| *Methane | $CH_4$ | Phosgene | $COCl_2$ |
| Nitrogen | $N_2$ | *Hydrogen Sulfide | $H_2S$ |
| *Ethylene | $C_2H_4$ | *Butadiene | $C_4H_6$ |
| *Flammable Gas | | | |

**Figure 3.1**

Water from the booster tanks on fire apparatus can be 70°F or higher. The water, if applied to the tank surface to cool it, could actually be heating the liquids rather than cooling them! Cryogenic liquids are already colder than the water at any temperature and the water will act as a super-heated material, causing the cryogenic to heat up and vaporize faster. This difference in temperatures can cause problems for responders unless the containers are handled properly. Care should be taken when applying water to "cool" containers.

The compressed gas class is divided into three subclasses: 2.1 gases are flammable, 2.2 gases are nonflammable, and 2.3 gases are poisons (formerly class

## CRITICAL TEMPERATURE AND PRESSURE

| Gas | Boiling Point | Critical Temp F° | Critical Pressure PSI |
|---|---|---|---|
| Ammonia | - 28 | 266 | 1691 |
| Butane | 31 | 306 | 555 |
| Carbon Dioxide | - 110 | 88 | 1073 |
| Hydrogen | - 422 | - 390 | 294 |
| Nitrogen | - 320 | - 231 | 485 |
| Oxygen | - 297 | - 180 | 735 |
| Propane | - 44 | 206 | 617 |

**Figure 3.2**

**LARGE EXPANSION RATIO - Liquid to Gas**

270 X
AND GREATER!

**Figure 3.3**

A poisons). Each of the compressed gas categories presents its own special hazards, in addition to the hazard of being under pressure, in a specially designed and regulated pressure container. Container pressures range from around 5 psi to as much as 6000 psi (Figure 3.4). The higher the pressure, the more substantial the container must be to contain the pressure; the higher the pressure, the greater the danger when the pressure is released or the container fails.

## FLAMMABLE GASES

Class 2.1 compressed gases are flammable. Flammable gases may be shipped and stored as liquefied gases, cryogenic liquids, or compressed gases. The DOT defines a flammable gas as "any material which is a gas at 68°F or less and 14.7 psi of pressure or above, which is ignitable when in a mixture of 13 percent or less by volume with air, or has a flammable range with air, of at least 12 percent, regardless of the lower limit."

Certain physical conditions must be present for flammable gases to burn. These physical characteristics are sometimes referred to as "parameters of combustion." All materials must be in the vapor state (gas state) in order to burn; solids and liquids do not burn. Subclass 2.1 materials are already gases; some of them may be found liquefied under pressure, others are pressurized or nonpressurized cryogenics and are shipped and stored in the liquid state. Many of these liquefied and cryogenic gases are already above their boiling point. The only

**CONTAINER PRESSURES**

| | |
|---|---|
| Atmospheric Pressure | 0 - 5 psi |
| Low Pressure | 5 - 100 psi |
| High Pressure | 100 - 3,000 psi |
| Ultra-high Pressure | 3,000 - 6,000 psi |

**Figure 3.4**

High-pressure gas pipeline marker.

thing keeping the compressed gases liquid is the pressure they are under. If the pressure is released from the container of a liquefied gas, such as when a boiling liquid expanding vapor explosion (BLEVE) occurs, all of the liquid turns into a gas instantly. This occurs because the liquid in the tank is already above its boiling point. Cryogenics are kept liquid by virtue of the fact that they are very cold and are shipped and stored in insulated containers. Cryogenics, on the other hand, are not usually under pressure in transportation. They may be pressurized or self-refrigerated in fixed storage. If a cryogenic liquid is spilled, the gas will vaporize, but only to the extent that it is heated and vaporized by the heat of the surrounding air. The vaporization will occur more slowly than liquefied compressed gases. A flammable vapor may ignite if an ignition source is present. Ignition must occur within the flammable range of that particular material.

## FLAMMABLE RANGE

Flammable range is represented by a scale, numbered from 0 to 100% (Figure 3.5). Flammable range is the point at which there is enough oxygen and fuel present for combustion to occur. There are two common terms applied to flammable range to express where the mixture is located within the range: upper explosive limit (UEL) and lower explosive limit (LEL). When a flammable gas is above the UEL, it is considered too rich to burn, which means there is plenty of fuel for combustion to occur, but not enough oxygen. When a flammable gas is below its LEL, there is plenty of oxygen for combustion to occur, but not enough fuel. For combustion to take place, a flammable gas must be between its upper and lower explosive limits, with just the right mixture of oxygen and fuel.

## FLAMMABLE RANGE
## (EXPLOSIVE LIMITS)

**The percentage of fuel in
the air that will burn.**

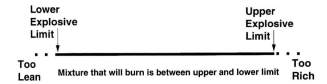

**Figure 3.5**

Different flammable gases have different flammable ranges (Figure 3.6). Most common hydrocarbon fuels have ranges from 1 to 13%. There are flammable materials that have wide flammable ranges, such as hydrogen gas, acetylene, ethers, aldehydes, and alcohols. The only way to determine if a flammable vapor is within its flammable range is to use monitoring instruments. Monitoring instruments check for a percentage of the lower explosive limit. The rule of thumb, according to the Environmental Protection Agency (EPA), is when you reach 10% of the lower explosive limit, it becomes too dangerous for personnel to proceed any further.

## VAPOR DENSITY

Another important physical characteristic of gases and vapors is vapor density. Vapor density is the relationship between the molecular weight of the gas and the air. The molecular weight of an average air molecule is 29 atomic mass units (AMU). Therefore, gases that have a molecular weight less than 29 AMU will be lighter than air, and those with a weight greater than 29 will be heavier. Air is given a value weight of 1 when compared to other gases; any gas that has a vapor density greater than 1, will be heavier than air. Propane has a vapor density of 1.56, therefore propane is heavier than air. If a gas has a vapor density less than 1, it will be lighter than air. Natural gas (methane) has a vapor density

### FLAMMABLE RANGES OF COMMON MATERIALS

| | |
|---|---|
| Acetylene | 2.5% - 80% |
| Ethyl Ether | 1.85% - 48% |
| Methyl Alcohol | 6% - 36.5% |
| Formaldehyde | 7% - 73% |
| Propane | 2.4% - 9.5% |
| Ammonia | 16% - 25% |
| Methane | 5% - 15% |

**Figure 3.6**

of 0.554, therefore natural gas is lighter than air. A propane explosion in a structure will cause damage near the foundation, because it is heavier than air and stays low to the ground. A natural gas explosion will cause damage in the upper part of the structure, because natural gas is lighter than air and will rise to the upper parts of the structure.

The DOT (Department of Transportation) has established criteria for a material to be classified as a flammable gas in transportation. First of all, its LEL must be below 13%. If it is, it is classified as a flammable gas. Some materials have wide flammable ranges that make them much more dangerous than materials with narrow ranges. The DOT says if a material has a flammable range greater than 12 percentage points, regardless of what the lower limit is, it is classified as a flammable gas.

Propane and butane are two very common flammable liquefied compressed gases. Propane has a flammable range of 2.4 to 9.5% in air, and butane 1.9 to 8.5% in air. Propane and butane have boiling points of –44°F and 31°F, respectively. Both materials are above their boiling points under ambient temperature conditions in many parts of the country year round. This makes the materials very dangerous when a leak or fire occurs, especially if there is flame impingement on the container. The vapor density of propane is 1.56 and of butane 2.07. Both propane and butane vapors are heavier than air. The vapors will tend to hug the ground and seek low places, such as basements and confined spaces.

Propane and butane are shipped on the highway in MC/DOT 331 uninsulated containers. On the rail, they are shipped in insulated pressure rail tank cars. The railcars are insulated to keep flame impingement from the tank surface. They are stored in uninsulated bulk pressure containers shaped much like the MC/DOT 331. They may also be found in varying sizes of portable pressure containers. Because the materials are already above their boiling points, flame impingement, radiant heat transfer, or increases in ambient temperature can cause the materials to boil faster. Faster boiling causes an increase in pressure within the container. Even though the containers are specially designed to withstand pressure and have relief valves provided to release excess pressure, there are limits to the pressure they can tolerate. If the pressure build-up in the container exceeds the ability of the tank to hold the pressure or of the relief valve to relieve the pressure, the container will fail.

**Hydrogen, H$_2$,** is a nonmetallic, diatomic, elemental flammable gas. Hydrogen is one of the most flammable materials known. It burns very clean, without smoke or visible flame. The only way to detect a hydrogen fire is from the radiant heat. Hydrogen may be found in transportation or storage as a cryogenic liquid (refrigerated liquid), a compressed gas, or a liquefied compressed gas. The flammable range of hydrogen is very wide at 4 to 75%, its boiling point is –423°F, its ignition temperature is 1075°F. Hydrogen gas is slightly soluble in water and is noncorrosive. It is an asphyxiant gas and can displace the oxygen in the air or in a confined space. There are traces of hydrogen in the atmosphere, and it is abundant in the sun and stars. It is the most abundant element in the universe. Hydrogen has a vapor density of 0.069; therefore it is lighter than air. As a cryogenic liquid, hydrogen has a four-digit UN identification number of 1066; as a compressed gas, its four-digit UN identification number is 1049. The NFPA 704 classification for hydrogen is health 3, flammability 4, and reactivity 0. It is usually shipped in insulated containers, insulated MC 338 tank trucks, tank cars, and tube trailers. Hydrogen is used in the production of ammonia, hydrogenation of vegetable oils, fuel for nuclear engines, hydrofining of petroleum, and cryogenic research.

## HYDROCARBON FAMILIES

Three of the hydrocarbon families of hazardous materials have compounds that are gases: alkanes, alkenes, and alkynes. Hydrocarbon families are compounds made up of carbon and hydrogen. Carbon has a unique ability to bond with itself almost indefinitely. Carbon forms compounds with varying numbers of carbons in chains with hydrogen filling the remaining bonds. The names of the compounds are based on the number of carbons bonded together in the compound and the type of bond between the carbons. All three of the families use the same prefixes to indicate the number of carbons. For example, a one-carbon prefix is "meth", two carbons "eth", three carbons "prop", and four carbons "but". The hypothetical number of carbons and prefixes is endless. For the purposes of this book, only the prefixes from 1- to 10-carbon chains will be considered. Those compounds that have more than four carbons are usually liquids and will be discussed in Chapter 4, Flammable Liquids. A listing of other prefixes above 10 carbons is located in the Appendix.

### Alkanes

The first hydrocarbon family to be considered is the alkanes. The alkane family has all single bonds between the carbons. The smallest alkane contains one carbon. The prefix for a one-carbon compound is "meth". The ending for the alkane family indicating all single bonds is "ane". Therefore, a one-carbon alkane is called methane, with a molecular formula of $CH_4$. A two-carbon compound has the prefix "eth" and the ending "ane", indicating the alkane family and all single bonds. The compound is called ethane, with a molecular formula of $C_2H_6$. A three-carbon compound has the prefix "prop" and the ending "ane", again indicating the alkane family and all single bonds. The compound is called propane, with a molecular formula of $C_3H_8$. A four-carbon compound is named butane, with the molecular formula of $C_4H_{10}$. The structures, molecular formulas, and some of the physical characteristics of these alkane compressed gas compounds follow. Notice the differences in boiling points and flammable ranges. As the carbon content increases, the boiling point of the compounds increases. In addition to carbon content, polarity and branching of compounds will affect the physical characteristics. This is also true of the flash point. This concept will be discussed in more detail in Chapter 4. For now, just be aware that there are relationships between the physical characteristics of flammable materials, which, in addition to boiling point and flash point, include ignition temperature, heat output, vapor content, and vapor pressure.

| Methane | Ethane | Propane | Butane |
|---------|--------|---------|--------|
| $CH_4$ | $C_2H_6$ | $C_3H_8$ | $C_4H_{10}$ |
| BP: $-259°$ F | BP: $-128°$ F | BP: $-44°$ F | BP: $31°$ F |
| FR: 5%–15% | FR: 3%–12.5% | FR: 2.1%–9.5% | FR: 1.9%–8.5% |

The alkane hydrocarbons occur naturally and their major hazard is flammability. The gases may also act as asphyxiants by displacing oxygen in the air. The alkanes are all considered to be saturated hydrocarbons because all of the single bonds are full. No other elements can be added to the compound without physically removing one or more of the hydrogens attached to the carbons. These hydrogens are removed, forming radicals, when the alkanes are used in the making of hydrocarbon derivatives.

The type of hydrocarbon family can be estimated from the molecular formula by looking at the ratio of carbons to hydrogens. In the case of the alkanes, there are twice as many hydrogens as carbons plus two hydrogens. For example, ethane has two carbons: $2 \times 2 = 4 + 2 = 6$, so there are two carbons and six hydrogens in ethane. Butane has four carbons: $4 \times 2 = 8 + 2 = 10$, so there are four carbons and ten hydrogens in butane. The molecular formula for butane is $C_4H_{10}$.

Fixed-facility tube bank with hydrogen gas.

## Isomers

Hydrocarbon structures may be altered so that their physical characteristics make them more economically valuable. One such alteration is called branching, or isomers. An **isomer** is a hydrocarbon that has the same molecular formula, i.e., the same number of hydrogens and carbons, but a different structure. The molecular formula for butane is $C_4H_{10}$; the molecular formula for the isomer of butane would be $i$-$C_4H_{10}$. The formula stays the same, but the structure is different. In order to determine the difference a prefix has to be added to the molecular formula and the name. When a structure is without a branch, it is sometimes referred to as normal or straight chained, i.e., all the carbons are connected together end to end in a chain. In both of the following examples, all of the carbons are connected together end to end. In the second example, the end carbon has been placed below the chain, but is still connected end to end with the rest of the carbons. So the second example is not a branch or an isomer, even though it might appear to have a "branch." The fact is it is just the same as the first example except for the arrangement of the carbons in the chain. To have a branched hydrocarbon, the branch cannot be on the end carbons, but must be on the carbons between the ends.

$$-C-C-C-C-C-C-C- \qquad \begin{matrix} -C-C-C-C-C-C \\ | \\ C \\ | \end{matrix}$$

The straight-chained hydrocarbon is sometimes referred to as the "normal" configuration. When a material is listed in a reference book you may see a small "n-" in front of the name or molecular formula; this indicates that it is the "normal" or straight-chained form of the compound. If the material is the isomer or branched form, a small "i" will be placed in front of the molecular formula and the prefix "iso" in front of the name. For example, butane can be normal, which would be written as normal butane or n-butane. The branched form of butane would be written isobutane. The molecular formula would be $C_4H_{10}$ or n-$C_4H_{10}$ for the normal, and i-$C_4H_{10}$ for the branched compound. Normal is not used all of the time with the normal form of the compound; it is found mostly in reference books and on laboratory containers. If no prefixes appear in the name, it is understood to be the normal form. However, "iso" or the small "i" must be used to designate the branched form of the compound. Without this designation there is no way to determine the branching from the name or the molecular formula.

The physical effect that branching has on a hydrocarbon is the lowering of the boiling point of that material. For example, butane is a liquefied compressed gas and has a boiling point of 31°F. One of the primary uses of butane is as a fuel for household and industrial purposes. Propane and butane tanks are usually located outdoors, and the liquefied gas takes on whatever the ambient temperature happens to be. In many parts of the country, the ambient temperature is below 31°F much of the winter. Because of the low ambient temperatures, butane would not be above its boiling point, and therefore would not be producing enough vapor to be used as a fuel. However, by changing the structure of butane and making it the branched isobutane, the boiling point becomes 10°F. Thus, with the lower boiling point, isobutane can be used as a fuel at lower ambient temperatures. The following is an illustration of the structures of normal butane and isobutane.

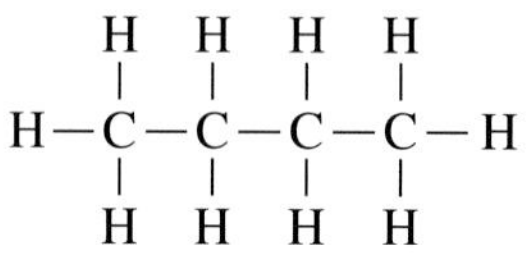

Butane
$C_4H_{10}$
BP: 31°F

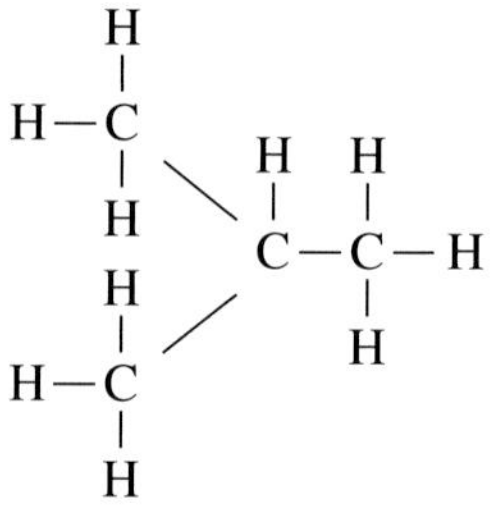

Isobutane
i-$C_4H_{10}$
BP: 11°F

One way of determining if the structure is branched or straight chained is to try to draw a line through all of the carbons connected together in a chain, without lifting the pencil or having to backtrack to reach another carbon. In the following examples the structure on the left is normal or straight chained because the line can be drawn through all of the carbons without backtracking. The structure on the right, however, requires lifting the pencil or backtracking to draw the line to the branched carbon. Therefore, the compound on the right is the isomer or branched compound.

Straight chained

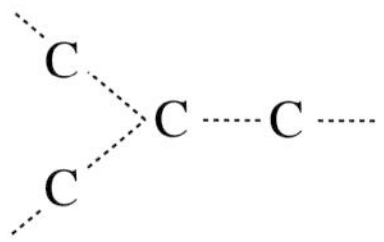

Branched

Branching can also occur in the alkenes and alkynes. In order for branching to occur in hydrocarbons, there must be at least a four-carbon compound. Propane cannot be branched until the hydrocarbon derivatives, when elements other than carbon and hydrogen are added to the structure of the compound. Other types of branching will be discussed in the Hydrocarbon Derivatives section of Chapter 4.

## Alkenes

The alkene family has one or more double bonds between the carbons in the chain. Alkene compounds do not occur naturally, they are man-made. They are considered to be unsaturated because the double bond can be broken by heat or oxygen. The double bond is actually out-of-plane electrons between the carbons.

The charges on the four electrons between the carbons are negative. Charges that are the same will repel each other. This forces the electrons out of plane and makes them vulnerable to the oxygen from the air. The following illustration shows the double bonds between two carbons. The structure on the left is how the double bond is usually represented in a structure. The structure on the right shows how the out-of-plane electrons really appear.

The double bond is
usually shown in
this manner.

This illustration more
correctly shows the
out-of-plane electrons.

When the double bonds are broken, heat is created and other elements, including atmospheric oxygen, attach to the compound. The major hazard of the alkenes is flammability. Some of the compounds may be toxic or irritants, and some are suspected of being carcinogenic. Double-bonded compounds are usually unstable and reactive. Oxygen from the air can react with the double bonds and break them, creating heat and forming other compounds. Alkenes are man-made and may also be used to make hydrocarbon derivatives. The same prefixes are used for the alkenes as for the alkanes. However, since there must be at least one double bond between two carbons to have an alkene, there are no single-carbon alkenes. The smallest alkene would be two carbons. The prefix for a two-carbon compound is "eth". The ending that indicates at least one double bond is "ene". Therefore, a two-carbon alkene is called ethene, with the molecular formula of $C_2H_4$.

Many times there is more than one way to name compounds in chemistry. In the case of the alkene family, sometimes a "yl" is inserted between the prefix and the ending "ene". Therefore, ethene may sometimes be called ethylene. A three-carbon compound has the prefix "prop" and the ending "ene". The compound is called propene, or propylene, with the molecular formula $C_3H_6$. A four-carbon compound would have the prefix "but" and the ending "ene". The compound is named butene, or butylene, with the molecular formula of $C_4H_8$. The structures for these alkene compounds follow:

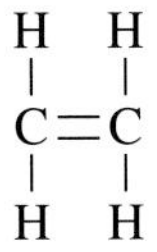

```
  H   H              H   H   H                H   H   H   H
  |   |              |   |   |                |   |   |   |
  C = C              C = C — C — H            C = C — C — C — H
  |   |              |       |                |       |   |
  H   H              H       H                H       H   H

  Ethene               Propene                    Butene
  Ethylene             Propylene                  Butylene
```

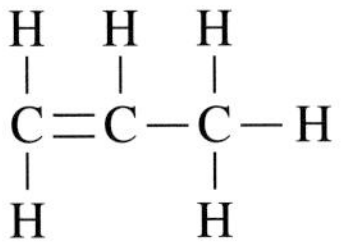

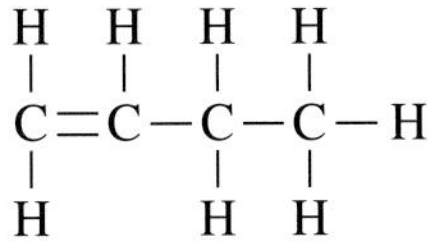

Some compounds of alkenes have more than one double bond in the structure. The naming of the prefix for the number of carbons is the same as with the other alkenes. The ending "ene" is still used to indicate double bonds and alkene family. There is, however, a prefix used to indicate the number of double bonds in the compound. The prefix "di" is inserted before the "ene" in the name to indicate two double bonds. For example, butene with two double bonds is called butadiene. The prefix "tri" is inserted to indicate three double bonds. Hexene with three double bonds is called hexatriene. There have to be at least four carbons before two double bonds are found. Two double bonds next to each other are very unstable and will not hold together long. Following are examples of two and three double bonds in compounds.

```
H   H   H   H              H   H   H   H   H   H
|   |   |   |              |   |   |   |   |   |
C = C - C = C              C = C - C = C - C = C
|           |              |                   |
H           H              H                   H
```

Butadiene                        Hexatriene

When trying to estimate the hydrocarbon family from the molecular formula, there is a ratio that indicates the alkenes. With the alkene family there are twice as many hydrogens as carbons. Propene has three carbons; $3 \times 2 = 6$, so there are three carbons and six hydrogens in propene ($C_3H_6$). In the case of alkenes that have more than one double bond, the ratio does not work. It is better to look at a molecular formula and draw out the structure rather than to try to guess from the ratio of the molecular formula. While the ratios work most of the time, there are exceptions, such as the two and three double-bonded compounds.

## Alkynes

The alkyne hydrocarbons have at least one triple bond between the carbons in the chain. The ending for the alkyne family is "yne". Alkynes are unsaturated, with the triple bond(s) being reactive to heat and the oxygen in the air. As with the alkene family, there are no one-carbon alkynes. The most commercially valuable alkyne is the two-carbon compound. The prefix for two carbons is "eth", so the chemical name for the two-carbon triple bond compound is ethyne, with the molecular formula of $C_2H_2$. This is probably the only alkyne that will ever be encountered by emergency responders. Ethyne, however, is known by the commercial name **acetylene.** This is a trade name, so it is not derived from any of the naming rules of the hydrocarbon families. Acetylene is a highly flammable, colorless gas with a flammable range of 2.5 to 80%. Pure acetylene is odorless; however, the ordinary commercial purity has a distinct garlic-like odor. Acetylene can be liquefied and solidified; however, both forms are very unstable. The vapor density is 0.91, so it is slightly lighter than air. Acetylene is produced when water is reacted with calcium carbide and other binary carbide salts. When these salts

Fixed-facility propane tank. Note condensation on side, indicating liquid level.

come in contact with water, acetylene gas is released. Acetylene is very unstable and burns very rich, with a smoky flame. The material may burn within its container. Acetylene is so unstable that it can detonate under pressure. It is dissolved in a solvent, such as acetone, to keep it stable within a specially designed container. The container has a "honeycomb mesh" of ceramic material inside to help keep the acetylene dissolved in the acetone. Acetylene is non-toxic and has no chronic harmful effects even in high concentrations. In fact, it has been used as an anesthetic. Like most gases, acetylene can be a simple asphyxiant if present in high concentrations that displace the oxygen in the air. The LEL of acetylene is reached well before asphyxiation can occur, and the danger of explosion is reached before any other health hazard is present. When fighting fires involving acetylene containers, the fire should be extinguished before closing the valve to the container. This is because the acetylene has such a wide flammable range that it can burn inside the container. Acetylene is incompatible with bromine, chlorine, fluorine, copper, silver, mercury, and their compounds. Acetylene has a four-digit UN identification number of 1001. The NFPA 704 designation is health 1, flammability 4, and reactivity 3. Reactivity is reduced to 2 when the acetylene is dissolved in acetone.

There are other alkyne compounds, but they do not have much commercial value, and will not be commonly encountered. A three-carbon compound with one triple bond has "prop" as a prefix for three carbons and is called propyne, with a molecular formula of $C_3H_4$. It is listed in the *Condensed Chemical Dictionary* as propyne, but you are referred to methylacetylene for information. It is listed as a dangerous fire risk, and it is toxic by inhalation. Propyne is used as a specialty fuel and as a chemical intermediate. A four-carbon alkyne has the prefix "but", and the compound is called butyne, with the molecular formula of $C_4H_6$. The chemical listing is under the name ethylacetylene, and it is designated as a dangerous fire risk. It is also used as a specialty fuel and as a chemical intermediate. Following are the structures for ethyne, propyne, and butyne:

$$H-C\equiv C-H \qquad H-\overset{\displaystyle H}{\underset{\displaystyle H}{C}}-C\equiv C-H \qquad H-\overset{\displaystyle H}{\underset{\displaystyle H}{C}}-C\equiv C-\overset{\displaystyle H}{\underset{\displaystyle H}{C}}-H$$

|     Ethyne     |     Propyne     |     Butyne     |
|:--------------:|:---------------:|:--------------:|
|    $C_2H_2$    |     $C_3H_4$    |    $C_4H_6$    |

While there are no commercially valuable two or three triple-bonded compounds, the same rules for naming them would apply as in the alkenes: the prefixes "di" for two and "tri" for three would be used.

There is a ratio of carbons to hydrogens that can be used to identify the compound from the molecular formula. With the alkyne family there are twice as many hydrogens as carbons –2. Ethyne has two carbons: $2 \times 2 = 4 - 2 = 2$. So there are two carbons and two hydrogens in the compound ethyne, with a molecular formula of $C_2H_2$.

## HYDROCARBON DERIVATIVES

There are some hydrocarbon derivative functional groups that have flammable gas compounds in their families. Alkyl halides are listed with toxicity as a primary hazard. However, there are some flammable alkyl halides. Vinyl chloride and methyl chloride are alkyl halides. Vinyl chloride is listed as one of the Top 50 industrial chemicals and will be detailed there. Vinyl chloride and methyl chloride are extremely flammable gases. The amines are also primarily toxic as a group; there are, however, some flammable amine gases. Methylamine, dimethylamine, ethylamine, and propylamine are all flammable gases. There are a few ether flammable gases, most of which do not use the trivial naming system and may not be recognized as ethers. For example, propylene oxide is an ether that is in the Top 50 industrial chemicals. Methyl ether may be found as a compressed gas or a liquid. In the aldehyde family, most compounds are liquids, except for the one-carbon aldehyde, formaldehyde. Formaldehyde is also one of the Top 50 industrial chemicals and is detailed in that section. These hydrocarbon derivative functional groups will be discussed in detail in Chapter 4, Flammable Liquids.

**Methyl chloride, $CH_3Cl$,** is an alkyl halide hydrocarbon derivative. It is a colorless compressed gas or liquid with a faintly sweet, ether-like odor. It is a dangerous fire risk, with a flammable range of 10.7 to 17% in air. The critical temperature is approximately 225°F, and the critical pressure is 970 psi. It is slightly soluble in water. The vapor density is 1.8, which is heavier than air. The

Highway transportation tube trailer with compressed hydrogen gas.

boiling point is −11°F, and the flash point is 32°F. The ignition temperature is 1170°F. It is a narcotic, producing psychic effects. The TLV (threshold limit value) is 50 ppm in air. The four-digit UN identification number is 1063. The NFPA 704 designation is health 1, flammability 4, and reactivity 0. The primary uses are as a catalyst in low-temperature polymerization, a refrigerant, a low-temperature solvent, an herbicide, and a topical anesthetic. The structure for methyl chloride is shown in the following illustration:

$$
\begin{array}{c}
H \\
| \\
H-C-Cl \\
| \\
H
\end{array}
$$

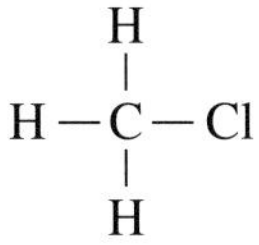

Methyl chloride
$CH_3Cl$

**Dimethylamine, $(CH_3)_2NH$,** an amine hydrocarbon derivative, is a gas with an ammonia-like odor. It is a dangerous fire risk, with a flammable range of 2.8 to 14% in air. It is insoluble in water. The vapor density is 1.55, which is heavier than air. The boiling point is 44°F, and the ignition temperature is 806°F. Dimethylamine is an irritant, with a TLV of 10 ppm in air. The four-digit UN identification number is 1032. The NFPA 704 designation is health 3, flammability 4, and reactivity 0. The primary uses are in electroplating and as gasoline stabilizers, pharmaceuticals, missile fuels, pesticides, and rocket propellants. The structure for dimethylamine is shown in the following illustration:

$$
\begin{array}{c}
\text{H} \\
| \\
\text{H}-\text{C}-\text{H} \\
\text{H} \quad | \\
| \quad | \\
\text{H}-\text{C}-\text{N}-\text{H} \\
| \\
\text{H}
\end{array}
$$

Dimethylamine
$(CH_3)_2NH$

## TOP 50 INDUSTRIAL CHEMICALS

There are five flammable gases in the Top 50 industrial chemicals: ethylene, propylene, formaldehyde, vinyl chloride, and butadiene. The fourth highest-volume industrial chemical is **ethylene, $C_2H_4$** (ethene), with 46.97 billion lbs. produced in 1995. Ethylene has a boiling point of –155°F and is a dangerous fire and explosion risk. The flammable range is fairly wide, with an LEL of 3% and an UEL of 36%. The vapor density is 0.975, which is slightly lighter than air. The critical pressure is 744 psi, and the critical temperature 9.5°C. Ethylene is not toxic, but can be an asphyxiant gas. The UN/DOT designation number for ethylene is 1962 as a compressed gas. The NFPA 704 designation is health 3, flammability 4, and reactivity 2. As a cryogenic liquid the UN designation number is 1038. The NFPA 704 designation is health 1, flammability 4, and reactivity 2. It is usually shipped in steel pressure cylinders and tank barges. Ethylene is used in the production of other chemicals, as a refrigerant, in welding and cutting of metals, as an anesthetic, and in orchard sprays to accelerate fruit ripening. The structure for ethylene is shown in the alkene section of this chapter.

**Propylene, $C_3H_6$** (propene), is the ninth most produced industrial chemical at 25.69 billion lbs. in 1995. Propylene has a boiling point of –53°F. The flammable range of propylene is 2 to 11%. The vapor density is 1.46, which is heavier than air. The four-digit UN identification number is 1077. The NFPA 704 designation is health 1, flammability 4, reactivity 1. It is not toxic, but can be an asphyxiant gas by displacing the oxygen in the air. It is usually shipped as a pressurized liquid in cylinders, tank cars, and tank barges. The structure for propylene is shown in the alkene section of this chapter.

**Vinyl chloride, $C_2H_3Cl$,** is the eighteenth most produced industrial chemical, with 14.98 billion lbs. in 1995. It is an alkyl halide hydrocarbon derivative. It is a compressed gas that is easily liquefied. Vinyl chloride is the most important vinyl monomer and may polymerize if exposed to heat. It has an ether-like odor, and phenol is added as an inhibitor during shipment and storage. It is highly flammable with a flash point of –108°F, and a boiling point of 7°F. The flammable range is 3.6 to 33% in air, with an ignition temperature of 882°F. Vinyl chloride is insoluble in water and has a specific gravity of 0.91, which is lighter than water. The vapor density is 2.16, which is heavier than air. It is toxic by inhalation,

MC 331 used to transport liquefied compressed gases.

ingestion, and skin absorption. Vinyl chloride is a known human carcinogen. The TLV is 5 ppm in air. The four-digit UN identification number is 1086. The NFPA 704 designation is health 2, flammability 4, and reactivity 2; uninhibited, the values would be higher for reactivity. The primary uses are in making polyvinyl chloride and as an additive in plastics. The structure for vinyl chloride is shown in the following example:

$$
\begin{array}{cc}
\text{H} & \text{H} \\
| & | \\
\text{C} & = \text{C} - \text{Cl} \\
| & \\
\text{H} &
\end{array}
$$

Vinyl chloride
$C_2H_3Cl$

**Formaldehyde, HCHO,** is the twenty-fourth most produced industrial chemical at 8.11 billion lbs. in 1995. It is an aldehyde hydrocarbon derivative. Formaldehyde is a gas with a strong, pungent odor. It readily polymerizes. Commercially it is offered as a 37 to 50% solution, which may contain up to 15% methanol to inhibit polymerization. The boiling points for the solutions range from 206 to 212°F. These commercial solutions have the trade name **formalin.** It is flammable, with a wide flammable range of 7 to 73% in air for formaldehyde gas. The boiling point is –3°F, and the flash point is 185°F. The ignition temperature for the gas is 572°F. It is water soluble. The vapor density is 1, which is the same as air. Formaldehyde is also toxic by inhalation, a strong irritant, and a carcinogen. The

TLV is 1 ppm in air. Nonflammable solutions are Class 9 miscellaneous hazardous materials with a four-digit UN identification number of 2209. The NFPA 704 designation is health 3, flammability 2, and reactivity 0. The flammable solutions are Class 3 flammable liquids with a four-digit UN identification number of 1198. The NFPA 704 designation is health 3, flammability 4, and reactivity 0. The following structure is for formaldehyde:

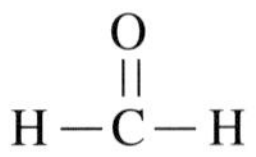

$$
\begin{array}{c}
O \\
\parallel \\
H-C-H
\end{array}
$$

Formaldehyde
HCHO

**Butadiene, $C_4H_6$,** is the thirty-sixth most produced chemical, with 3.68 billion lbs. in 1995. It has a boiling point of 24°F, and a flammable range of 2 to 11%. The vapor density is 1.93 psi, which means it is much heavier than air. It is highly flammable and may polymerize. Butadiene may form explosive peroxides in contact with air. It has a four-digit UN identification number of 1010. The NFPA 704 designation is health 2, flammability 4 and reactivity 2. Butadiene must be inhibited during transportation and storage. It is usually shipped in steel pressure cylinders, tank cars, and tank barges. The structure is shown in the alkene section of this chapter.

## INCIDENTS

There have been numerous incidents over the years involving pressure containers and flammable gases. One of the most recent incidents happened in August 1994 in White Plains, NY. A propane tanker crashed into a house and exploded, killing two people.

In Kingman, AZ, a 33,500-gal railroad tank car containing liquefied petroleum gas was being off-loaded at a gas distribution plant. As the liquid lines were attached to the tank, a leak was detected. During attempts by workers to tighten the fittings to stop the leak, a fire occurred. Both workers were severely burned and one later died from the injuries. Shortly after the fire department arrived, the liquid line failed and flame impingement began on the vapor space of the tank car. Nineteen minutes after the flames began contacting the surface of the tank car a BLEVE occurred. A large section of the 20-ton tank was propelled 1200 feet by the explosion. Railroad conductor Hank Graham, who took many of the now-famous Kingman photographs, had the hair burned off his arms as he took the pictures. Twelve Kingman firefighters were killed in the explosion. Only one firefighter from Kingman who was close to the scene survived. Ninety-five spectators on nearby Highway 66 were injured by the blast. The explosion set fire to lumber stored nearby, a tire company, and other businesses within 900 feet of the blast.

Kingman, AZ: railcar fire just before explosion. Note flame impingement on vapor space of tank. Courtesy of Hank Graham.

Kingman, AZ: propane explosion. Note pieces of tank car being propelled by the explosion. Courtesy of Hank Graham.

Waverly, TN: propane tank car UTLX 83013 after BLEVE. Courtesy of Waverly Fire Department.

Another quite different type of incident occurred in Waverly, TN, involving two 28,000-gal propane tank cars. A 24-car train derailment occurred just six blocks from the town square in this town of 6,000 residents. The derailment created a mass of piled railcars that damaged at least one of the tank cars of propane. Initially 50 to 100 people were evacuated within one quarter mile of the accident site as a precaution, although no leaks, fires, or explosions occurred when the train derailed. The accident occurred around 5:10 a.m. on a cold winter day. There were about 2 in. of snow on the ground, it was cloudy, and the temperatures had been below freezing for several days. The task of cleaning up the crash site began about mid-day on Wednesday. By Friday, the situation was felt to be well under control, security was relaxed, and people were allowed to return to their homes. Spectators and nonessential personnel were allowed into the immediate area of the derailment. Workmen smoked and used acetylene cutting torches at will in the process of cleaning up the site. A tank truck was brought in to off-load the damaged propane tank cars. Friday afternoon, the temperature began to rise, causing an increase in pressure inside the damaged tank. Normally the pressure-relief valve would function to release pressure created by increases in ambient temperature. However, in this instance, the damaged portion of the tank was weaker than the relief-valve pressure setting. At approximately 3:00 p.m., the tank could no longer withstand the increased pressure and opened up, releasing into the environment liquid propane, which instantly vaporized. The vapors quickly found an ignition source and flames shot 1000 feet into

Waverly, TN: aerial view of train derailment site after BLEVE, showing damage to structures from fires following explosion. Courtesy of Waverly Fire Department.

Waverley, TN: propane tank car UTLX 83013 before BLEVE. Courtesy of Waverly Fire Department.

Crescent City, IL: fireball from BLEVE of propane tank. Courtesy of Irma Hill.

the air. The giant fireball was visible for 30 miles around Waverly. The resulting explosion and fire killed 16 people, including the Waverly fire and police chiefs and five firefighters. Fifty-four people were injured, many severely burned.

Crescent City, IL, was the site of still another accident involving propane in transportation. Sixteen cars of an eastbound freight train derailed in the center of town at approximately 6:40 a.m., including 10 cars each containing 34,000 gal of liquid propane. Two additional propane tanks remained on the tracks. During the derailment, one of the propane tank cars was punctured by a coupler of another car, causing a leak that ignited almost immediately. Flames reached several hundred feet into the air. A nearby house and business were set on fire from the radiant heat, injuring several residents. Relief valves on the other tank cars began to open as the pressure built up from the surrounding fires. The first explosion (BLEVE) occurred around 7:33 a.m., almost one hour after the derailment. In that first blast, several firefighters and bystanders were injured and some fire equipment was damaged. Additional explosions occurred at 9:20, 9:30, 9:45, 9:55, and 10:10 a.m. Parts of tank cars were propelled all over town, setting fires and damaging structures. There were no injuries to civilians from the explosions, because of a quick evacuation after the derailment; 73 firefighters, police officers, and press personnel were injured by the explosions, but there were no fatalities. Twenty-four living quarters were destroyed by fire and three homes destroyed by "flying" tank cars; numerous other homes received damage. Eighteen businesses were destroyed. The remaining propane tanks were allowed to burn, which took some 56 hours after the derailment.

The city of Weyauwega, WI, population 1700, was evacuated for 20 days following a train derailment in the middle of town on March 4, 1996 at 5:55 a.m. Thirty-four cars were derailed, including seven containing LPG, seven propane, and two sodium hydroxide. The resulting fires from leaking propane damaged a feed mill and a storage building. There were no explosions as a result of the derailment and resulting fires. The incident commander, Jim Baehnman, Assistant Chief of the Weyauwega Fire Department, said that "the tone of the response from the beginning was not time driven, but rather safety driven." This may have very well accounted for the fact that not a single death or injury occurred as a direct result of this incident.

When dealing with emergencies involving pressure containers and flammable gases, great caution should be taken. Flame impingement on the vapor space of a container is a "no-win" situation. If a BLEVE is going to occur, it is just a matter of time. To try to fight a fire under those conditions is to play Russian roulette. The NFPA *Fire Protection Guide* says that BLEVE times range from 8 to 30 min with the average time being 15 min. There is usually no way to know how long the flame impingement has been going on prior to the fire department arrival, and no way to know exactly when the BLEVE will occur. If the only threat is to the lives of the emergency responders, there is little reason to risk their lives needlessly. If the impingement is on the liquid space, the liquid will absorb the heat for a period of time, and will boil faster as it does. There will be an increase in pressure within the tank as the liquid boils faster. This can still be a dangerous situation if not handled properly. Conditions involving the tank must

Weyauwega, WI: train derailment burning feed mill from propane fires. Photograph by Robert Ehrenberg of the Weyauwega Fire Department.

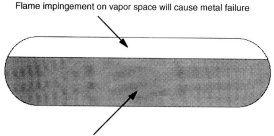

Flame impingement on vapor space will cause metal failure

The heat from flame impingement on the liquid level is absorbed
by the liquid causing increased vapor production but no metal failure

be monitored constantly for changes, including liquid level, pressure increases, and signs of tank failure.

Precautions should be taken to ensure personnel safety when fighting fires involving flammable gases. Flammable gas fires should not be extinguished until the source of the gas has been shut off. It is much safer to have the gas on fire and know where it is than to have the gas leaking and going where it wants to go. Flammable gases are more dangerous than flammable liquids. Flammable gases do not have a flash point. Many flammable gases have wide flammable ranges. Flammable gases can be heavier or lighter than air.

## NONFLAMMABLE GASES

Class 2.2 includes materials that are nonflammable, nonpoisonous, compressed gases. The materials in this class can be compressed gases, liquefied gases, pressurized cryogenic gases, and gases in solution. Though these materials are nonflammable, the containers can still BLEVE, under flame impingement conditions on the tank, from a fire involving other materials. BLEVE can also occur from a damaged or weakened container, or from an overpressure of the container caused by overfilling. Increases in pressure caused by increases in ambient temperature can also cause container failure. Pressure containers can be very dangerous under accident conditions; ambient temperature changes, flame impingement, and damage to containers can cause BLEVEs.

The DOT definition of a nonflammable gas is "a material that exerts in the packaging an absolute pressure of 41 psi or greater at 20°C, and does not meet the definition of Division 2.1 or 2.3." If the pressure is less than 41 psi, a gas does not belong in this category. Many cryogenic materials are shipped at atmospheric pressure and so are not considered compressed gases. Except when under pressure or some other hazard, cryogenics are not considered a DOT hazard class as a group. If they are shipped above 41 psi, cryogenics are considered compressed gas. Cryogenics may carry other placards, such as flammable gas, poison gas, or oxidizer. If cryogenics do not have any other hazard, they are not required to be placarded under DOT regulations. Materials under this class that are shipped as liquefied gases, such as cryogenics, exhibit other hazards not indicated by the

## BOILING POINTS OF CRYOGENIC LIQUIDS

| | | | |
|---|---|---|---|
| Helium | - 452 °F | Air | -318 °F |
| Neon | - 411 °F | Fluorine | -307 °F |
| Nitrogen | - 321 °F | Hydrogen | -423 °F |
| Argon | - 303 °F | Methane | -257 °F |
| Oxygen | - 297 °F | Nitric Oxide | -241 °F |
| Krypton | - 244 °F | CO | -312°F |
| Xenon | - 162 °F | $NF_3$ | -200°F |

**Figure 3.7**

placard. Liquefied refrigerated gases, such as cryogenics, are very cold materials; boiling points are –130°F or greater (Figure 3.7). Liquid helium has a boiling point of –452°F; it is the coldest material known. It is also the only material on earth that never exists as a solid, only as a cryogenic liquid and a gas.

Gases are liquefied into cryogenic liquids by a process of pressurization, cooling, and ultimate release of pressure. Therefore, they do not require pressure to keep them in the liquid state, unless they will be in the container for a long period of time; then they are pressurized. They are kept cold by the temperature of the liquid and the insulation in the tanks. The cryogenic liquefaction process begins when gases are placed into a large processing container. They are pressurized to 1500 psi. The process of pressurizing a gas causes the molecules to move faster, causing more collisions with each other and the walls of the container. This causes heat to be generated. The top of a self-contained breathing apparatus (SCBA) bottle or an oxygen bottle feels hot when being filled. Once the pressure of 1500 psi is reached, the material is cooled to 32°F by using ice water. Once cooled, the pressure is once again increased, this time up to 2000 psi, again with an increase in temperature. The material is then cooled to –40°F with liquid ammonia. Once the material is cooled, all of the pressure is released and the resulting temperature decrease turns the gases into cryogenic liquids.

Many of the gases found on the Periodic Table are extracted from the air and turned into cryogenic liquids. These include neon, argon, krypton, xenon, oxygen, and nitrogen. All but oxygen are considered inert, i.e., they are nontoxic, nonflammable, and nonreactive. To extract these materials from the air, the air is first turned into a cryogenic liquid. Then the liquid is run through a type of distillation tower, where each component gas is extracted as it reaches its boiling point. Those gases are then liquefied by the same process mentioned above. Oxygen, though nontoxic, is very reactive with hydrocarbon-based materials and is an oxidizer. Liquid oxygen in contact with an asphalt surface, such as a parking lot or highway, can create a contact explosive; dropping an object, driving or even walking on the area, can cause an explosion to occur.

**Helium, He,** is a gaseous, nonmetallic element from the noble gas family, family eight on the Periodic Table. Helium is a colorless, odorless, and tasteless gas. It is nonflammable, nontoxic, and nonreactive. Helium has a boiling point of –452°F. It is slightly soluble in water. Even though it is an inert gas, helium can still displace oxygen and cause asphyxiation. The vapor density is 0.1785, which is lighter than air. Helium is derived from natural gas by liquefaction of all other components. Helium has a four-digit UN identification number of 1046 as a compressed gas and 1963 as a cryogenic liquid.

**Neon, Ne,** is a gaseous, nonmetallic element from the noble gas family. It is colorless, odorless, and tasteless, and is present in the Earth's atmosphere at 0.0012% of normal air. It is nonflammable, nontoxic, and nonreactive, and does not form chemical compounds with any other chemicals. It is, however, an asphyxiant gas and will displace oxygen in the air. The boiling point of neon is –410°F. It is slightly soluble in water. Neon has a vapor density of 0.6964, which is lighter than air. The four-digit UN identification number is 1065 when compressed and 1913 as a cryogenic liquid. Its primary uses are in luminescent electric tubes and photoelectric bulbs. It is also used in high-voltage indicators, lasers (liquid), and cryogenic research.

**Argon, Ar,** is a gaseous, nonmetallic element of family eight. It is present in the Earth's atmosphere to 0.94% by volume. It is a colorless, odorless, and tasteless gas. It does not combine with any other chemicals to form compounds. The boiling point is –302°F. It is slightly soluble in water. The vapor density is 1.38, so it is heavier than air. The four-digit UN identification number is 1006 as a compressed gas and 1951 as a cryogenic liquid. It is used as an inert shield in arc welding, electric and specialized lightbulbs (neon, fluorescent, and sodium vapor), in Geiger-counter tubes, and lasers.

Fixed-facility cryogenic liquid oxygen container outside of a hospital with heat exchanger.

**Krypton, Kr,** is a gaseous, nonmetallic element of family eight. It is present in the Earth's atmosphere to 0.000108% by volume. It is a colorless, odorless gas. It is nonflammable, nontoxic, and nonreactive. It is, however, an asphyxiant gas and can displace oxygen in the air. At cryogenic temperatures, krypton exists as a white, crystalline substance with a melting point of 116°K. The boiling point of krypton is –243°F. Krypton is known to combine with fluorine at liquid nitrogen temperature by means of electric discharges or ionizing radiation to form $KrF_2$ or $KrF_4$. These materials decompose at room temperature. Krypton is slightly water soluble. The vapor density is 2.818, which is heavier than air. The four-digit UN identification number is 1056 for the compressed gas, and 1970 for the cryogenic liquid. It is used in incandescent bulbs, fluorescent light tubes, lasers, and high-speed photography.

**Xenon, Xe,** is a gaseous, nonmetallic element from family eight. It is a colorless, odorless gas or liquid. It is a nonflammable, nontoxic gas at standard temperatures and pressures, but is an asphyxiant and will displace oxygen in the air. The boiling point is –162°F, the vapor density is 05.987, which is heavier than air. It is chemically unreactive; however, it is not completely inert. The four-digit UN identification number is 2036 for the compressed gas and 2591 for the cryogenic liquid. Xenon is used in luminescent tubes, flash lamps in photography, lasers, and as an anesthesia.

**Xenon compounds.** Xenon combines with fluorine through a process of mixing the gases, heating in a nickel vessel to 400°C, and cooling. The resulting compound is xenon tetrafluoride, $XeF_4$, composed of large, colorless crystals. Compounds of xenon difluoride, $XeF_2$, and hexafluoride, $XeF_6$, can also be formed in a similar manner. The hexafluoride compound melts to a yellow liquid at 122°F and boils at 168°F. Xenon and fluorine compounds will also combine with oxygen to form oxytetrafluoride, $XeOF_4$, which is a volatile liquid at room temperature. These compounds with fluorine must be protected from moisture to prevent the formation of xenon trioxide, $XeO_3$, which is a dangerous explosive when dried out. The solution of xenon trioxide is a stable weak acid that is a strong oxidizing agent.

Cryogenic liquids have very large expansion ratios, some as much as 900 or more to 1 (Figure 3.8). Because of this expansion ratio, if the material is flammable or toxic these hazards are intensified because of the potential of large gas

## EXPANSION RATIOS FOR CRYOGENICS

| | |
|---|---|
| Argon | 841/1 |
| Ethane | 487/1 |
| Fluorine | 981/1 |
| Helium | 754/1 |
| Hydrogen | 840/1 |
| *LNG | 637/1 |
| Nitrogen | 697/1 |
| Oxygen | 862/1 |

*Liquefied Natural Gas

**Figure 3.8**

Trailer used to haul anhydrous ammonia to be used as a fertilizer in farming.

cloud production from a very small amount of liquid. As the size of the leak increases, so does the size of the vapor cloud. This means that 1 gal of a material can produce as much as 900 gal of a gas. Those materials that are not under pressure above 41 psi or are not flammable or poisonous are not required to display placards. These materials can still pose a serious danger to responders. Due to the large expansion ratios, these materials can displace oxygen in the air, which can harm responders by asphyxiation. Asphyxiation is not poisoning; it is simply not enough oxygen to breathe. Normal atmospheric oxygen content is about 21%. When the oxygen in the lungs and ultimately the blood is reduced, unoxygenated blood reaches the brain, and the brain shuts down. It may only be a few seconds between the first breath and collapse. Being very cold, these materials can cause frostbite and solidification of body parts. When the parts thaw out, the tissue is irreparably damaged.

Cryogenic liquids are shipped and stored in special containers. On the highway, the MC 338 tanker is used to transport cryogenic liquids. The tank is usually not pressurized, but is heavily insulated to keep the materials cold. There is a heat exchanger underneath the belly of the tank truck to facilitate the off-loading of product as a gas. Railcars are also specially designed to keep the cryogenic liquids cold inside the containers to minimize the boiling off of the gas. Fixed storage containers of cryogenic liquids are usually very tall, small-diameter tanks. These are insulated and resemble large vacuum bottles that keep the liquid cold. These containers are also under pressure to keep the material liquefied. The heat exchanger is a series of metal tubes with fins around the outside. The liquid runs through the tubes, is warmed, and turns into a gas. The gas is then used in the facility for whatever purpose it was intended. Other gases, such as hydrogen, are liquefied, sometimes made into cryogenics, and placarded as flammable gases. Liquid oxygen is placarded as an oxidizer or nonflammable compressed gas.

There are some "foolers" in the nonflammable gases. Anhydrous ammonia, for example, is regulated by the DOT as a nonflammable compressed gas. This is the only country in the world that placards anhydrous ammonia in this manner. This placarding is a result of lobbying efforts by the agricultural fertilizer industry. Everywhere else, anhydrous ammonia is placarded as a poison gas, not to mention that it is also flammable under the right conditions. If it is inside a building or in a confined space, it may very well be within its flammable range and burn if an ignition source is present. If anhydrous ammonia met the DOT definition of a flammable gas, which is a LEL of less than 13% and/or a flammable range of greater than 12 percentage points, it would be placarded as such. It does not, however, meet the DOT definition; it has a LEL of 16% and a flammable range of 16 to 25%.

## HYDROCARBON DERIVATIVES

There are a few hydrocarbon derivatives from the alkyl halide family that are 2.2 nonflammable compressed gases. This illustrates the wide range of hazards of the alkyl halides as a group. Some are flammable, some are toxic, and some are nonflammable and nontoxic. They can still act as asphyxiants and displace the oxygen in the air. It is important to remember that the primary hazard of the alkyl halides is toxicity. Some of them are also flammable; therefore, all must be assumed to be toxic and flammable until the individual chemical is researched. It is interesting to note that while the DOT lists tetrafluoromethane as a nonflammable, nonpoisonous gas, the *Condensed Chemical Dictionary* lists the compound as toxic by inhalation. The *NIOSH Pocket Guide to Chemical Hazards* does not list the compound. The best source of information on this compound and others may be the MSDS (material safety data sheet). Examples of nonflammable Class 2.2 alkyl halides are tetrafluoromethane and trifluoromethane.

**Tetrafluoromethane, CF$_4$,** also known as carbon tetrafluoride and fluorocarbon 14, is a colorless gas that is slightly soluble in water. It is nonflammable, but is listed as toxic by inhalation by the *Condensed Chemical Dictionary*. The four-digit UN identification number is 3159. The primary uses are as a refrigerant and gaseous insulator. The structure is illustrated in the following example:

Tetrafluoromethane
CF$_4$

**Trifluoromethane, $CHF_3$,** also known as fluoroform, propellant 23, and refrigerant 23, is a colorless gas that is nonflammable. There are no hazards listed for trifluoromethane. It may be an asphyxiant gas and displace oxygen in the air and in confined spaces. The four-digit UN identification number is 2035 for the compressed gas and 3136 for the cryogenic liquid. The primary uses for trifluoromethane are as a refrigerant, a direct coolant for inferred detector cells, and a blowing agent for urethane foams. The structure is shown in the following illustration:

$$
\begin{array}{c}
F \\
| \\
F - C - F \\
| \\
H
\end{array}
$$

Trifluoromethane
$CHF_3$

## TOP 50 INDUSTRIAL CHEMICALS

**Nitrogen, $N_2$,** is a gaseous, nonmetallic element, the second most produced industrial chemical, with 68.04 billion lbs. in 1995. It is a colorless, odorless, tasteless gas that makes up 78% of the air that is breathed. The boiling point of nitrogen is –320°F. It is slightly soluble in water. Nitrogen does not burn and is nontoxic. It may, however, displace oxygen and be an asphyxiant gas. The vapor density of nitrogen is 0.96737, which makes it slightly lighter than air. The four-digit UN identification number is 1066 as a compressed gas and 1977 as a cryogenic liquid. As a cryogenic liquid, the NFPA 704 designation is health 3, flammability 0, and reactivity 0. Nitrogen is used in the production of ammonia, cyanides, and explosives, as an inert purging agent, and as a component in fertilizers. It is usually shipped in insulated containers, insulated MC 338 tank trucks, and tank cars.

**Oxygen, $O_2$,** like nitrogen, is a nonmetallic elemental gas, the third largest volume industrial chemical, with 53.48 billion lbs. produced in 1995. Oxygen makes up approximately 21% of the air breathed. The boiling point of oxygen is –297°F. It is nonflammable, but supports combustion. Oxygen can explode when exposed to heat or organic materials. The vapor density of oxygen is 1.105, which makes it slightly heavier than air. Oxygen is incompatible with oils, grease, hydrogen, flammable liquids, solids, and gases. The four-digit UN identification number for oxygen is 1072 as a compressed gas and 1073 as a cryogenic liquid. The NFPA 704 designation for liquid oxygen is health 3, flammability 0, and reactivity 0. Liquid oxygen is shipped in Dewier flasks and MC 338 tank trucks. It may also be encountered in cryogenic railcars.

**Anhydrous ammonia, $NH_3$,** is the sixth highest volume industrial chemical produced, with 35.60 billion lbs. in 1995. Anhydrous means without water, so this type of material seeks water. This can be particularly dangerous to responders

MC 338 used for the transportation of cryogenic liquids.

because ammonia can seek water in the eyes, lungs, and other moist parts of the body. Ammonia is toxic, with a TLV of 25 ppm; the inhalation of concentrated fumes can be fatal above 2000 ppm. Ammonia odor is detectable at 1 to 50 ppm. It has a boiling point of –28°F and a flammable range of 16 to 25% in air, although it does not meet the DOT definition of a flammable gas. The ignition temperature is 1204°F. The vapor density of ammonia is 0.6819, which makes ammonia gas lighter than air. It is also water soluble. Hose streams can be used to control vapor clouds of ammonia gas. The run-off created, however, is ammonium hydroxide, which is corrosive so the run-off should be contained. Ammonia is incompatible with mercury, hydrogen fluoride, calcium hypochlorite, chlorine, and bromine. Mixing ammonia and chlorine will yield phosgene gas. This type of accident has occurred numerous times in toilet bowls where chlorine bleach is mixed with household ammonia. The four-digit UN identification number is 1005, and the NFPA 704 designation is health 3, flammability 1, and reactivity 0. Ammonia maybe shipped as a cryogenic liquid or a liquefied compressed gas. Ammonia is used as an agricultural fertilizer and as a coolant in cold storage buildings and food lockers. It is usually shipped in MC 331 tank trucks, railcars, barges, and steel cylinders.

Ammonia and propane are often shipped in the same type of container. In the late winter and early spring, the containers are purged of the propane and used for ammonia. In the late summer and early fall, the containers are purged of ammonia and used for propane. One of the primary uses of ammonia is as a fertilizer. The tanks used for farm application of ammonia can also be used for propane in the winter for heating and grain drying purposes. Ammonia will attack copper, zinc, and their alloys of brass and bronze. Propane valves, fittings, and

Dewier container of cryogenic nitrogen.

piping are often made of these materials. If the valves and other fixtures are not changed to steel before being used for ammonia, or if the ammonia is not completely purged from the containers before propane use, serious accidents can occur. The ammonia can damage the fittings and cause leaks of the very flammable propane, which may result in fires. There have also been situations where propane has been transported in ammonia containers and the placards for ammonia have been left on the containers.

**Carbon dioxide, $CO_2$,** is the twenty-second highest volume industrial chemical, with 10.89 billion lbs. produced in 1995. Carbon dioxide is a colorless, odorless gas. It can also be a solid (dry ice), which will undergo sublimation and turn back into carbon dioxide gas or a cryogenic liquid. It is miscible with water. It is not flammable or toxic, but can be an asphyxiant gas and displace oxygen. In 1993, two workers were killed aboard a cargo ship when a carbon dioxide fire-extinguishing system discharged. The oxygen in the area was displaced by the carbon dioxide and the men were asphyxiated. Carbon dioxide has a vapor density of 1.53, which is heavier than air. It may be shipped as a cryogenic or liquefied compressed gas. It has a four-digit UN number of 2187 as a cryogenic

Heat exchanger underneath MC 338 allows the cryogenic liquid to be off-loaded as a gas.

and 1013 as a compressed gas. The NFPA 704 designation is health 3, flammability 0, and reactivity 0. It is used primarily in carbonated beverages and fire-extinguishing systems.

## INCIDENTS

According to NFPA studies on ammonia incidents between 1929 and 1969, there were 36 incidents in which released ammonia gas was ignited; 28 resulted in a combustion explosion. All of the explosions occurred indoors. In Shreveport, LA, one firefighter was killed and one badly burned in a fire involving anhydrous ammonia. A leak developed inside a cold-storage plant. The firefighters donned Level A chemical protective clothing and went inside to try to stop the leak. Something caused a spark and the anhydrous ammonia caught fire.

In Verdigris, OK, a tank car was being filled with anhydrous ammonia. It was unknown that there was a weakened place on the tank car, which gave way from the pressure of the ammonia and resulted in a BLEVE. There was no fire, just a vapor cloud that traveled downwind, defoliating trees and turning other vegetation brown. One worker who was filling the tank car was killed in the incident.

In Delaware County, PA, ammonia was being removed from an abandoned cold-storage facility when a leak occurred. Firefighters were exposed to ammonia and complained of irritation and burning of the face and other exposed skin surfaces. Several were transported to local hospitals for treatment after going through decontamination at the scene.

In Ortanna, PA, two workers were killed while doing routine maintenance on an ammonia system in a cold-storage building used for storing fruit. One of the workers was an assistant chief with the local volunteer fire company. Several firefighters were injured by the ammonia vapors while trying to rescue the workers.

Anhydrous ammonia is a very common but dangerous material when not properly handled. Firefighters should wear full Level A chemical protective clothing when exposed to ammonia vapors. Be aware, however, that ammonia is also flammable, and Level A protective clothing provides little thermal protection. When responding to incidents involving ammonia, use caution. Firefighter turnouts do not provide adequate protection from ammonia vapors. If there are victims exposed to ammonia for any length of time, the chance of rescue is slim. Do not expose unprotected rescue personnel to ammonia vapor.

## POISON GASES

Subclass 2.3 materials are an inhalation hazard, and some may also be absorbed through the skin. Firefighters are exposed to certain types of toxic materials whenever they fight a fire. These toxic materials are byproducts of the combustion process. Toxic fire gases include carbon monoxide (CO), hydrogen chloride (HCl), hydrogen cyanide (HCN), sulfur dioxide ($SO_2$), nitrogen dioxide ($NO_2$), ammonia ($NH_3$), hydrogen sulfide ($H_2S$), and phosgene ($COCl_2$). These toxic materials kill thousands of persons each year. There are, however, poisons in every community that are potentially more dangerous, toxic materials capable of killing tens of thousands of people in a matter of minutes.

Ton containers of chlorine, similar to the ton containers used for mustard agents.

The DOT definition of poison gas is "a material that is a gas at 68°F or less at 14.7 psi and is so toxic to humans as to pose a hazard to health during transportation, or in the absence of adequate data on human toxicity, is presumed to be toxic to humans because when tested on laboratory animals it has an $LC_{50}$ value of not more than 5000 ml/m³." These materials are considered so toxic that, when transported, the vehicle must be placarded regardless of the quantity. The potential exists for 2.3 materials to affect large populations by creating toxic gas clouds.

In order to understand the toxic effects of poisons, it is necessary to know some toxicological terminology. One thing to remember about toxicological data is that the tests that were conducted to gain the information were not conducted on humans. The data are the result of tests on laboratory animals. The toxicity for humans is really nothing more than an educated guess. Most of the terms mentioned here are applied to workplace exposures. Acceptable exposures in many cases are 8 hours a day, 40 hours a week. The concentrations encountered on the scene of an incident will be much higher than any ordinary workplace exposure, but for a shorter period of time.

**TLV-TWA** is the threshold limit value-time weighted average concentration for a normal 8-hour workday and a 40-hour workweek, to which nearly all workers may be repeatedly exposed, day after day, without adverse effect.

**IDLH** is immediately dangerous to life and health. IDLH determines the highest concentration that a person can be exposed to for a maximum of 30 min and still escape without any irreversible health effects.

**$LC_{50}$** is the lethal concentration by inhalation for 50% of the laboratory animals tested.

**STEL** is the short-term exposure limit, defined as a 15-min TWA exposure, which should not be exceeded at any time during a workday even if the 8-hour TWA is within the TLV-TWA. Exposures above the TLV-TWA up to the STEL should not be longer than 15 min and should not occur more than four times a day. There should be at least 60 min between successive exposures in this range.

**Concentration** is the amount of one substance found in a given volume of another substance. Depending on the materials involved, there are many different ways of expressing concentration. Two of the most common ways are ppm (parts per million) and milligrams per kilogram (mg/kg).

Toxicology will be discussed further in Chapter 7. Examples of some 2.3 poison gases are fluorine, chlorine, carbon monoxide, hydrogen sulfide, phosgene, phosphine, and chloropicrin–methyl bromide mixture.

**Carbon monoxide, CO,** is an odorless, colorless, tasteless gas that is toxic by inhalation, with a TLV of 50 ppm. A 1% concentration is lethal to adults if inhaled for 1 min. Carbon monoxide binds to the blood hemoglobin 220 times tighter than oxygen. The more carbon monoxide that binds to the blood, the less oxygen can be carried. In addition to its primary hazard of toxicity, it is also highly flammable and is a dangerous fire and explosion risk. The boiling point is −313°F, and the ignition temperature is 1292°F. Carbon monoxide has a flammable range from 12 to 75%. Its vapor density is 0.967, which is slightly less than that of air. It is slightly soluble in water. Carbon monoxide will prevent oxygen from being taken into the blood, thus causing a type of chemical asphyxiation. The four-digit UN identification number is 1016 as a compressed gas and 9202 as a cryogenic liquid. The NFPA 704 designation for carbon monoxide is health 3, flammability 4, and reactivity 0. The primary uses are in the synthesis of organic compounds, such as aldehydes, acrylates, alcohols, and in metallurgy.

**Hydrogen sulfide, $H_2S$,** is a colorless gas, with an odor like rotten egg. It is the only common material that can halt respiration. It is toxic by inhalation with a TLV of 10 ppm. The minimal perceptible odor is found at concentrations of 0.13 ppm; at 4.60 ppm, the odor is moderate; at 10 ppm, tearing begins; at 27 ppm, there is a strong, unpleasant but not intolerable odor. When TLV reaches 100 ppm, coughing begins, eye irritation occurs, and loss of sense of smell begins after 2 to 15 min. Marked eye irritation occurs at 200 to 300 ppm and respiratory irritation after 1 hour of exposure. Loss of consciousness takes place at 500 to 700 ppm with the possibility of death in 30 min to 1 hour. Concentrations of 700 to 1000 ppm cause unconsciousness, cessation of respiration, and death. Instant unconsciousness occurs at 1000 to 2000 ppm, with cessation of respiration and death in a few minutes. Death may occur even if the victim is removed to fresh air at once. Hydrogen sulfide is highly flammable and a dangerous fire and explosion risk. The boiling point is –76°F. The flammable range is 4.3 to 46%. The ignition temperature is 500°F. Its vapor density is 1.189, which is heavier than air. It is soluble in water. Hydrogen sulfide is incompatible with oxidizing gases and fuming nitric acid. The four-digit UN identification number is 1053. The NFPA 704 designation for hydrogen sulfide is health 4, flammability 4, and reactivity 0. It is used in the purification of hydrochloric and sulfuric acids and is a source of hydrogen and sulfur. Hydrogen sulfide is usually shipped in steel pressure cylinders.

**Fluorine, $F_2$,** is a nonmetallic elemental gas from the halogens, which is family seven on the Periodic Table. It is the most electronegative and powerful oxidizing agent known. It reacts vigorously with most oxidizible substances at room temperature, frequently causing combustion. Fluoride compounds form with all elements except helium, neon, and argon. Fluorine is a pale yellow gas with a pungent odor. It is nonflammable, but will support combustion because it is an oxidizer. The boiling point is –307°F. The vapor density is 1.31, which is heavier than air. Fluorine is water reactive. The primary hazard is toxicity; fluorine is toxic by inhalation and extremely irritating to tissue. The TLV is 1 ppm, and the IDLH is 25 ppm in air. Fluorine is incompatible with and should be isolated from everything! The four-digit UN identification number is 1045. The NFPA 704 designation is health 4, flammability 0, and reactivity 4. The white section at the bottom of the diamond contains a W with a slash through it, indicating water reactivity. Because of the strong reactivity with other materials, it is shipped in special steel containers. The primary uses are in the production of metallic and other fluorides, fluorocarbons, fluoridation of drinking water, and in toothpaste.

**Boron trifluoride, $BF_3$,** is a colorless gas with a vapor density of 2.34, which is heavier than air. It is water soluble and does not support combustion. It is also water reactive, toxic by inhalation, and corrosive to skin and tissue. The TLV is

Pressure railcar of chlorine.

1 ppm, and the IDLH is 100 ppm in air. The boiling point is −148°F. The four-digit UN identification number is 1008. The NFPA 704 designation is health 4, flammability 0, and reactivity 1. The primary uses are as a catalyst in organic synthesis, in instruments for measuring neutron intensity, in soldering fluxes, and in gas brazing.

**Dichlorosilane, $H_2SiCl_2$,** is a pyrophoric, water-reactive gas. It is flammable, with a wide flammable range of 4.1 to 99% in air. The boiling point is 47°F, and the flash point is −35°F. The ignition temperature is 136°F. The vapor density is 3.48, which is heavier than air. It is immiscible in water and very water reactive. Contact with water releases hydrogen chloride gas. It is toxic by inhalation and skin absorption. Hydrogen chloride causes severe eye and skin burns and is irritating to the skin, eyes, and respiratory system. The four-digit UN identification number is 2189. The NFPA 704 designation is health 4, flammability 4, and reactivity 2. The white area at the bottom of the diamond contains a W with a slash through it, indicating water reactivity. It is shipped in carbon steel cylinders.

**Phosgene, $COCl_2$,** is a clear to colorless gas or fuming liquid, with a strong stifling or musty hay-type odor. It is slightly soluble in water. The vapor density is 3.41, which is heavier than air. Phosgene is a strong irritant to the eyes, is highly toxic by inhalation, and may be fatal if inhaled. The TLV is 0.1 ppm, and the IDLH is 2 ppm in air. The boiling point is 46°F, and it is noncombustible. When carbon tetrachloride comes in contact with a hot surface, phosgene gas is evolved, which is one of the main reasons that carbon tetrachloride fire extinguishers are no longer approved. The four-digit UN identification number is 1076. The NFPA 704 designation is health 4, flammability 0, and reactivity 1. It is shipped in steel cylinders, special tank cars, and tank trucks. The primary uses are in organic synthesis, including isocyanates, polyurethane, and polycarbonate resins; in carbamate, organic carbonates, and chloroformates pesticides; and in herbicides. The structure for phosgene follows:

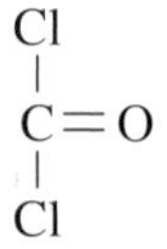

$$
\begin{array}{c}
Cl \\
| \\
C = O \\
| \\
Cl
\end{array}
$$

Phosgene
$COCl_2$

**Phosphine, $PH_3$,** a nonmetallic compound, is the gas evolved when binary phosphide salts come in contact with water. It is colorless, with a disagreeable, garlic-like or "decaying fish" odor. It is toxic by inhalation and is a strong irritant. It has a TLV of 0.3 ppm, and an IDLH of 200 ppm in air. It is also highly flammable (pyrophoric), and will spontaneously ignite in air. The flammable range is extremely wide at 1.6 to 98% in air. It is slightly soluble in cold water. The vapor density is 1.17, which is heavier than air. The four-digit UN identification number is 2199. The NFPA 704 designation is health 4, flammability 4, and reactivity 2. It is shipped in steel cylinders. The primary uses are in organic compounds, as a polymerization initiator, and as a synthetic dye. The structure follows:

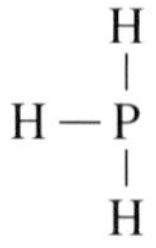

$$
\begin{array}{c}
H \\
| \\
H - P \\
| \\
H
\end{array}
$$

Phosphine
$PH_3$

**Diborane, $B_2H_6$,** is a colorless gas with a nauseating sweet odor. It decomposes in water and is highly reactive with oxidizing materials, including chlorine. It is toxic by inhalation and a strong irritant, with a TLV of 0.1 ppm in air. The IDLH is 40 ppm. In addition to being toxic, diborane is also a dangerous fire risk. It is pyrophoric and will ignite upon exposure to air. The boiling point is −135°F and the flammable range is 0.8 to 88% in air. The ignition temperature is 100° to 140°F, and the flash point is 130°F. Diborane will react violently with halogenated fire-extinguishing agents, such as the halons. The four-digit UN identification number is 1911. The NFPA 704 designation is health 4, flammability 4, and reactivity 3. The white section of the diamond has a W with a slash through it, indicating water reactivity. The primary uses are as a polymerization catalyst, fuel for air-breathing engines and rockets, a reducing agent, and a doping agent for p-type semiconductors.

$$
\begin{array}{ccc}
& H & H \\
& | & | \\
H- & B-B & -H \\
& | & | \\
& H & H \\
\end{array}
$$

Diborane

$B_2H_6$

Poison gases may be encountered as gases, liquefied gases, or cryogenics. The placard will indicate poison gas; it will not tell you the material has been liquefied or turned into a cryogenic liquid. The container type will help determine the physical state of the materials.

## TOP 50 INDUSTRIAL CHEMICALS

**Chlorine, Cl₂,** an elemental gas, is one of the most common poison gases transported and stored. Chlorine does not occur freely in nature. Chlorine is derived from the minerals halite (rock salt), sylvite, and carnallite, and is found as a chloride ion in sea water (Figure 3.9). It is the tenth highest volume industrial chemical, with 25.09 billion lbs. produced in 1995. Chlorine is nonflammable. Its vapor density is about 2.45, which makes it heavier than air. Chlorine is toxic, with a TLV of 0.5 ppm in air, and is also a very strong oxidizer. Chlorine will behave much the same way as oxygen in accelerating combustion during a fire. It is also corrosive. The liquid density of chlorine is 1.56, which makes it heavier than water, and it is only slightly soluble in cold water. Chlorine is incompatible with ammonia, petroleum gases, acetylene, butane, butadiene, hydrogen, sodium, benzene, and finely divided metals. The four-digit UN identification number is 1017. The NFPA 704 designation for chlorine is health 3, flammability 0, and reactivity 0. It may be encountered in 150-lb. cylinders, 1-ton containers, and rail tankcar quantities. Chlorine is nonflammable, but a BLEVE is possible if the container is exposed to flame because it is a liquefied gas. However, the likelihood of container failure is low. The NFPA has never recorded an incident of a BLEVE involving chlorine.

## <u>COMPONENTS OF SEAWATER</u>

| | |
|---|---|
| Oxygen | 91% |
| Hydrogen | 5.7% |
| Chlorine | 1.9% |
| Sodium | 1.1% |
| Others | 0.3% |

**Figure 3.9**

Level A chemical protection is necessary when dealing with poison gases.

Because chlorine is so common, its hazards are sometimes taken for granted. There was a time when firefighters handled chlorine leaks with turnouts and SCBAs (self-contained breathing apparatus); that is no longer an acceptable practice. Poison gases pose a large threat not only to the public, but also to emergency responders. To properly protect responders, full Level A chemical protective clothing and SCBAs must be worn for protection. Chlorine is used as a swimming pool chlorinator, a water-treatment chemical, and for many other industrial uses.

**Ethylene oxide, $CH_2OCH_2$,** is the twenty-sixth most produced industrial chemical, with 7.62 billion lbs. in 1995. It is a colorless gas at room temperature. It is miscible with water and has a specific gravity of 0.9, which is lighter than water. It is an irritant to the skin and eyes, with a TLV of 1 ppm in air. Ethylene oxide is a suspected human carcinogen. In addition to toxicity, it is highly flammable, with a wide flammable range of 3 to 100% in air. The flash point is –20°F and the boiling point is 120°F. The ignition temperature is 1058°F. The vapor density is 1.5, which is heavier than air. The four-digit UN identification number is 1040. The NFPA 704 designation is health 3, flammability 4, and reactivity 3. The primary uses are in the manufacture of ethylene glycol and acrylonitrile, as a fumigant, and as a rocket propellant.

$$
\begin{array}{cc}
\text{H} & \text{H} \\
| & | \\
\text{H}-\text{C}-\text{C}-\text{H} \\
\diagdown \diagup \\
\text{O}
\end{array}
$$

Ethylene oxide
$CH_2OCH_2$

## INCIDENTS

In Atlanta, a small pressurized cylinder fell from a truck in the garage of the Hilton Hotel. The resulting leak of chlorine, a 2.3 poison gas, sent 33 people to the hospital, including 6 firefighters and 4 police officers.

On December 3, 1984, Bhopal, India experienced a release of methyl isocyanate (MIC) at the Union Carbide pesticide plant. Thousands were injured, over 3000 people were killed, and many more are likely to die from the long-term effects. The accident occurred around 12:40 a.m. local time, when most of the victims were sleeping. The dead included large numbers of infants, children, and older men and women. These age groups are often adversely impacted by toxic exposures. It was this incident that led to the Emergency Planning and Community Right-To-Know Act of 1986 (EPCRA). This same chemical has been released on several occasions from the Union Carbide plant in Institute, WV, shortly after Bhopal and as recently as 1996. Fortunately, these releases did not affect the surrounding community.

Eight people were killed and 88 others were injured as the result of leaking chlorine from a railroad tank car during a derailment in Youngstown, FL. The liquid chlorine car ruptured, releasing a toxic cloud of chlorine. Chlorine is 2.5 times heavier than air, stays close to the ground, and has an expansion ratio of 460:1, which means that 1 gal of liquid chlorine will vaporize into 460 gal of chlorine gas. The chlorine gas settled into a low area on a nearby highway. As cars passed through the chlorine cloud, they stalled. Drivers were overcome by the chlorine and eight of them died. A few breaths of chlorine at 1000 ppm concentration can be fatal. The concentrations at the accident scene were estimated by environmental personnel to be 10,000 to 100,000 ppm.

## CHEMICAL RELEASE STATISTICS

Anhydrous ammonia is the second most released chemical from fixed facilities on the EPA listing, with 3586 accidents resulting in the release of over 19 million lbs. of ammonia. Anhydrous ammonia is consistently among the seven highest volume hazardous materials shipped on the railroads, accounting for over 59,000 carloads in 1994. In 1995, anhydrous ammonia was involved in 93 transportation incidents, including all modes except pipeline, according to the DOT. It is ranked thirty-sixth in number of incidents compared to all other hazardous materials, and twenty-second in terms of serious injuries during transportation releases.

Chlorine is the fourth most released chemical from fixed facilities on the EPA listing, with 2099 accidents resulting in the release of over 84 million lbs. of chlorine. Chlorine is also consistently among the seven highest volume hazardous materials shipped on the railroads, accounting for over 49,000 carloads in 1994. Chlorine is not listed among the Top 50 transportation incidents; however, it is number 16 in terms of total injuries. Chlorine is not often involved in transportation incidents, but when it is, it causes serious injuries.

Liquefied petroleum gases (LPG), which include propane, butane, isobutane propylene, and mixtures, are ranked twenty-sixth in number of incidents in all modes of transportation during 1995. LPG was involved in 123 transportation incidents. These gases, however, accounted for the largest number of serious injuries of all hazardous materials incidents and were second in the number of deaths. LPG is consistently one of the highest volume hazardous materials shipped by rail. In 1994, LPG was ranked number 2, with over 150,000 tank carloads being shipped. LPG shipments accounted for 22% of all tank car shipments by rail and 6% of the leaks.

## SUMMARY

Particular attention must be paid to compressed gases in emergency response situations. Compressed gases present responders with multiple hazards, including poisons, flammables, oxidizers, cryogenics, and the hazard of the pressure in the container. If the container fails or opens up, it can become a projectile or throw pieces of the container over a mile from the incident scene. Learn to recognize pressure containers and be very cautious when there is flame impingement on a pressure container.

## REVIEW QUESTIONS

1. Which of the following is true of cryogenic liquids?
   A. They have wide flammable ranges.
   B. They are very cold materials.
   C. They have large expansion ratios.
   D. They are atmospheric temperature materials.
   E. Both B and C.

2. Hazard Class 2 is composed of compressed gases that may have which of the following hazards?
   A. Flammability
   B. Elevated temperature
   C. Sublimation
   D. Air reactivity

3. The flammable range of a 2.1 compressed gas occurs between the _____ and the _____.

4. A compressed gas with a vapor density greater than 1 will be _____ than air.

5. Indicate whether the following formulas are alkanes, alkenes, or alkynes.

   $CH_4$     $C_2H_2$     $C_3H_6$     $C_2H_6$     $C_3H_4$

6. Provide the names and formulas for the compounds represented by the following structures.

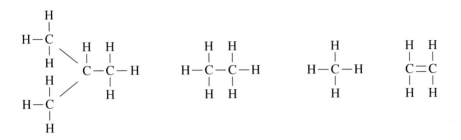

7. Match the following endings with the appropriate hydrocarbon family.

A. -ane          _____ Alkyne
B. -ene          _____ Alkane
C. -yne          _____ Not a hydrocarbon ending
D. -one          _____ Alkene

# 4 FLAMMABLE LIQUIDS

Class 3 materials are liquids that are flammable or combustible. Flammable liquids cause more fires than flammable gases because they are more abundant. The vapors of many flammable liquids are heavier than air. Most flammable liquids float in water. Flammable liquids may be incompatible with ammonium nitrate, chromic acid, hydrogen peroxide, sodium peroxide, nitric acid, and the halogens. According to the Department of Transportation (DOT), flammable liquids, "have a flash point of not more than 141°F, or any material in a liquid phase with a flash point at or above 100°F, that is intentionally heated and offered for transportation or transported at or above its flash point in bulk packaging." There is an exception to this definition that involves flammable liquids with a flash point between 100 and 140°F. Those liquids may be reclassified as combustible liquids, and the combustible or fuel oil placards may be used. Liquids between 100 and 140°F, as an option of the shipper, may be placarded flammable or combustible. Even though the DOT wanted all liquids up to 140° to be placarded flammable, the exception was made because of public comments, particularly from the fuel oil industry. Combustible liquids are defined as "materials that do not meet the definition of any other hazard class specified in the DOT flammable liquid regulations and have flash points above 141°F and below 200°F."

The National Fire Protection Association (NFPA) has a classification system for flammable and combustible liquids in fixed storage facilities (Figure 4.1). This system is part of the consensus standard NFPA 30, the Flammable and Combustible Liquids Code. The NFPA system further divides the flammable and combustible liquid categories into subdivisions based upon the flash points and boiling points of the liquids. The NFPA classification system does not apply to transportation of hazardous materials, since DOT regulations supersede NFPA 30. Examples of liquids in the various classification categories are listed in Figure 4.2.

As previously mentioned, all of the DOT hazard classes identify only the most severe hazard of that material. All of the classes have hidden hazards that are both chemical and physical in nature. Flammability is not the only hazard

## NFPA 30
## Flammable and Combustible Liquid
## Classification

### Flammable

Class IA = FP < 73° F - BP < 100° F
Class IB = FP < 73° F - BP > 100° F
Class IC = FP > 73° F - BP < 100° F

### Combustible

Class II = FP > 100° F < 140° F
Class IIIA = FP > 140° F < 200° F
Class IIIB = FP > 200° F

**Figure 4.1**

associated with Class 3 flammable liquids. They may also be poisons or corrosives. For all general purposes, there are no UN/DOT subclasses of flammable liquids. Emergency responders should realize that all materials with red placards will burn under certain conditions. Appropriate precautions should be taken when dealing with flammable liquids. The dividing line for flammable and combustible liquids is 140°F in the DOT regulations and 100°F in the NFPA standard. Those liquids with flash points below 100 and 140°F, respectively, are considered flammable; those above are considered combustible. The problem with classifying flammable and combustible liquids in emergency response situations is ambient temperature and radiant heat. Ambient temperatures near or above 100°F are common in many parts of the country. The radiant heat from the sun or an exposure fire can reach well above 100°F. Many of the liquids classified as combustible have flash points at or near 100°F. Surfaces such as roadways may have temperatures well above the flash points of combustible liquids. When a combustible liquid is spilled on a surface, it may be heated above its flash point and the combustible liquid will produce flammable vapor much like a flammable liquid. Because of the uncertainty of the potential flammability of combustible liquids, they will be referred to as flammable liquids throughout this book. It is highly recommended that they be treated the same way on the incident scene for the purpose of responder and public safety. Flash point will be discussed in detail

## NFPA Flammable and Combustible Liquids

| Class I | Class II | Class III |
|---------|----------|-----------|
| Acetone | Acetic Acid | Benzyl Chloride |
| Benzene | Fuel Oil | Corn Oil |
| Carbon Disulfide | Kerosene | Linseed Oil |
| Gasoline | Decane | Nitro Benzene |
| Methanol | Pentanol | Parathion |

**Figure 4.2**

later in this chapter. It is important, however, to note at this point that the flash point temperature is the most critical factor in determining if a flammable liquid will burn. The most important precautions are to control ignition sources at the incident scene and keep personnel from contacting the liquid without proper protective clothing.

According to DOT and Environmental Protection Agency (EPA) statistics, flammable liquids are involved in over 52% of all hazardous materials incidents. This should not be surprising since flammable liquids are used as motor fuels for highway vehicles, railroad locomotives, marine vessels, and aircraft. Additionally, many flammable liquids are used to heat homes and businesses. Effective handling of flammable liquids at an incident scene requires that emergency responders have a basic understanding of the physical characteristics of flammable liquids.

## EFFECTS OF TEMPERATURE ON FLAMMABLE LIQUIDS

Many of the physical characteristics of flammable liquids involve temperature. It is important to understand that there are different temperature scales listed in reference books; make sure of which scale is used when materials are researched. There is a big difference between the temperatures of the Fahrenheit and centigrade (Celsius) scales. The Fahrenheit scale is familiar to most emergency responders because it is the temperature used most commonly in the United States. The centigrade scale is used predominantly throughout the rest of the world and within the scientific and technical community of the United States, as well as in many of the reference books used by emergency responders to obtain information on hazardous materials. Shown in the following examples are temperature conversion formulas used to convert centigrade to Fahrenheit and Fahrenheit to centigrade.

$$\text{From } C° \text{ to } F°: \quad F° = \frac{9 \times C°}{5} + 32$$

Example:

$$C \text{ Temp} = 40°F = \frac{9 \times 40}{5} + 32 = \frac{360}{5} = 72 + 32 = 104°F$$

$$\text{From } F° \text{ to } C°: \quad C° = \frac{5(F° - 32)}{9}$$

Example:

$$F \text{ Temp} = 104°C = \frac{5(104° - 32)}{9} = \frac{5 \times 72}{9} = \frac{360}{9} = 40°C$$

Another temperature scale that may be encountered on a less frequent basis is the Kelvin scale, also known as absolute temperature. This scale is used principally in theoretical physics and chemistry, and in some engineering calculations. Absolute temperatures are expressed either in degrees Kelvin or in degrees Rankine, corresponding respectively to the centigrade and Fahrenheit scales. Temperatures in Kelvin are obtained by adding 273 degrees to the centigrade temperature (if above 0°C), or subtracting the centigrade temperature from 273 (if below 0°C). Degrees Rankine are obtained by subtracting 460 from the Fahrenheit temperature.

Absolute zero is the temperature at which the volume of a perfect gas theoretically becomes zero and all thermal motion ceases, which occurs at –273.13°C, or –459.4°F.

## BOILING POINT

The first physical characteristic discussed is the boiling point of a flammable liquid. Boiling point is a physical characteristic that is affected by temperature of the liquid and atmospheric pressure. Boiling point is defined as the temperature at which the vapor pressure of a liquid equals the atmospheric pressure of the air. Atmospheric pressure is 14.7 psi at sea level. Liquids want to become gases; it is atmospheric pressure that keeps a liquid from becoming a gas at normal pressures. Atmospheric pressure decreases as altitude increases. The higher the altitude, the lower the boiling point of any given material (Figure 4.3). For example, water boils at 212°F at sea level. In Denver, CO, the altitude is 1 mile (5280 ft) above sea level; water boils at approximately 203°F. At Pikes Peak, CO, the altitude is more than 14,000 feet above sea level, and water boils at approximately 186°F. The atmospheric pressure decreases as altitude increases because the air is thinner at higher altitudes.

Atmospheric pressure is pushing down on a liquid in an open tank or in a spill. With atmospheric pressure pushing down on a liquid there is not much vapor moving away from the surface of the liquid. There is a direct relationship between the boiling point of a liquid and the amount of vapor present in a spill or in a container. If a flammable liquid is above its boiling point, vapor will be produced and it will move away from the surface of the liquid. The more heat that is applied, the more vapor that is produced, the further the vapor will travel

### Boiling Point of Water and Atmospheric Pressure (psi) vs. Elevation

| Elevation | Atmospheric Pressure | Boiling Point |
|---|---|---|
| Sea Level | 14.7 | 212° F |
| 3300 Feet | 13.03 | 203° F |
| 5280 Feet (1 mile) | 12.26 | 201° F |
| 10,000 Feet | 10.17 | 194° F |
| 13,000 Feet | 9.00 | 188° F |

(Values are approximate)

**Figure 4.3**

away from the spill. If the flammable liquid is in a closed container, the vapor pressure will increase inside the container as the temperature of the liquid increases. This increase in temperature can come from many different sources. Increases in ambient temperature, radiant heat from the sun or a nearby fire can increase the vapor pressure in the container. As the pressure increases in the container, it will reach the setting on the pressure-relief valve and the relief valve will function. If this pressure increase occurs in a container that does not have a relief valve, the container may rupture. Rupture may also occur in a container with a relief valve if the pressure rises too fast for the relief valve to vent the material into the air or if the relief valve is not working properly. In either case, the rupture may be violent, with a fireball and flying pieces of tank that can travel over a mile from the blast site.

## FACTORS AFFECTING BOILING POINT

A number of outside factors can determine the relative boiling point of a flammable liquid. The first consideration is the molecular weight of a compound. Each element in a compound has an atomic weight, listed on the Periodic Table. Many flammable liquids are made up of hydrogen and carbon, and are referred to as hydrocarbon compounds. Carbon has an atomic weight of 12, so each carbon atom in a compound will weigh 12 atomic mass units (AMU). Hydrogen has an atomic weight of 1, so each hydrogen atom in a compound will weigh 1 AMU. Therefore, in the compound butane, there are 4 carbon atoms each weighing 12 AMU (which equals 48) and 10 hydrogen atoms, each weighing 1 AMU (which equals 10). Butane therefore weighs 58 AMU. The molecular weight of a compound would be the sum of all the weights of all the atoms. The more carbons in a hydrocarbon compound, the heavier it will be. The heavier it is, the more energy it will take to get the liquid to boil and overcome atmospheric pressure. Therefore, the heavier a compound is, the higher the boiling point it will have. In the following example are three hydrocarbon compounds. Compare the molecular weights of each and look at their boiling points. You can readily see the relationship between heavy compounds and high boiling points.

```
      H  H  H  H            H  H  H  H  H           H  H  H  H  H  H
      |  |  |  |            |  |  |  |  |           |  |  |  |  |  |
  H — C — C — C — C — H  H — C — C — C — C — C — H  H — C — C — C — C — C — C — H
      |  |  |  |            |  |  |  |  |           |  |  |  |  |  |
      H  H  H  H            H  H  H  H  H           H  H  H  H  H  H
        Butane                  Pentane                  Hexane
        58 AMU                  72 AMU                   86 AMU
        BP: 31°F                BP: 97°F                 BP: 156°F
```

Hydrocarbon derivatives are compounds with other elements in addition to hydrogen and carbon. Weight will still determine boiling points when comparing hydrocarbon derivatives within the same family. However, when comparing different families, the concept of polarity has to be considered with some of the compounds. There is a rule in chemistry that says "like materials dissolve like materials." Another term used in this case would be miscibility. If a material is miscible, it will mix with another. If a material is immiscible, it will not mix with another. All polar compounds are alike in terms of being polar. Therefore, polar compounds are soluble in polar compounds. One of the main reasons polarity is discussed in terms of emergency response is because of foam used for firefighting with flammable liquid fires. When fighting flammable liquid fires, two general types of foam are used: hydrocarbon foam and polar-solvent foam, which is sometimes referred to as alcohol-type foam. The reason that two different types of foam are necessary for flammable liquid fires is polarity.

## Polarity

Water is a polar compound. Because like dissolves like, water is miscible with most polar solvents. The main ingredient of firefighting foam is water. Foam is a mixture of foam concentrate, water, and air. If regular foam is put on a polar-solvent liquid such as alcohol, the water will be removed from the foam by the polar solvent and the foam blanket will break down. To effectively extinguish fires with polar solvents, it is necessary to use polar-solvent foam. Polarity is discussed here to familiarize you with the types of materials that are polar solvents and the effects that polarity has on flammable liquids. Be aware that polar solvents, such as alcohols, aldehydes, organic acids, and ketones, require a special foam to extinguish fires. Choosing the right type of foam is part of the process of effectively managing a flammable-liquid incident.

Polar liquids tend to have higher boiling points than nonpolar liquids. Polarity is said to have the effect of raising the boiling point of a liquid. There are two types of structure that represent polarity in hydrocarbon derivative compounds: the carbonyl structure and hydrogen bonding.

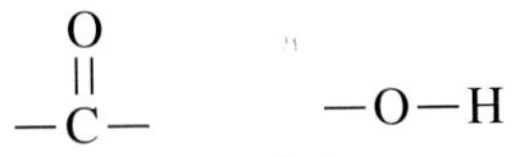

$$
\begin{array}{cc}
\underset{\text{Carbonyl}}{-\overset{\overset{\textstyle O}{\|}}{C}-} & \underset{\text{Hydrogen bond}}{-O-H}
\end{array}
$$

A carbonyl contains a carbon-to-oxygen double bond. Most double bonds are reactive; however, with the carbonyl families, except for aldehyde, the double bond is protected by hydrocarbon radicals on either side of the carbonyl. This prevents oxygen from getting to the bond and breaking it. Double bonds are stable with the carbonyl families, except for aldehydes. The carbonyl structure is shown in the following illustration, first by itself, and then in the hydrocarbon functional groups: ketone, aldehyde, organic acid, and ester. All four compounds are polar because of the carbonyl, and the organic acid also has a hydrogen bond. The amount of polarity is generally the same between the ketone, ester, and aldehyde. The polarity of an acid is higher because of the hydrogen bond and the carbonyl in the same compound, like a double dose of polarity.

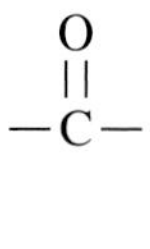

$$\begin{array}{c} O \\ || \\ -C- \end{array}$$

Carbonyl

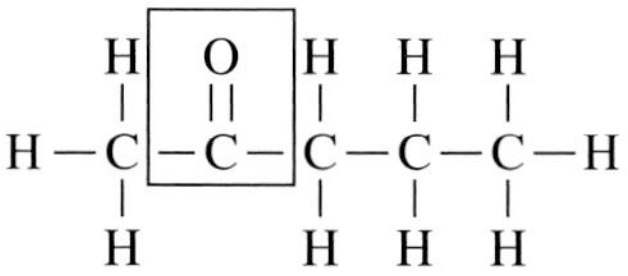

$$\begin{array}{ccccccc} & H & O & H & H & H \\ & | & || & | & | & | \\ H- & C- & C- & C- & C- & C- & H \\ & | & & | & | & | \\ & H & & H & H & H \end{array}$$

Ketone

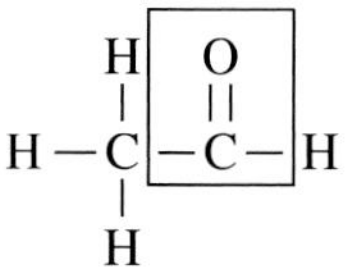

$$\begin{array}{ccc} & H & O \\ & | & || \\ H- & C- & C- & H \\ & | & \\ & H & \end{array}$$

Aldehyde

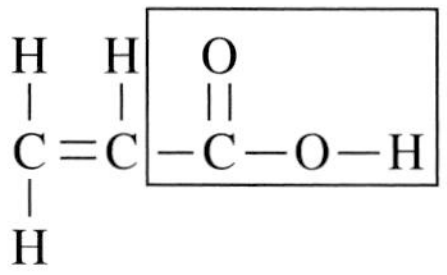

$$\begin{array}{ccccc} H & H & O \\ | & | & || \\ C= & C- & C- & O- & H \\ | & & \\ H & & \end{array}$$

Organic acid

$$\begin{array}{ccccccc} H & H & O & H & H \\ | & | & || & | & | \\ C= & C- & C- & O- & C- & C- & H \\ | & & & | & | \\ H & & & H & H \end{array}$$

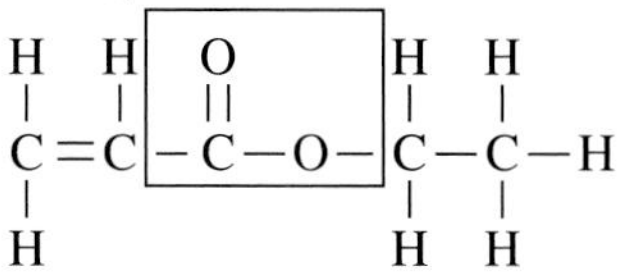

Ester

When oxygen and hydrogen covalently bond, the bond is very polar. Water has a hydrogen–oxygen bond that gives water polarity. It is well known that water is a liquid between 32 and 212°F. Water has a molecular weight of 18 AMU. The molecular weight of an average air molecule is 29 AMU. The water molecule is lighter than air! Water should be a gas at normal temperatures and pressures, but it is a liquid, due to polarity. Even though water has a molecular weight of 18 AMU, it has a boiling point of 212°F! If water was not polar, there would be no life on earth as we know it. The hydrogen bond is a very polar bond.

There are two hydrocarbon derivative families presented in this book that have hydrogen bonding: alcohols and organic acids. The organic acid also has a carbonyl bond. This double polarity makes organic acid the most polar material among the hydrocarbon derivatives. Many materials exhibit some degree of polarity. However, only the polarity of carbonyl and hydroxyl groups of hydrocarbon derivatives will be discussed here. All other materials mentioned will be considered nonpolar. The hydrocarbon and hydrocarbon derivative compounds are listed in Figure 4.4. They are ranked in order of descending degree of polarity. Organic acids are the most polar of all the materials listed, here referred to as "super-duper polar." Alcohols are the second most polar, "super polar." The remaining carbonyls are polar with the exception of amines, which are "slightly polar." All other compounds discussed will be considered nonpolar. In the following structures, compare the polarity of the compounds listed and notice the effect that polarity has on boiling point.

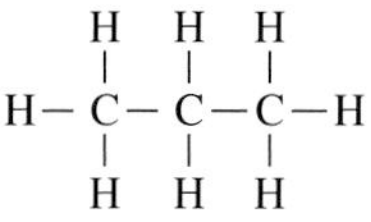

$$\begin{array}{ccc} H & H & H \\ | & | & | \\ H- C- & C- & C- H \\ | & | & | \\ H & H & H \end{array}$$

Propane
44 AMU
BP: -40°F

Non-Polar

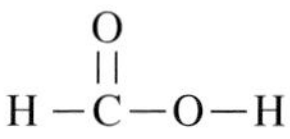

$$\begin{array}{c} O \\ || \\ H- C- O- H \end{array}$$

Formic Acid
46 AMU
BP: 213°F

Super Duper Polar

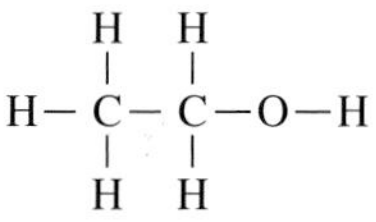

$$\begin{array}{cc} H & H \\ | & | \\ H- C- & C- O- H \\ | & | \\ H & H \end{array}$$

Ethyl Alcohol
46 AMU
BP: 170°F

Super Polar

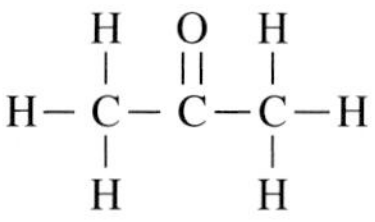

$$\begin{array}{ccc} H & O & H \\ | & || & | \\ H- C- & C- & C- H \\ | & & | \\ H & & H \end{array}$$

Di-Methyl Ketone
56 AMU
BP: 133°F

Polar

## POLARITY OF HYDROCARBON DERIVATIVES

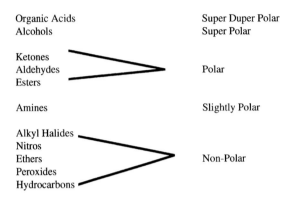

**Figure 4.4**

Formic acid has the highest boiling point even though it weighs less than dimethyl ketone. Methyl alcohol has the next highest boiling point; it also weighs less than dimethyl ketone. The weights of methyl alcohol and ethane are about the same, yet the difference in boiling point is 275°F! Polarity has a great effect on the boiling point of liquids.

To fully understand the concept of polarity, we must revisit the structure of the atom presented in Chapter 1. The nucleus contains positive protons and the energy levels outside the nucleus contain negative electrons. There is normally an equal number of negative electrons and positive protons in an element and the elements of a compound. Hydrogen has only one electron in its outer energy level. Oxygen has 16 electrons in its outer energy levels. As hydrogen and oxygen bond together, oxygen has a tendency to draw the one electron from hydrogen toward the oxygen side of the covalent bond. This exposes the positive nucleus of hydrogen, creating a slightly positive side to the hydrogen end of the molecule. Oxygen has drawn the electron from hydrogen toward its nucleus. In doing so, oxygen has more negative electrons near the nucleus than positive protons inside. This creates a slightly negative field around the oxygen side of the molecule. This is illustrated in the following illustration.

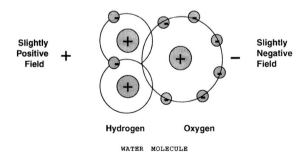

These hydrogen–oxygen bonded molecules are then attracted to other hydrogen–oxygen molecules because of the rule in chemistry that "opposite charges attract." The positive ends of the hydrogen–oxygen molecules are attracted to the negative ends of other hydrogen–oxygen molecules. This attraction holds the molecules together so that it takes more energy to break them apart and cause the compound to boil. This attraction of molecules is illustrated in the following example:

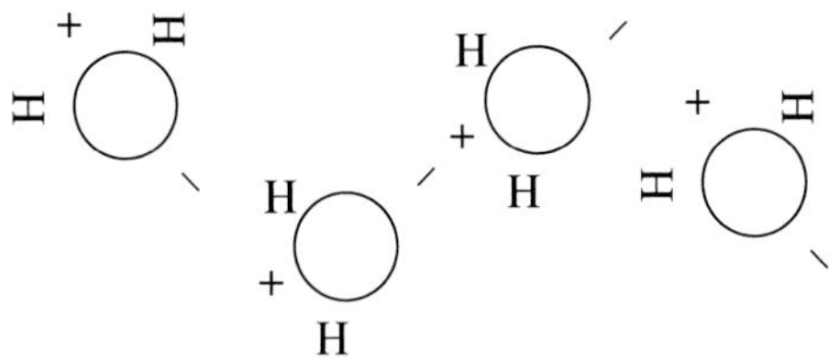

Water molecules with polar attraction holding them together.

## Branching

The last factor that affects the boiling point of a flammable liquid is branching. (Branching was discussed in Chapter 3 under Isomers in the Hydrocarbon section.) Branched compounds are all man-made. Since branching does not occur naturally, branching of a compound is done for a particular purpose. When a compound is branched, it has a lower boiling point than the unbranched liquid. In the following illustration, the structures for butane and isobutane are shown with the corresponding boiling points. The effect that branching has on the boiling point of a liquid is clearly visible.

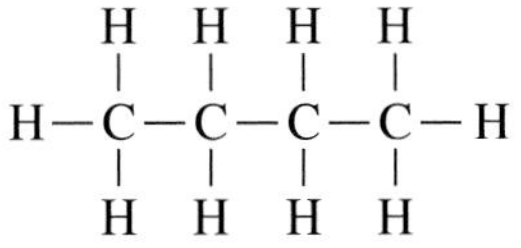

Butane
$C_4H_9$
BP: 31°F

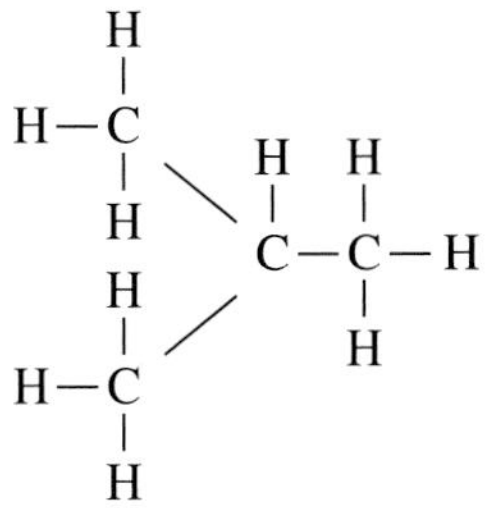

Isobutane
i-$C_4H_9$
BP: 11°F

There is another type of structure that has an effect on the boiling point of a liquid, called a cyclic compound. Cyclic compounds with five, six, or seven carbon atoms are very stable, and materials that are cyclic tend to have higher boiling points than straight-chained compounds with the same number of carbons. Those with less than five carbons are reactive and come apart very easily. Cyclic compounds with more than seven carbon atoms tend to fragment. The cyclic compounds will be discussed later in this chapter under Hydrocarbons and the Aromatic Compounds. In the following example, hexane is compared to cyclohexane. Notice the difference that the cyclic structure has on the boiling point.

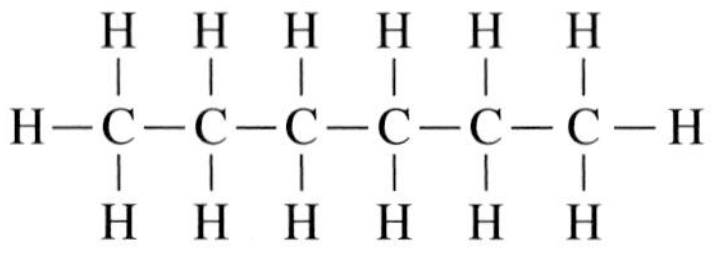

Hexane
$C_6H_{14}$
BP: 156°F

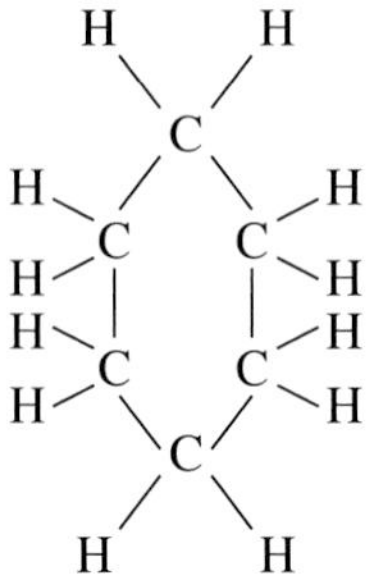

Cyclohexane
c-$C_6H_{12}$
BP: 179°F

## FLASH POINT

The next and probably the most important physical characteristic of a flammable liquid is its flash point. Flash point is the most important information for emergency responders to know about a flammable liquid. Flash point, more than any other characteristic, helps to define the flammability hazard of a liquid. If a flammable liquid is not at its flash point temperature, **it will not burn.** Flash point is defined as the minimum temperature to which a liquid must be heated to produce enough vapor to allow a vapor flash to occur (if an ignition source is present). After all, it is the vapor that burns, not the liquid, so the amount of vapor present is critical in determining whether the vapor will burn. Flash point is a measurement of the temperature of the liquid. Therefore, even if the ambient temperature is not at the flash point temperature, the liquid may have been heated to its flash point by some external heat source. For example, the radiant heat of the sun, heat from a fire, or heat from a chemical process may heat the liquid to its flash point. If an ignition source is present, ignition can and probably will occur!

Reference books used to research chemical characteristics in hazardous materials emergency response may show different flash point values. There are different tests used to determine the flash point of a liquid. Tests use open-cup and closed-cup testing apparatus, which often produce somewhat different temperatures. Open-cup flash-point tests try to simulate conditions of a flammable liquid in the open, such as a spill from a container to the ground. The open-cup test usually results in a higher flash-point temperature for the same flammable liquid than the closed-cup method, which produces a flash-point temperature that is higher than the actual flash point when tested in a vertical tube using the upper-flame propagation principle. The flash point of a liquid varies with the oxygen

Flammable liquids must be at their flash point before combustion can occur if an ignition source is present and the material is within its flammable range.

content of the air, pressure, purity of the liquid, and the method of testing. If reference books give conflicting flash point temperatures, *use the lowest flash point value given.*

Flash point should not be confused with fire point. Fire point is the temperature at which the liquid is heated to produce enough vapor for ignition to occur after the vapor flash occurs. The fire-point temperature is 1 to 3 degrees above the flash-point temperature. The fact that the fire point is so close to the flash point really does not give it much significance to emergency responders. If a liquid is at its flash point, prepare for a fire!

The author is familiar with only one occasion when there was a vapor flash from vapors being ignited and no fire occurred after the vapors burned off. In that situation, there may not have been a fire because all of the vapor was burned off and there was nothing else to burn. A gasoline tanker was filling the underground storage tanks of a service station early one morning. The vent pipes for the underground tanks were not up to code and were just 3 ft above the ground. NFPA 30, the Flammable and Combustible Liquids code, requires that the vents be a minimum of 3 ft above the roof line of the nearest building. As a result of the improper vents and the vapors being heavier than air, the vapors collected near the ground. When a soda machine compressor turned on near the vapor spill, it provided the ignition source and a vapor flash occurred. Other than a very surprised delivery man, there was not any injury or damage from the flash fire. Do not be concerned about fire point; it is the flash point that is the important temperature to look up and be aware of at the scene of an incident.

There is a direct, parallel relationship between boiling point and flash point. Generally speaking, a liquid that has a low boiling-point temperature will have a low flash-point temperature. If a flammable liquid has a high boiling point, it will have a high flash point.

Boiling point is also related to vapor pressure and vapor content. Materials with low boiling points and flash points will have high vapor pressure and high vapor content. The relationship is opposite in nature. The vapor pressure of a liquid is defined in the *Condensed Chemical Dictionary* as "the characteristic at any given temperature of a vapor in equilibrium with its liquid or solid form. This pressure is often expressed in millimeters of mercury, mm Hg." In simple terms, vapor pressure is the pressure being exerted by the liquid against atmospheric pressure. When the pressure of the vapor is greater than atmospheric pressure, the vapor will spread beyond an open container or an open spill. If the liquid is in a container, the vapor pressure is the pressure being exerted by the liquid vapors on the container. When a gas has been liquefied, the only thing keeping it a liquid is the pressure in the tank. The liquid is already above its boiling point. The pressure inside the container is the atmospheric pressure in that container. It can be much higher than outside atmospheric pressure. For example, if the pressure in the tank is 50 psi, the atmospheric pressure in that tank is 50 psi.

Vapor content is the amount of vapor that is present in a spill or open container. The lower the boiling point and the flash point of a liquid, the more vapor there will be. The parallel relationship of boiling point and flash point is comparable to the opposite relationship of vapor pressure and vapor content in the diagram below, using a seesaw to illustrate the up-and-down and opposing relationship.

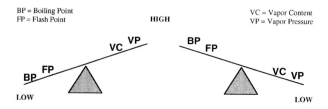

As shown in the previous illustration, when the boiling point and flash point are low, vapor content and vapor pressure are high. When boiling point and flash point are high, vapor pressure and vapor content are low. The lower the boiling point or flash point of a liquid, the higher the vapor content at a spill and the higher the vapor pressure inside a container. If a liquid is above its boiling-point temperature, there is likely to be more vapor moving farther away from a spill. If the liquid is below its boiling-point temperature, there will be some vapor above the surface of the liquid, but it will not travel very far. If the flammable liquid in a container is below its boiling point, the vapor content and vapor pressure in the container will be low. If the flammable liquid in a container is above its boiling point, the vapor content and vapor pressure in the container will be high.

## VAPOR DENSITY

Vapor density is a physical characteristic that affects the travel of vapor; it is the weight of a vapor compared to the weight of air (Figure 4.5). Vapor density

## VAPOR DENSITY

**Density of a gas or vapor compared to air.  Air = 1**

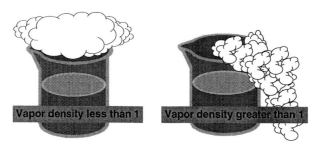

**Figure 4.5**

is usually determined in the reference books by dividing the molecular weight of a compound by 29, which is the assumed molecular weight of air. Air is given a weight value of 1, which is used to compare the vapor density of a material. If the vapor of a material has a density greater than 1, it is considered to be heavier than air. Heavier-than-air vapor will lie low to the ground and collect in confined spaces and basements. This can cause problems because many ignition sources are in basements, such as hot-water heaters and furnace pilot lights. If the vapor density is less than 1, the vapor is considered to be lighter than air, so it will move up and travel farther from the spill.

Another term associated with vapor is volatility. It is the tendency of a solid or a liquid to pass into the vapor state easily. A volatile liquid or solid will produce significant amounts of vapor at normal temperatures, creating an additional flammability hazard. The volatility of a liquid is affected by wind, vapor pressure, temperature, and surface area. Temperature always causes an increase in vapor pressure and vapor content in an incident.

## SPECIFIC GRAVITY

Specific gravity is to water what vapor density is to air. Specific gravity is the relationship of the weight of a liquid to water or another liquid (Figure 4.6). Like air, water is given a weight value of 1. If a flammable liquid has a specific gravity greater than 1, it is heavier than water and will sink to the bottom in a water spill. If a flammable liquid has a specific gravity less than 1, it will float on top of water. The specific gravity of a flammable liquid is very important in a water spill because it will determine what tactics are necessary to contain the spill. Specific gravity is the theory behind the construction of overflow and underflow dams, which are used to stop the flow of hazardous materials in water spills. Overflow dams are constructed for liquids heavier than water. The liquid sinks to the bottom of the water, and the water flows over the dam, while the hazardous liquid is stopped by the dam. Underflow dams are constructed for liquids lighter than water. The liquid floats on the surface of the water and is

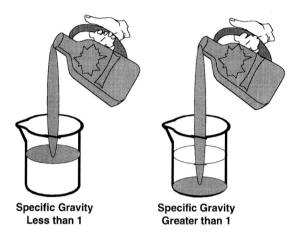

**Specific Gravity
Less than 1**          **Specific Gravity
Greater than 1**

**Figure 4.6**

stopped by the top of the dam, while the water continues to flow through a pipe at the bottom of the dam.

A term often associated with flammable and other types of liquids is miscibility. If a flammable liquid is miscible, it will mix with water, which could make clean-up very difficult. If the flammable liquid is not miscible with water, it will form a separate layer. The layer will form on top or on the bottom of the water, depending on the specific gravity of the liquid.

## POLYMERIZATION/PLASTICS

Some flammable liquids may undergo a chemical reaction called polymerization, in which a large number of simple molecules combine to form a large-chained molecule. This process is used under controlled conditions to create plastics (Figure 4.7). This large-chained molecule is referred to as a polymer, and the simple molecules used to make up a polymer are called monomers. Hydrocarbon derivatives, such as aldehydes, alkyl halides, and esters, and the aromatic hydrocarbon styrene may undergo polymerization. There are other monomers that are flammable and can polymerize, but their primary hazard is poison. Monomers can be flammable liquids, flammable gases, and poisons.

When a monomer such as styrene is transported or stored, an inhibitor is included in solution to keep the styrene from polymerizing. An inhibitor, usually an organic compound, retards or stops an unwanted chemical reaction, such as corrosion, oxidation, or polymerization. If an accident should occur, this inhibitor can become separated from the monomer and a runaway polymerization may occur. Phenol is used as an inhibitor for vinyl chloride. Dibutylamine is used as an inhibitor for butadiene. During a normal chemical reaction to create a particular polymer from a monomer, a catalyst is used to control the reaction. A catalyst is any substance that in a small amount noticeably affects the rate of a chemical

## TYPES OF PLASTICS

| Thermoplastics | Thermosets |
| --- | --- |
| ABS | Polyurethane |
| Acrylics | Amino Resins |
| Nylons | Epoxy Resins |
| Polycarbonate | Phenolic Resins |
| Polyesters | Polyesters |
| Polyethylene | |
| Polypropylene | |
| Polystyrene | |
| Polyurethane | |
| Polyvinyl Chloride | |

**Figure 4.7**

reaction, without itself being consumed or undergoing a chemical change. For example, phosphoric acid is used as a catalyst in some polymerization reactions. Once an uncontrolled polymerization starts at an incident scene, it will not be stopped until it has completed its reaction, no matter what responders may try to do. If the polymerization occurs inside a tank, the tank may rupture violently. If a container of a monomer is exposed to fire, it is important to keep the container cool. Heat from an exposure fire may start the polymerization reaction. In the following example, the monomer vinyl chloride is shown along with the process of polymerization of the vinyl chloride molecules.

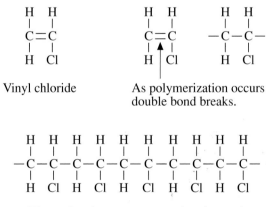

Vinyl chloride

As polymerization occurs double bond breaks.

The molecules or monomers bond to each other to form a long-chain polymer.

This has been a very abbreviated explanation of polymers and plastics, which are fairly complicated subjects. An entire book could be written on them. The most important thing for responders to understand about monomers and polymers is the danger they present in the uncontrolled conditions of the incident scene.

## IGNITION TEMPERATURES OF
## COMMON COMBUSTIBLES

| | |
|---|---|
| Wood | 392° F |
| #1 Fuel Oil | 444° F |
| Paper | 446° F |
| 60 Octane Gas | 536° F |
| Wheat Flour | 748° F |
| Acetylene Gas | 571° F |
| Corn | 752° F |
| Propane Gas | 871° F |

**Figure 4.8**

## IGNITION TEMPERATURE

An often-misunderstood physical characteristic associated with flammable liquids is ignition temperature (Figure 4.8). The definition of ignition temperature (also known as autoignition temperature) is "the minimum temperature to which a material must be heated to cause autoignition without the need for an ignition source." In other words, the material autoignites by being heated to its ignition temperature. For example, consider a pan of cooking oil on a stove. Cooking oils are animal or vegetable oils. They are combustible liquids with high boiling points and flash points. If the heat is turned up too high on the stove, the oil catches fire. Many kitchen fires occur because of cooking oils or grease being overheated. The reason for this is ignition temperature. Liquids that have high boiling and flash points have low ignition temperatures. Corn oil, commonly used as a cooking oil, has an ignition temperature of 460°F. If corn oil is heated on a stove to 460°F, it will autoignite. Gasoline has an ignition temperature around 800°F, depending on the blend. If gasoline were placed on a stove without an ignition source, the stove would not produce a temperature high enough to autoignite the gasoline. It would just boil away into vapor. This can be a real "fooler", particularly when dealing with combustible liquids where the boiling points and flash points are high. Responders sometimes become complacent when dealing with combustible liquids. They think that because the boiling and flash points are high the danger of fire is low. If, however, ignition sources are not controlled, what little vapor is present above the liquid can ignite if the temperature of the ignition source is above the ignition temperature of the liquid (Figure 4.9). For example, a lighted cigarette, with no drafts, has a surface temperature of 550°F and number 1 fuel oil has an ignition temperature of 444°F. A cigarette can be an ignition source for a combustible liquid because of the low ignition temperatures. A cigarette cannot be an ignition source for gasoline because the temperature of the lighted cigarette, without a draft, is below gasoline's ignition temperature.

Referring back to the seesaw, there is an opposite relationship between boiling point/flash point and ignition temperature. Flammable liquids that have

## TEMPERATURES OF COMMON
## IGNITION SOURCES

| | |
|---|---|
| Lighted Cigarette, No drafts | 550° F |
| Lighted Cigarette, With drafts | 1,350° F |
| Struck Match | 2,000° F + |
| Electric Arc | 2,000° F + |

**Figure 4.9**

low boiling points and flash points have high ignition temperatures. Liquids that have high boiling points and flash points have low ignition temperatures.

Combustible liquids generally are more difficult to ignite than flammable liquids. However, once they are ignited they have much higher heat output than flammable liquids. Because of this high heat output, combustible liquid fires are much more difficult to extinguish than flammable liquid fires. There is a parallel relationship between boiling point and heat output. Materials that have low boiling points have low heat outputs. This relationship between boiling point, flash point, ignition temperature, vapor content, vapor pressure, and heat output is illustrated in the following example.

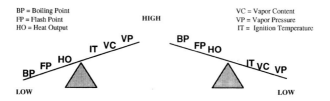

## FLAMMABLE RANGE

The last physical characteristic to be presented is flammable range. Flammable range is defined as the percent of vapor in air necessary for combustion to occur and is referred to as the explosive limit. It is expressed on a scale from 0 to 100% (Figure 4.10). There is an upper explosive limit and a lower explosive limit; between the two there is a proper mixture of vapor and air for combustion to occur. Above the upper explosive limit, there is too much vapor and not enough air, in other words, the mixture is too rich to burn. Below the lower explosive limit, there is enough air but too little vapor, therefore, the mixture is too lean to burn. Most flammable liquids have explosive limits between 1 and 12%. They have a very narrow flammable range; all of the conditions must be just right for combustion to occur. The liquid must be at its flash point, the air–vapor mixture must be within its flammable range, and the ignition–source temperature must be above the ignition temperature of the liquid.

## TYPICAL FLAMMABLE RANGES
## OF FLAMMABLE LIQUID FAMILIES

| | |
|---|---|
| Fuel Family | 1 to 8% |
| Aromatic Hydrocarbons | 1 to 7% |
| Ketones | 2 to 12% |
| Esters | 1 to 9% |
| Amines | 2 to 14% |
| Alcohols | 1 to 36% |
| Ethers | 2 to 48% |
| Aldehydes | 3 to 55% |
| Acetylene | 2 to 85% |

**Figure 4.10**

Wide flammable range materials can burn inside a container because they burn very rich. Alcohols, ethers, and aldehydes are families of flammable liquids that have very wide flammable ranges and should be addressed with extreme caution (Figure 4.11).

## ANIMAL AND VEGETABLE OILS

Some combustible liquids, such as animal and vegetable oils, have a hidden hazard: they may burn spontaneously when improperly handled. They have high boiling and flash points, narrow flammable ranges, low ignition temperatures, and are nonpolar. Examples of these liquids are linseed oil, cottonseed oil, corn oil, soybean oil, lard, and margarine. These unsaturated materials can be dangerous when rags containing residue are not properly disposed of or they come in contact with other combustible materials. There is a double bond in the chemical make-up of animal and vegetable oils that reacts with oxygen in the air. This reaction causes the breakage of the double bond, which creates heat. If the heat is allowed to build up in a pile of rags for example, spontaneous combustion will occur over a period of hours.

In Verdigris, OK, a fire occurred in an aircraft hangar at a small airport. The owner's living quarters were on the second level of the hangar. Workers had been polishing wooden parts of an airplane in the afternoon. The rags used to apply linseed oil were placed in a plastic container in a storage room in the hangar, just below the living quarters. Around 2 a.m., the rags with the linseed oil spontaneously ignited and the fire traveled up the wall into the living quarters. Fortunately the owner had smoke detectors; the family was awakened and the fire department was called promptly. The fire was quickly extinguished with a minimum of damage. The V-pattern on the wall led right back to the box where the linseed oil–soaked rags had been placed. There was little doubt what had happened; the confinement of the pile allowed the heat to build up as the double bonds were broken in the linseed oil and spontaneous combustion occurred.

# FLAMMABLE RANGE SCALE

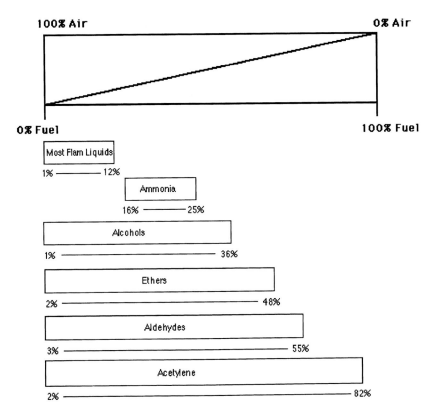

Figure 4.11

Ordinary petroleum products, such as motor oil, grease, diesel fuel, and gasoline, to name a few, do not have a double bond in their chemical make-up. Therefore, those materials **do not undergo spontaneous combustion!** This fact may come as a surprise to some people because the author knows there have been numerous fires blamed on soiled rags with those products on them. The fact is that those types of flammable liquids do not spontaneously ignite and cannot start to burn without some other ignition source.

## FIRE-EXTINGUISHING AGENTS

The theory behind fire extinguishing was first represented by the fire triangle and more recently by the fire tetrahedron. The triangle (Figure 4.12) represents the three components that were thought to be necessary for fire to occur: heat, fuel, and oxygen. If any of the components were removed, the fire would go out. The current theory (Figure 4.13), uses a four-sided geometric figure called a

**THE FIRE TRIANGLE**

**OXYGEN**

**Figure 4.12**

tetrahedron, representing the four components necessary for fire to occur: the original three plus chemical chain reaction. It is believed that fire is a chemical chain reaction; certain extinguishing agents work by interrupting this chemical chain reaction.

Extinguishing agents, such as foam, act to eliminate one of the four components. Foam excludes oxygen from the fuel by blanketing the surface of the liquid, and can also cool the material (removing heat). There are two general categories of firefighting foam: chemical and mechanical.

Chemical foam was developed in the late 1800s. The foam bubble was produced by a chemical reaction of sodium bicarbonate powder, aluminum sulfate powder, and water. The reaction produces carbon dioxide, which is then encapsulated in the interior of a bubble. Chemical foams are expensive to produce and use, and create a very rigid foam blanket that does not reseal well when disrupted. Presently chemical foams have been replaced by more effective and economical mechanical foams.

There are three types of mechanical firefighting foams: protein, fluoroprotein, and aqueous film forming (AFFF). Different foams have different physical characteristics that affect their ability to form a foam blanket and extinguish a fire (Figure 4.14). The type of foam selected will depend on the type of foam available and the firefighting tactics chosen to extinguish the fire. Make sure that whatever

**THE FIRE TETRAHEDRON**

**HEAT**

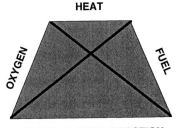

**CHEMICAL CHAIN REACTION**

**Figure 4.13**

**FOAM TYPE COMPARISON**

|                                 | Protein | Fluoroprotein | AFFF                    |
|---------------------------------|---------|---------------|-------------------------|
| **Drain Time**                  | Long    | Shorter       | Rapid                   |
| **Viscosity**                   | High    | Low           | Very Low                |
| **Flow Rate**                   | Low     | Moderate      | Very Fast               |
| **Polar Solubility Compat.**    | None    | None          | Varies by Manufact.     |

**Figure 4.14**

extinguishing agent is used, it is the proper agent for the type of fire. Be sure there is enough agent on-scene to extinguish the fire before fire suppression efforts are started. Fires require a certain volume of water to effect extinguishment. Flammable liquid fires also require a certain volume of foam to extinguish the fire. If that amount of foam is not available, putting a lesser amount of foam on the fire is not going to extinguish it and is just a waste of time and foam.

In addition to foam, fires involving flammable liquids can be extinguished by using dry chemical, dry powder, halon, Purple K™, carbon dioxide, and water. Fire-extinguishing agents are rated according to the class of fire that they are effective in extinguishing (Figure 4.15). Dry powder acts primarily to exclude atmospheric oxygen. Dry chemical, dry powder, halon, and Purple K™ interrupt

## CLASSES OF FIRES

**Class A**
Ordinary Combustibles

Water
All Purpose Dry Chemical

**Class B**
Flammable Liquids

All Purpose Dry Chemical
Foam
Purple K™
Halon
Carbon Dioxide
Sodium Bicarbonate

**Class C**
Electrical

All Purpose Dry Chemical
Carbon Dioxide
Halon

**Class D**
Combustible Metals

Dry Powder
Graphite
Sodium Chloride

**Figure 4.15**

the chemical chain reaction. Dry chemical fire extinguishers use sodium bicarbonate and monoammonium phosphate as agents. Purple K™ extinguishers use potassium bicarbonate, which is where the purple color comes from. Dry-powder extinguishers use sodium chloride and graphite as agents. Water acts by cooling the fire or, in other words, removing the heat. Water may be ineffective against some flammable liquid fires. This precaution usually applies to materials with flash points below 100°F. When water is used, it should be applied in the form of a water spray. Halon and carbon dioxide displace the oxygen needed for the fire to burn, so if there is not enough oxygen the fire goes out. Because of their effect on the ozone layer above the earth, halon fire-extinguishing agents are being phased out, along with other gaseous halogenated chemicals.

Carbon tetrachloride was once used as a fire-extinguishing agent. In fact it was one of the first halons, halon 1040. It was discovered, however, that carbon tetrachloride has a hidden hazard. When carbon tetrachloride contacts fire or a hot surface, it gives off phosgene gas. As a result, carbon tetrachloride is no longer an approved extinguishing agent, but this does not mean that you will not see carbon tetrachloride extinguishers still in use. As recently as 1987 during routine fire inspections, they were found in use in a rural school, an apartment building laundry room, and in the basement of a drug store. They may also be found in antique stores and homes of collectors.

Soda acid fire extinguishers are also obsolete and no longer approved in fire extinguishment. They are the familiar copper, sometimes brass extinguishers that are now collectors' items. They have a screw-on cap that exposes a glass bottle of sulfuric acid when removed. The rest of the tank is filled with soda water. When the extinguisher is turned upside down, the acid and soda water mix, expelling the mixture through a rubber hose with a nozzle on the end. After 10 to 15 years of service, these extinguishers became dangerous. The pressure that builds up to expel the acid–soda water mix is around 100 psi; however, if the hose or nozzle is plugged or the hose is kinked, the pressure can reach 300 psi. Many of these extinguishers failed when used, causing injury to operators. As recently as the late 1980s this type of extinguisher was found still in use in rural schools, old hotels, and main street businesses. If these obsolete extinguishers are encountered, they should promptly be removed from service and properly disposed of.

## HYDROCARBONS

The hydrocarbon families of flammable materials were introduced in Chapter 3 in the Flammable Gases section. The first group is the alkane family. The lighter alkanes (methane, ethane, propane, and butane) are all gases at normal conditions. All of these gases, however, may be encountered as liquefied gases. The alkane family is recognized by the "ane" suffix of the compound names and the single bonds between the carbons in the compounds. Starting with pentane, a five-carbon alkane, the remaining alkanes are flammable liquids. They are all naturally occurring and are byproducts of crude oil. As a group, they have low flash points, narrow flammable ranges, high ignition temperatures, and are nonpolar. The structures and molecular formulas of pentane through decane are shown in the following illustration:

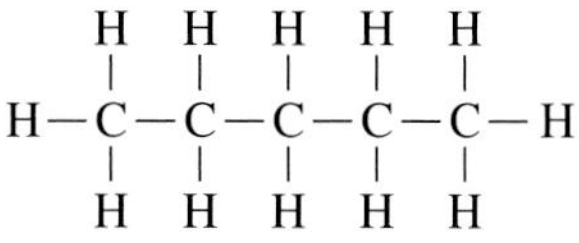

H   H   H   H   H
|   |   |   |   |
H—C—C—C—C—C—H
|   |   |   |   |
H   H   H   H   H

Pentane $C_5H_{12}$

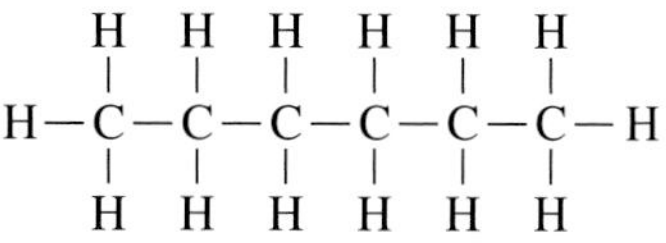

H   H   H   H   H   H
|   |   |   |   |   |
H—C—C—C—C—C—C—H
|   |   |   |   |   |
H   H   H   H   H   H

Hexane $C_6H_{14}$

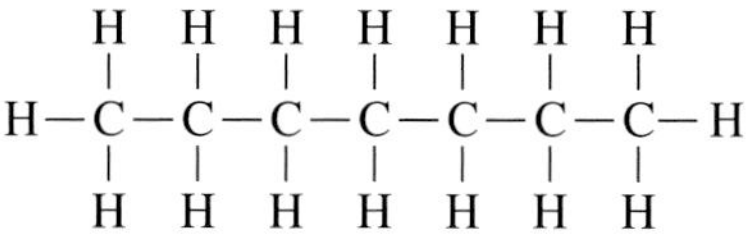

H   H   H   H   H   H   H
|   |   |   |   |   |   |
H—C—C—C—C—C—C—C—H
|   |   |   |   |   |   |
H   H   H   H   H   H   H

Heptane $C_7H_{16}$

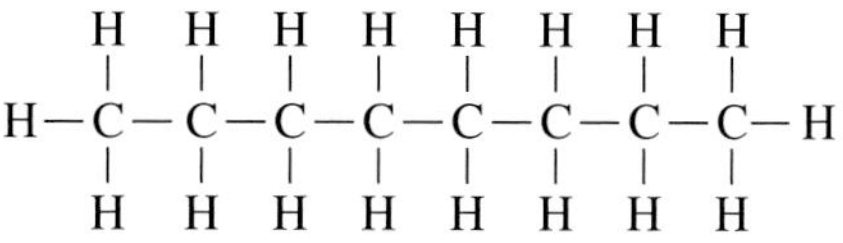

H   H   H   H   H   H   H   H
|   |   |   |   |   |   |   |
H—C—C—C—C—C—C—C—C—H
|   |   |   |   |   |   |   |
H   H   H   H   H   H   H   H

Octane $C_8H_{18}$

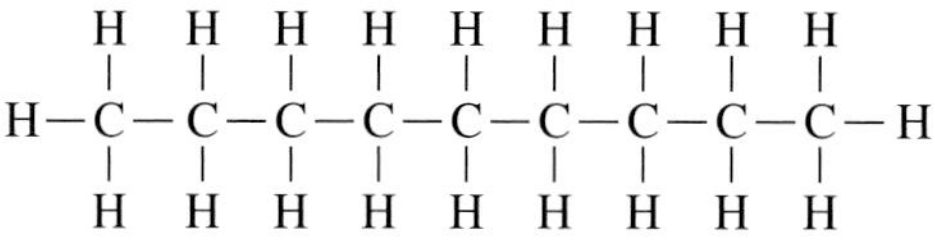

H   H   H   H   H   H   H   H   H
|   |   |   |   |   |   |   |   |
H—C—C—C—C—C—C—C—C—C—H
|   |   |   |   |   |   |   |   |
H   H   H   H   H   H   H   H   H

Nonane $C_9H_{20}$

H   H   H   H   H   H   H   H   H   H
|   |   |   |   |   |   |   |   |   |
H—C—C—C—C—C—C—C—C—C—C—H
|   |   |   |   |   |   |   |   |   |
H   H   H   H   H   H   H   H   H   H

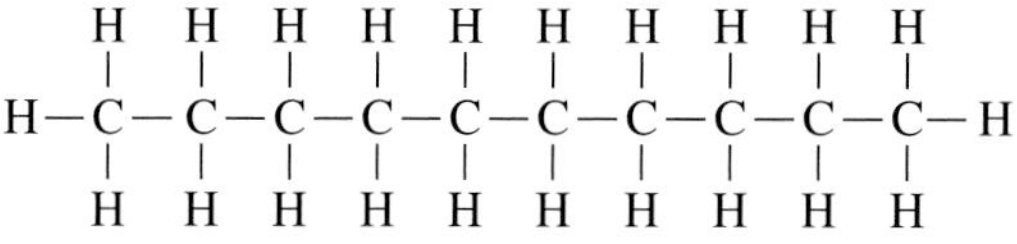

Decane $C_{10}H_{22}$

Pentane has a boiling point of 97°F, a flash point of –40°F, a flammable range of 1.5 to 7.8%, and an ignition temperature of 500°F. As the carbon content of a compound increases, so do the boiling point and flash point. Hexane has a boiling point of 156°F, a flash point of –7°F, a flammable range of 1.1 to 7.5%, and an ignition temperature of 437°F. There is usually an opposite relationship between boiling point and ignition temperature. Compounds that have low boiling points generally have high ignition temperatures. Compounds with high boiling points generally have low ignition temperatures. Decane has a boiling point of 345°F, a flash point of 115°F, a flammable range of 0.8 to 5.5%, and an ignition temperature of 410°F. While it is entirely possible to encounter these compounds in transportation and fixed facilities as individual chemicals, they will more likely be found in mixtures with other compounds. Mixtures do not involve chemical reactions or bonding. They are pure chemical compounds that have been mixed together to form a solution without losing their individual chemical make-up. In simple terms, a pure compound is like octane, pentane, or isooctane. It is possible

## PHYSICAL CHARACTERISTICS OF SOME COMMON FUELS

|  | Gasoline | Kerosene | Fuel Oil #4 | Jet Fuel |
|---|---|---|---|---|
| Boiling Point | 100° to 400° F | 338° to 572° F |  | 250° F |
| Flash Point | -45° F | 100° to 150° F | 130° F | 100° F |
| Ignition Temp | 536° to 853° F | 444° F | 505° F | 435° F |
| Vapor Density | 3.0 to 4.0 | 4.5 |  | 1 |
| Specific Gravity | 0.8 | 0.81 | <1 | 0.8 |
| LEL | 1.4% | 0.7% |  | 0.6% |
| UEL | 7.6% | 5% |  | 3.7% |

**Figure 4.16**

to draw structures or write formulas for these compounds. When pure compounds are mixed together you cannot draw a structure or write a formula for the mixture. These mixtures include gasoline, diesel fuel, jet fuel, fuel oil, petroleum ether, and others. Gasoline has a boiling point between 100°F and 400°F, depending on the contents of the mixture; its flash point is –36° to –45°F; the flammable range is 1.4 to 7.6%; and the ignition temperature ranges from 536° to 853°F. Diesel fuel has a flash point range of 100° to 130°F. Hazardous materials reference books do not list a boiling point or ignition temperature for diesel fuel. Figure 4.16 shows the physical characteristics of some common hydrocarbon fuels. Remember that the fuels are mixtures of hydrocarbons and other additives rather than pure compounds.

Physical characteristics of mixtures will vary depending on the components of the mixture. Many times mixtures are designed to have certain ranges of flash point so they will perform a particular function. If a high boiling and flash point liquid is mixed with a low boiling and flash point liquid, the boiling and flash point of the mixture will be somewhere between the two liquids that were mixed together. Mixtures will be discussed further in the Hydrocarbon Derivatives section.

The next hydrocarbon family is the alkene family. Alkenes have one or more double bonds between the carbons in the structure of the compound. The gases ethene (ethylene), propene (propylene), and butene (butylene) were discussed in Chapter 3. The first flammable liquid in the alkene family is pentene. Because it is an alkene, it has a double bond in the compound. Pentene has a boiling point of 86°F, a flash point of 0°F, a flammable range of 1.5 to 8.7%, and an ignition temperature of 527°F. Hexene has a boiling point of 146°F, a flash point of less than 20°F, and an ignition temperature of 487°F. As with the alkanes, as the carbon content increases, so do the boiling point and flash point temperatures.

There are alkene family flammable liquids that have more than one double bond. **Pentadiene,** also known as 1,3,pentadiene and piperylene is a five-carbon compound with two double bonds. Pentadiene is a highly flammable liquid with an NFPA 704 designation for flammability of 4. It has a boiling point of –45°F and a flash point of 112°F. **Hexadiene** is a six-carbon liquid compound that is highly flammable with a boiling point of 147°F, a flash point of –6°F, and a flammable range of 2 to 6.1%. The structures and molecular formulas of pentadiene and hexadiene are illustrated in the following example.

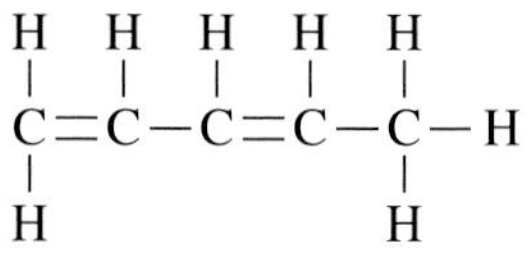

```
H   H   H   H   H                    H   H   H   H   H   H
|   |   |   |   |                    |   |   |   |   |   |
C = C — C = C — C — H                C = C — C = C — C — C — H
|           |                        |           |   |
H           H                        H           H   H
```

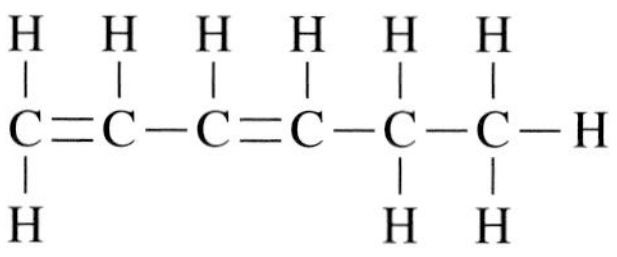

Pentadiene                          Hexadiene
$C_5H_8$                            $C_6H_9$

The last hydrocarbon family discussed is known as the aromatic hydrocarbons, sometimes referred to at the BTX fraction (benzene, toluene, xylene). All four of the aromatics to be discussed here are in the Top 50 industrial chemicals produced annually and are also components of several others. The aromatics as a group are toxic and flammable. They have low boiling and flash points, narrow flammable ranges, high ignition temperatures, and are nonpolar. Benzene is a known carcinogenic, while toluene and xylene are suspect. The parent member of the family is benzene, which has a molecular formula of $C_6H_6$. There is a ratio that can be used to recognize the benzene ring in a formula: there has to be a minimum of six carbons to have a benzene ring, and the number of carbons to hydrogens is almost in a 1:1 ratio. None of the other hydrocarbon families have that kind of carbon to hydrogen ratio. The structure of the compound is ringed with six carbons, sometimes referred to as the benzene ring. Aromatics were thought at one time to be unsaturated because the structure was thought to have double bonds. (See the First Theory benzene structure.) The structure appeared to have three double bonds to satisfy the octet rule of bonding. However, in reality, aromatics do not behave like unsaturated compounds. They burn with very incomplete combustion. They are unreactive, so it is theorized that instead of three double bonds, they have a unique structure where the six extra electrons are in a state of resonance within the benzene ring. They are not attached to any one of the carbons, but rather go from one to another at a speed faster than the speed of light, much the same way a rotor works inside a distributor in an automobile. The bond in the aromatic hydrocarbons is called a resonant bond. This resonant bonding is represented by a large circle within the benzene ring, shown in the following illustrations:

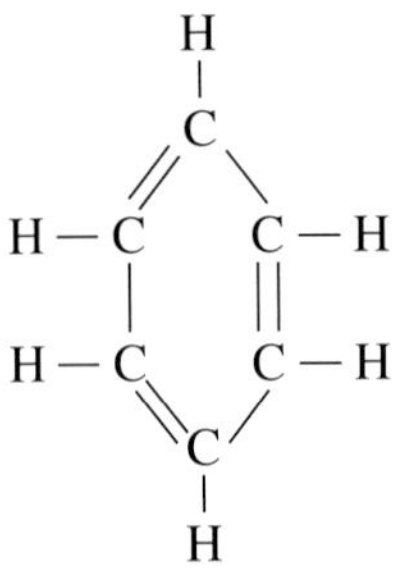

|              |                  |               |
| ------------ | ---------------- | ------------- |
| First theory | Benzene          | Resonant      |
| benzene structure | sometimes shown | bond benzene |
|              | without hydrogen | structure     |
|              | and carbon       |               |

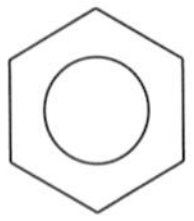

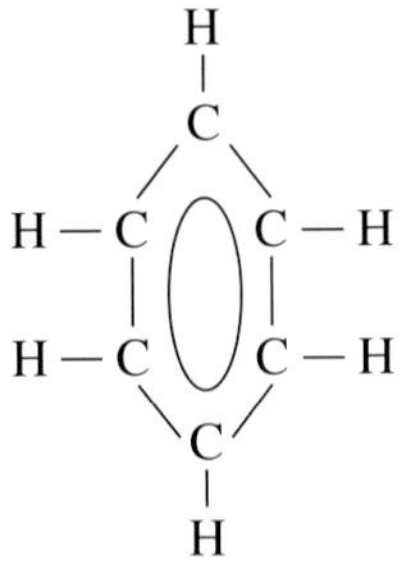

The benzene ring is the backbone of three other aromatic compounds that will be presented here. The first is toluene, with the molecular formula $C_6H_5CH_3$. Sometimes all of the carbons and hydrogens are combined and the formula is shown as $C_7H_8$. Toluene is a benzene ring with one hydrogen removed to attach a methyl radical $CH_3$. Xylene is the next aromatic, with a molecular formula of $CH_3C_6H_4CH_3$. If all of the carbons and hydrogens are added together, the formula would be $C_8H_{10}$. The benzene ring is the backbone for xylene with two hydrogens removed. Two methyl radicals replace the hydrogens in the ring to make xylene. The structures of benzene, toluene, and xylene are shown in the following illustration:

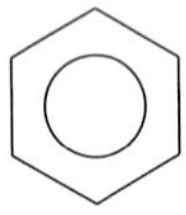

| Benzene | Toluene | Xylene |
| ------- | ------- | ------ |
| $C_6H_6$ | $C_7H_8$ | $C_8H_{10}$ |
|         | $C_6H_5CH_3$ | $CH_3C_6H_4CH_3$ |

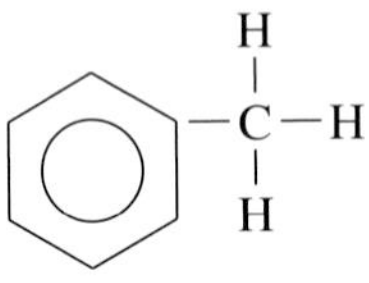

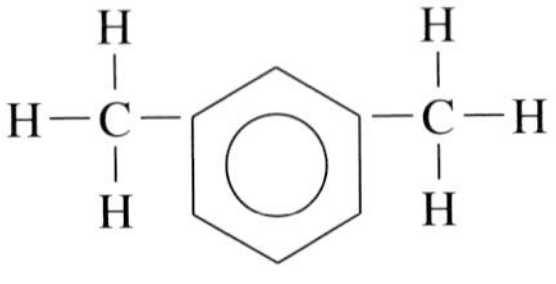

The positioning of the methyl radicals on the xylene ring are of particular importance. There are names for the different positions around the benzene ring. If the methyl radicals are placed on the top and first side position, it is referred to as the "ortho" position; "ortho" refers to "straight ahead." If the methyl radicals are placed on the top and on the second side position; it is referred to as the "meta" position, "meta" translates to "beyond." If the methyl radicals are on the top and bottom of the ring, it is referred to as the "para" position; "para" means "opposite." The effects that positioning have on a compound are changes in toxicity and some physical characteristics, such as melting point. The changes to boiling point, flash point, and ignition temperature are insignificant for these isomers of xylene. It is important when looking up these materials that you make sure which xylene is involved in an incident. The "para", "meta", and "ortho" structural isomers of xylene are often used in the making of pesticides; paraxylene is one of the Top 50 produced industrial chemicals. Shown in the following illustration are the different positions of the methyl radicals on the benzene ring for xylene.

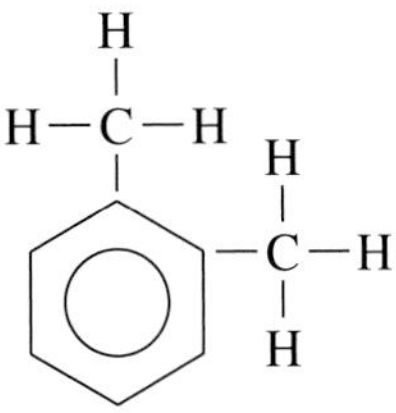

Orthoxylene
o-$C_8H_{10}$

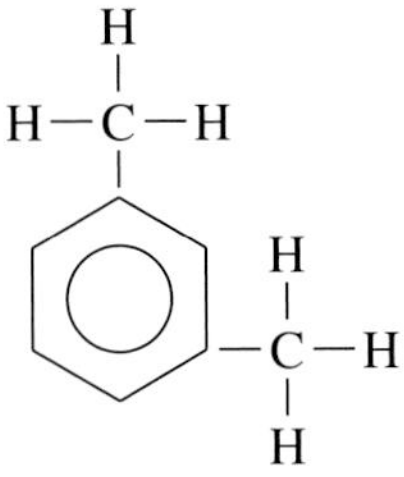

Metaxylene
m-$C_8H_{10}$

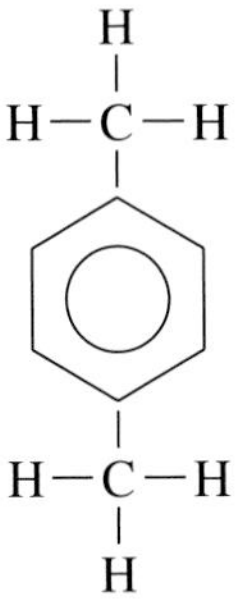

Paraxylene
p-$C_8H_{10}$

The final aromatic is styrene. It is, however, quite different than the other members of the aromatic family. Styrene is a monomer used in the manufacture of polystyrene. It has a vinyl radical attached to the benzene ring. The double bond in the vinyl radical is reactive. The reaction can occur with the oxygen in the air, with an oxidizer, or it can self-react in storage. The structure and molecular formula follow:

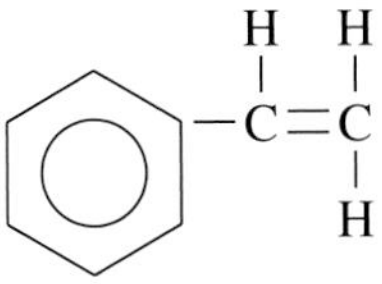

Styrene
$C_6H_5C_2H_3$

## HYDROCARBON DERIVATIVES

There are six hydrocarbon derivatives whose primary hazard is flammability (Figure 4.17): amines, ethers, alcohols, ketones, aldehydes, and esters. All but the amines and ethers are polar. The ethers, alcohols, and aldehydes have wide flammable ranges. Some organic acids are flammable. If an acid is going to burn, it will be an organic acid because inorganic acids do not burn. However, flammability is not the primary hazard of most organic acids. They will be discussed in detail in Chapter 9. Some of the alkyl halides are also flammable but their primary hazard is toxicity and they will be presented in Chapter 7. Hydrocarbon derivatives are man-made materials. They are made from hydrocarbons by removing hydrogens and adding some other element. The primary elements used in making hydrocarbon derivatives, in addition to carbon and hydrogen, are oxygen, nitrogen, chlorine, fluorine, bromine, and iodine.

Once a hydrogen is removed from a hydrocarbon, the hydrocarbon becomes a radical. The same prefixes are used for single-bonded hydrocarbon radicals as for the hydrocarbons; however, a "yl" is added to the prefix, indicating that it is a radical. For example, a one-carbon radical is called "methyl," two carbons "ethyl," three carbons "propyl," etc. Remember that at least one hydrogen has been removed, so the radical is not a complete compound. It must be attached to a hydrocarbon-derivative functional group to be complete. If a compound has more than one radical of the same type (except for ether), prefixes are used to indicate the number of radicals present. The prefixes are "di" for two, "tri" for three, and "tetra" for four. Radicals for single-bonded hydrocarbons are shown in the following illustration:

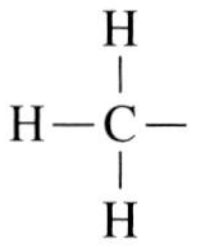

$$
\begin{array}{c}
\text{H} \\
| \\
\text{H}-\text{C}- \\
| \\
\text{H}
\end{array}
$$

Methyl/Form
$CH_3$

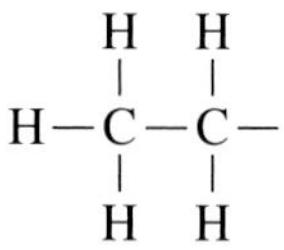

$$
\begin{array}{c}
\text{H}\quad\text{H} \\
|\quad\ | \\
\text{H}-\text{C}-\text{C}- \\
|\quad\ | \\
\text{H}\quad\text{H}
\end{array}
$$

Ethyl/Acet
$C_2H_5$

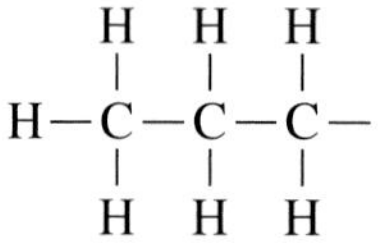

$$
\begin{array}{c}
\text{H}\quad\text{H}\quad\text{H} \\
|\quad\ |\quad\ | \\
\text{H}-\text{C}-\text{C}-\text{C}- \\
|\quad\ |\quad\ | \\
\text{H}\quad\text{H}\quad\text{H}
\end{array}
$$

Propyl
$C_3H_7$

$$
\begin{array}{c}
\text{H}\quad\text{H}\quad\text{H}\quad\text{H} \\
|\quad\ |\quad\ |\quad\ | \\
\text{H}-\text{C}-\text{C}-\text{C}-\text{C}- \\
|\quad\ |\quad\ |\quad\ | \\
\text{H}\quad\text{H}\quad\text{H}\quad\text{H}
\end{array}
$$

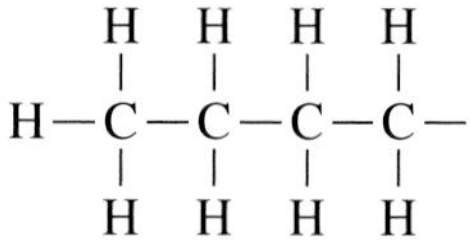

Butyl
$C_4H_9$

Hydrocarbon compounds with double bonds can also be made into radicals. In order to have a double bond, there must first be at least two carbons. There are no double-bonded radicals with only one carbon. Only two double-bonded hydrocarbon radicals are important here. They are two-carbon and three-carbon radicals with double bonds between the carbons. Since the prefixes for two and three carbons have been used up in the single-bonded compounds, the names for these double-bonded radicals are different from the others: a two-carbon compound with a double bond is called vinyl, which is actually a radical of ethene or ethylene; the three-carbon compound with one double bond is called "acryl", which is a radical of propene or propylene. The structures for the vinyl and acryl radicals are shown in the following example:

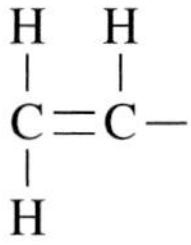

$$
\begin{array}{cc}
H & H \\
| & | \\
C & = C - \\
| & \\
H &
\end{array}
$$

Vinyl radical

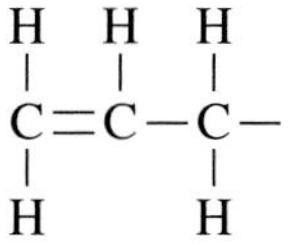

$$
\begin{array}{ccc}
H & H & H \\
| & | & | \\
C & = C - C - \\
| & & | \\
H & & H
\end{array}
$$

Acryl radical

A hydrogen can be removed from aromatic compounds to make radicals. If one carbon is removed from benzene, the radical is called "phenyl." If one carbon is removed from toluene, the radical is called "benzyl." The structures and molecular formulas are shown in the following illustration for the phenyl and benzyl radicals:

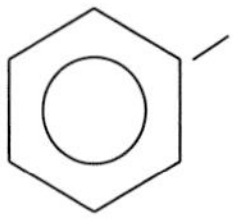

Phenyl radical
$C_6H_5$

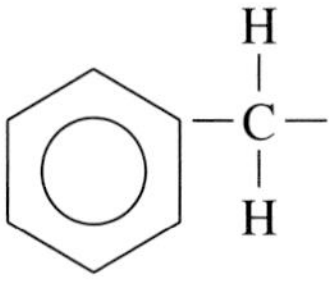

Benzyl radical
$C_6H_5CH_2$

With certain hydrocarbon-derivative functional groups, alternate names for one- and two-carbon single-bonded radicals are used. This occurs when the radicals are used with the aldehydes, esters, and organic acids. A one-carbon radical for aldehydes, esters, and organic acids is called "form". The two-carbon radical is called "acet". Additionally, when naming the radical for these compounds, the carbon in the functional group is counted as part of the total number of carbons when choosing the prefix. The following examples show the structures, molecular formulas, and names for the one- and two-carbon compounds of aldehydes, esters, and organic acids. (Just a hint for future reference: when naming ester functional groups, nothing is named ester. There are some alternate naming rules for esters based on which radical is attached to the carbon in the ester functional group.)

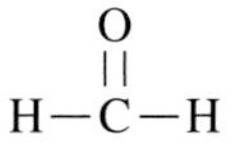

Formaldehyde
HCHO

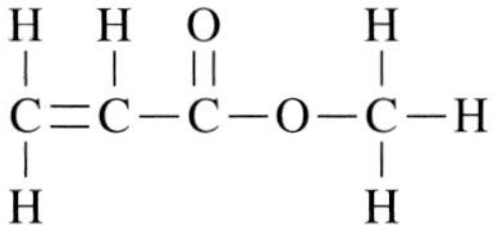

Methyl acrylate (ester)
$C_2H_3COOCH_3$

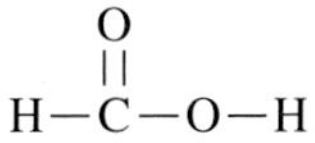

Formic acid
HCOOH

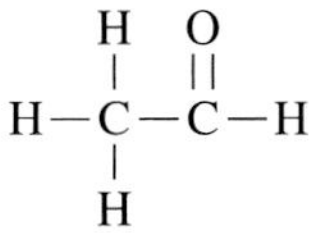

Acetaldehyde
$CH_3CHO$

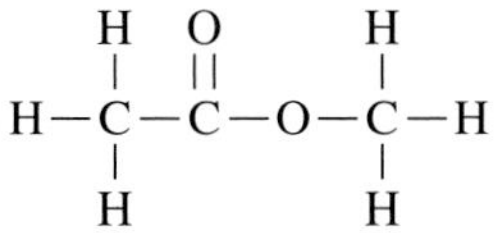

Methyl acetate
$CH_3COOCH_3$

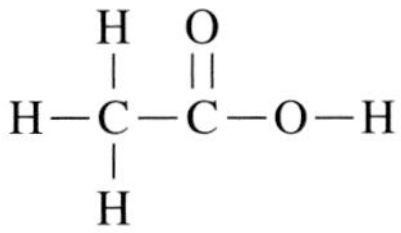

Acetic acid
$CH_3COOH$

## Amines

The first flammable liquid hydrocarbon-derivative functional group is the amines. They have low boiling points and flash points, narrow flammable ranges, and high ignition temperatures. In addition to being flammable, amines are toxic and irritants. They have a characteristic unpleasant odor, similar to the odor of the bowel or rotten flesh. The amines are considered slightly polar when compared to nonpolar materials. The amine functional group is represented by a single nitrogen surrounded by two or less hydrogens. Nitrogen requires three bonds to satisfy the octet rule of bonding. The general formulas for the amines are **R–NH₂, R₂NH,** and **R₃N.** Nitrogen identifies the amine group, not the number of hydrogens attached to the nitrogen. There may be one, two, or three radicals connected to the nitrogen. R–NH₂ indicates one radical and two hydrogens attached to the nitrogen R₂–NH indicates two radicals and one hydrogen attached to the nitrogen, R₃–N indicates three radicals and no hydrogens attached to the nitrogen. To name the amines, start with the smallest radical and proceed through however many more radicals there are, in order of size, and end with the word amine. The type of radical(s) attached to the nitrogen may be the same or different. In the following examples, structures, molecular formulas, and names are shown for amines with one, two, and three radicals attached. These are not the only radicals that can be attached, just examples.

Propylamine
$C_3H_7NH_2$

Butylamine
$C_4H_9NH_2$

Isopropylamine
$C_3H_7NH_2$

**Propylamine** is a colorless liquid that is slightly soluble in water. The specific gravity is 0.7, which is lighter than water. Propylamine is highly flammable, with a flammable range of 2 to 10% in air. The boiling point is 120°F and the flash point is –35°F. The ignition temperature is 604°F. The vapor density is 2, which is heavier than air. In addition to being flammable, propylamine is corrosive, and is a strong irritant to skin and tissue. The four-digit UN identification number is 1277. The NFPA 704 designation is health 3, flammability 3, and reactivity 0. It is shipped in glass bottles, cans, drums, and tank cars. The primary uses are as a chemical intermediate and as a lab reagent.

**Butylamine** is a colorless, volatile liquid with an amine-like odor. It is miscible with water and has a specific gravity of 0.8, which is lighter than water. It is a dangerous fire risk, with a flammable range of 1.7 to 9.8% in air. The flash point is 10°F, with a boiling point of 172°F. The ignition temperature is 594°F.

The vapor density is 2.5, which is heavier than air. Butylamine is also a skin irritant, with a TLV ceiling of 5 ppm in air. The four-digit UN identification number is 1125. The NFPA 704 designation is health 3, flammability 3, and reactivity 0. It is shipped in glass bottles, cans, drums, and tank cars. The primary uses are in the manufacture of pharmaceuticals, insecticides, dyes, and rubber chemicals.

**Isopropylamine** is a colorless, volatile liquid. It is highly flammable with a flammable range of 2 to 10.4% in air. The boiling point is 93°F and the flash point is –15°F, and the ignition temperature is 756°F. It is miscible in water with a specific gravity of 0.69, which is lighter than water. The vapor density is 2.04, which is heavier than air. In addition to flammability, isopropylamine is a strong irritant to tissue and has a TLV of 5 ppm in air. The four-digit UN identification number is 1221. The NFPA 704 designation for isopropylamine is health 3, flammability 4, and reactivity 0. The primary uses for isopropylamine are pharmaceuticals, dyes, insecticides, and a dehairing agent.

### Ethers

The next flammable liquid hydrocarbon derivative is ether. The primary hazard of ether is flammability. Ethers have low boiling and flash points, wide flammable ranges, low ignition temperatures, and are nonpolar. In addition to being flammable, ethers are anesthetic; they have wide flammable ranges, from 2 to 48% in air; and can form explosive peroxides as the ether ages. Ethers are nonpolar materials. When a container of ether is opened, oxygen from the air gets inside and bonds with the oxygen in the ether, forming a very unstable peroxide. Heat, friction, or shock can cause the peroxide to explode. Oxygen has been known to permeate the soldered seam in a metal container even without the container being opened. Ethers can be very dangerous in storage. Most ethers should not be stored longer than 6 months. If an aging container of ether is discovered, the nearest bomb squad should be called; the container should be treated as if it were a bomb.

Ether is composed of a single oxygen with two hydrocarbon radicals, one on either side of the oxygen. It is expressed by the general formula **R–O–R.** Ether names do not always follow the trivial naming system. However, the formula will have a single oxygen, which indicates the ether family. The radicals on either side of the oxygen may be the same or they may be different. An ether is named by identifying the hydrocarbon radicals and ending with the word ether. As a general rule, if the radicals are different, start with the smallest and name it, then name the second radical, and end with the word ether. For example, the following structure has a methyl radical and an ethyl radical on either side of the oxygen. The smallest radical is methyl, the second radical is ethyl, and the name ends with ether. The name of the compound is methyl ethyl ether. There may be occasions when a compound is looked up in a reference book, but it is not listed by the trivial naming system. When this happens, just look it up using the other radical name, such as ethyl methyl ether.

$$\begin{array}{c}
H \quad\quad H \quad H \\
| \quad\quad\; | \quad\; | \\
H-C-O-C-C-H \\
| \quad\quad\; | \quad\; | \\
H \quad\quad H \quad H
\end{array}$$

Ethyl methyl ether
$CH_3OC_2H_5$

When the radicals on each side of the oxygen are the same, the compound is named with just the one radical name, with no prefix to indicate two of the same radical. Because ether must have two radicals, and there is only one radical in the name, it is understood that there are two of the same radical. For example, methyl ether has two methyl radicals, one on each side of the oxygen. Ether is the only hydrocarbon derivative where the prefixes "di", "tri", and "tetra", indicating the number of radicals, are **not** used with the trivial naming system. However, even though uncommon, the "di" prefix is used and listed as a synonym for the common name of the ether compound. Therefore, it is not wrong to use "di", it is just not common. Shown in the following example are three common ethers, their structures, and molecular formulas:

$$\begin{array}{c}
H \quad H \quad\quad H \quad H \\
| \quad\; | \quad\quad | \quad\; | \\
H-C-C-O-C-C-H \\
| \quad\; | \quad\quad | \quad\; | \\
H \quad H \quad\quad H \quad H
\end{array}$$

Ethyl ether
$C_2H_5OC_2H_5$

$$\begin{array}{c}
H \quad\quad\quad\quad H \\
| \quad\quad\quad\quad\; | \\
H-C-H \quad H-C-H \\
| \quad\quad\quad\quad\quad | \\
H-C-O-C-H \\
| \quad\quad\quad\quad\quad | \\
H-C-H \quad H-C-H \\
| \quad\quad\quad\quad\quad | \\
H \quad\quad\quad\quad H
\end{array}$$

Isopropyl ether
$iC_3H_7OiC_3H_7$

$$\begin{array}{c}
H \quad H \quad H \quad H \quad\quad H \quad H \quad H \quad H \\
| \quad\; | \quad\; | \quad\; | \quad\quad | \quad\; | \quad\; | \quad\; | \\
H-C-C-C-C-O-C-C-C-C-H \\
| \quad\; | \quad\; | \quad\; | \quad\quad | \quad\; | \quad\; | \quad\; | \\
H \quad H \quad H \quad H \quad\quad H \quad H \quad H \quad H
\end{array}$$

Butyl ether
$C_4H_9OC_4H_9$

**Ethyl methyl ether** is a colorless liquid that is soluble in water. The specific gravity is 0.70, which is lighter than water. It is highly flammable, with a flammable range of 2 to 10.1% in air. The boiling point is 51°F, the flash point is −35°F, and the ignition temperature is 374°F. The vapor density is 2.07, which is heavier than air. In addition to flammability, ethyl methyl ether is anesthetic and can form explosive peroxides as it ages. The four-digit UN identification number is 1039. The NFPA 704 designation is health 1, flammability 4, and reactivity 1. The primary use is in medicine as an anesthetic.

**Ethyl ether** (diethyl ether) is a colorless, volatile, mobile liquid. It is slightly soluble in water with a specific gravity of 0.7, which is lighter than water. It is a severe fire and explosion risk when exposed to heat or flame. The compound forms explosive peroxides from the oxygen in the air as it ages. The flammable range is wide, from 1.85 to 48% in air. The boiling point is 95°F, the flash point is –49°F, and the ignition temperature is 356°F. The vapor density is 2.6, which is heavier than air. In addition to flammability, it is an anesthetic, which causes central nervous system depression by inhalation and skin absorption, with a TLV of 400 ppm in air. The four-digit UN identification number is 1155. The NFPA 704 designation is health 1, flammability 4, and reactivity 1. The primary uses are in the manufacture of smokeless powder, as an industrial solvent, in analytical chemistry, and as an anesthetic.

**Isopropyl ether** (diisopropyl ether) is a colorless, volatile liquid, that is slightly soluble in water. The specific gravity is 0.7, which is lighter than water. It is highly flammable with a wide flammable range of 1.4 to 21% in air. The boiling point is 156°F, the flash point is –18°F, and the ignition temperature is 830°F. The vapor density is 3.5, which is heavier than air. In addition to flammability, isopropyl ether is toxic by inhalation and a strong irritant, with a TLV of 250 ppm in air. The four-digit UN identification number is 1159. The NFPA 704 designation is health 1, flammability 3, and reactivity 1. The primary uses are as a solvent and in rubber cements.

**Butyl ether** (dibutyl ether), is a colorless, stable liquid, with a mild ether-like odor. It is immiscible in water, with a specific gravity of 0.8, which is lighter than water. Butyl ether is a moderate fire risk and will form explosive peroxides on aging. The flammable range is 1.5 to 7.6% in air, with a boiling point of 286°F, and a flash point of 77°F. The ignition temperature is 382°F, and the vapor density is 4.5, which is heavier than air. In addition to flammability, butyl ether is toxic on prolonged inhalation. The four-digit UN identification number is 1149. The NFPA 704 designation is health 2, flammability 3, and reactivity 1. The primary use is as a solvent.

## Alcohols

The next flammable liquid hydrocarbon derivative is alcohol. In addition to being flammable, alcohols have wide flammable ranges from 1 to 36% in air, and are toxic to some degree. They have high boiling points, moderate flash points, and high ignition temperatures. Small fires involving alcohols should be fought with dry-chemical fire extinguishers. Large fires should be fought with alcohol-type foam; water may be ineffective. Alcohols are very miscible with water. Water, as it mixes with alcohol, will at some point raise the boiling and flash points of the alcohol until the mixture of the water and the alcohol are no longer flammable.

The problem with this is that the container must be large enough to hold the mixture or this method of extinguishment will not work. Therefore, care must be taken in choosing the method of extinguishing alcohol fires. Methyl alcohol, or methanol as it is sometimes called, is toxic by ingestion and can cause blindness or death, with a TLV of 200 ppm. Ethyl alcohol or ethanol, also referred to as grain alcohol, is consumed in alcoholic beverages. It is classified as a depressant drug; too much of it can produce toxic effects and can lead to liver damage.

### Toxic Effects of Ethyl Alcohol in the Blood

| 0.08% | 2.0% | 3.0% | 4.0% | 5.0% |
|-------|------|------|------|------|
| Happy | Very happy | Drunk | Falling down drunk | Death |

The alcohol functional group is identified by the general formula **R–O–H**. There will be one radical attached to the oxygen. Alcohols are polar liquids because they have hydrogen bonding. Alcohol is the second-most polar functional group; the most polar are the organic acids. Because of polarity, alcohols are miscible with water and require the use of polar solvent or alcohol-type foam to extinguish fires. Because of polarity, alcohols as a family have high boiling and flash points. An alcohol is named by identifying the radical attached to the oxygen. The radical is named first and the compound ends in the word alcohol. There are often various ways of naming the same chemical compounds. With alcohol, an "ol" ending may be added to the radical, indicating that it is an alcohol. For example, the radical "methyl" is attached to the oxygen in the alcohol functional group. The name of the compound is methyl alcohol or may be called methanol. The structure and molecular formula for one-, three-, and four-carbon radicals attached to the alcohol functional group are shown in the following example:

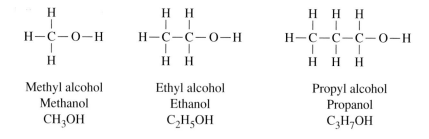

| Methyl alcohol | Ethyl alcohol | Propyl alcohol |
|---|---|---|
| Methanol | Ethanol | Propanol |
| $CH_3OH$ | $C_2H_5OH$ | $C_3H_7OH$ |

**Methyl alcohol** and **butyl alcohol** are in the Top 50 industrial chemicals and will be discussed in detail later in this chapter.

**Ethyl alcohol,** or ethanol, is a colorless, clear, volatile liquid. It is polar and soluble in water, with a specific gravity of 0.8, which is lighter than water. It is highly flammable with a flammable range of 3.3 to 19% in air. The boiling point is 173°F and the flash point is 55°F. The vapor density is 1.6, which is heavier than air. The ignition temperature is 685°F and ethanol is classified as a depressant drug with a TLV of 1000 ppm in air. The four-digit UN identification number is 1170. The NFPA 704 designation is health 0, flammability 3, reactivity 0. The primary uses are as a solvent, in beverages, antifreeze, gasohol, pharmaceuticals, and explosives.

**Propyl alcohol,** 1-propanol, is a colorless liquid with an odor similar to ethanol. It is polar, soluble in water, and has a specific gravity is 0.8, which is lighter than water. It is a dangerous fire risk with a flammable range of 2 to 13% in air. The vapor density is 2.1, which is heavier than air. The boiling point is 207°F, the flash point is 74°F, and the ignition temperature is 775°F. It is toxic by skin absorption with a TLV of 200 ppm in air. The four-digit UN identification number is 1274. The NFPA 704 designation is health 1, flammability 3, and reactivity 0. The primary uses are in brake fluid, as a solvent, and as an antiseptic.

Isomers were introduced in Chapter 3. An isomer is a compound with the same formula, but a different structure. Isomers are sometimes referred to as branched compounds. Branching of a compound has the effect of lowering the boiling point. In the Hydrocarbon section of Chapter 3, only the "iso" branch was discussed. With the hydrocarbon-derivative functional groups, there will also be an "iso" branch and, in addition, there will be secondary and tertiary branches. When dealing with the hydrocarbon compounds, the branch is a carbon attached to one of the center carbons of the chain. In the derivatives, the functional group is a part of that carbon chain and considered when determining branching of the compound. The types of branches will be shown in this section because they occur commonly with the alcohol compounds. However, branching can occur in any of the hydrocarbon derivative groups. Shown in the following illustration are examples of branches of a four-carbon alcohol. The branch is determined by the location of the functional group on the carbon chain. The first structure is the straight-chained compound butyl alcohol with a molecular formula of $C_4H_9OH$. Straight-chained compounds are sometimes referred to as the "normal" form. Normal butyl alcohol might be represented by a small "n" in front of the name and formula. The next structure is isobutyl alcohol. The "iso" branch is determined by locating the O–H of the alcohol functional group and using it as an entry point into the structure, then go to the first carbon that is attached to the O–H. See how many carbons are attached to the first carbon. In the case of the "iso" branch, only one carbon is attached. The third structure is secondary butyl alcohol. The O–H is attached to a carbon that is attached to two other carbons. In the final structure, the compound is called tertiary butyl alcohol. The functional group is attached to a carbon that is attached to three other carbons. Notice that all of the compounds have the same molecular formula. To distinguish between them, it is necessary to include a small letter indicating which branch is in the structure. A small "i" is used for "iso" branches, a small "s" for secondary, and a small "t" for tertiary branches.

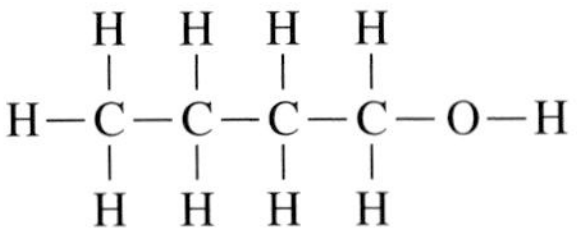

Butyl alcohol
Butanol
$C_4H_9OH$

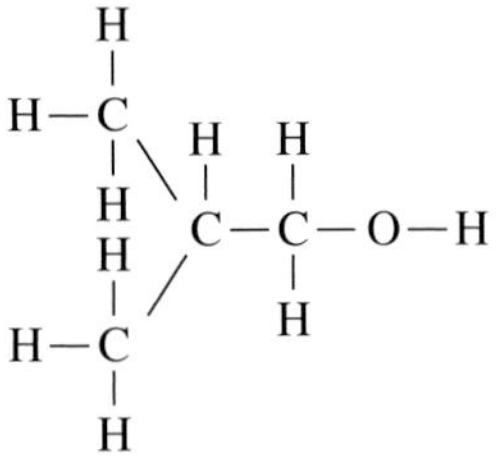

Isobutyl alcohol
Isobutanol
$i\text{-}C_4H_9OH$

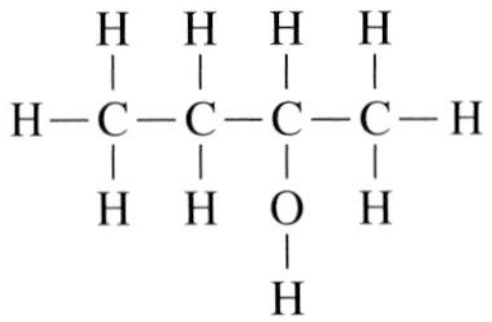

Secondary butyl alcohol
Secondary butanol
$s\text{-}C_4H_9OH$

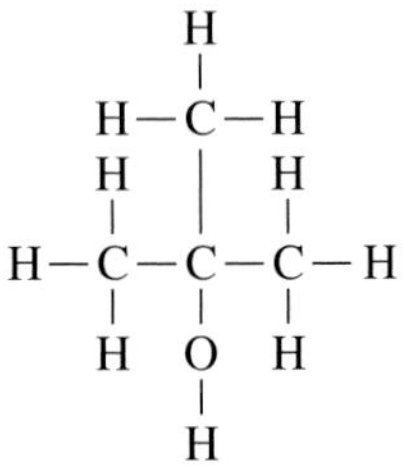

Tertiary butyl alcohol
Tertiary butanol
$t\text{-}C_4H_9OH$

There is one exception to the branching rules of hydrocarbon derivatives. Propane cannot be branched as a straight hydrocarbon because there are not enough carbons to create a branch. However, in the derivatives, the functional group becomes a part of the carbon chain for the purpose of determining branching. It is possible to put a functional group of the center carbon of propane. The structure formed appears to be secondary, according to the examples above. However, there is only one way to branch propane, so it is called the "iso" branch. That is something that has to be committed to memory as an exception to the branching rules. Another way to remember the exception is that secondary and tertiary are not used until there are four carbons in a compound. Illustrated in the following example is the structure for isopropyl alcohol, also known as isopropanol.

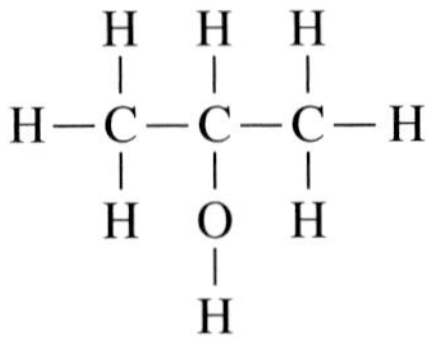

Isopropyl alcohol
Isopropanol
$i\text{-}C_3H_7OH$

Closed floating roof tank used to store polar-solvent flammable liquids, such as alcohols and ketones.

**Denatured alcohol** is ethyl alcohol or ethanol to which another liquid has been added to make it unfit to use as a beverage. The primary reason for denaturing is for tax purposes. There are approximately 50 formulations of denatured alcohol. The hazards are the same as for ethanol. The primary uses for denatured alcohol are in the manufacture of acetaldehyde, solvents, antifreeze, brake fluid, and fuels.

## Ketones

The next hydrocarbon derivatives are ketones. Ketones are flammable and narcotic. They have moderate boiling and flash points, narrow flammable ranges, and high ignition temperatures. Ketones are polar and fires should be fought with alcohol or polar-solvent type foams because water may be ineffective. They are made up of a carbon double-bonded to an oxygen, with a radical on each side. The general formula is **R–C–O–R.** The ketones are the first of several compounds that are part of the carbonyl family. Carbonyl compounds have a carbon double-bonded to an oxygen. Carbonyls are polar. The degree of polarity is less than that of the alcohols and organic acids. There are two radicals required in ketone compounds. The radicals may be the same, in which case the prefix "di" is used to indicate two, or the radicals may be different. When naming these compounds, the smallest is named first, then the second radical, ending in the word ketone.

Some ketones have trade names by which they are commonly known. There are no naming rules that can be used to determine trade names. There are some hints, such as with acetone, which is a three-carbon ketone, also called dimethyl ketone. The "one" ending indicates ketone, just as the "ol" ending indicates alcohol. The ending would be a tip-off that acetone may be a ketone. DMK is the trade name used for dimethyl ketone, which is also known as acetone. MEK is often used as a shortened name or trade name for methyl ethyl ketone. These are common ketones and the more familiar you are with hazardous materials, the more familiar you will become with alternate names and trade names. Shown in the following illustration are the structures, molecular formulas, and names for three common ketones.

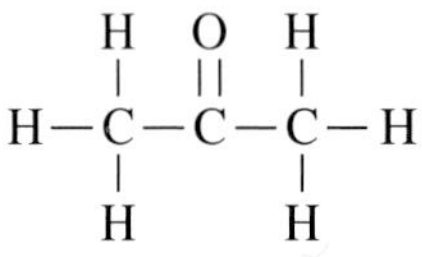

Dimethyl ketone
DMK
Acetone
CH₃COCH₃

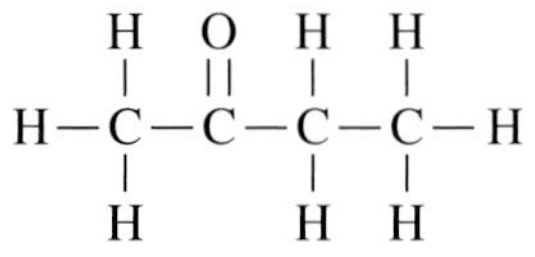

Methyl ethyl ketone
MEK
CH₃COC₂H₅

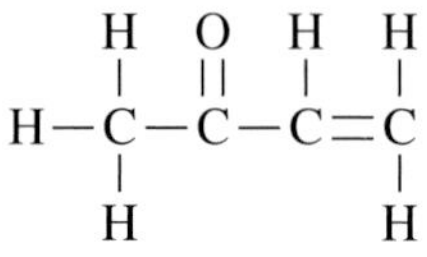

Methyl vinyl ketone
MVK
CH₃COC₂H₃

**Dimethyl ketone,** more commonly known as acetone, is the forty-first most produced industrial chemical. It will be discussed in detail under the Top 50 industrial chemicals later in this chapter.

**Methyl ethyl ketone, MEK,** is a colorless liquid with an acetone-like odor. MEK is polar, soluble in water, and has a specific gravity of 0.8, which is lighter than water. It is highly flammable, with a flammable range of 2 to 10% in air, a flash point of 16°F, and a boiling point of 176°F. The ignition temperature of MEK is 759°F, and the vapor density is 2.5, which is heavier than air. It is toxic by inhalation, with a TLV of 200 ppm in air. The four-digit UN identification number is 1193. The NFPA 704 designation for MEK is health 1, flammability 3, and reactivity 0. The primary uses of MEK are as a solvent, in the manufacture of smokeless powder, cleaning fluids, in printing, and acrylic coatings.

**Methyl vinyl ketone, MVK,** or vinyl methyl ketone, is a colorless liquid that is soluble in water. It is polar, with a specific gravity of 0.8636, which is lighter than water. It is flammable, with a flammable range of 2.1 to 15.6% in air, and the vapor density is 2.4, which is heavier than air. The boiling point is 177°F, the flash point is 20°F, and the ignition temperature is 915°F. MVK is a skin and eye irritant. The four-digit UN identification number is 1251. The NFPA 704 designation is health 4, flammability 3, and reactivity 2. The primary uses are as a monomer for vinyl resins and as an intermediate in steroid and vitamin A synthesis.

**Methyl isobutyl ketone** is a colorless, stable liquid with a pleasant odor. It is slightly soluble in water, with a specific gravity of 0.8, which is lighter than water. The vapor density is 3.5, which is heavier than air. It is highly flammable,

with a flammable range of 1.2 to 8% in air. The boiling point is 244°F, the flash
point is 64°F, and the ignition temperature is 840°F. It is toxic by inhalation,
ingestion, and skin absorption, with a TLV of 50 ppm in air. The four-digit UN
identification number is 1245. The NFPA 704 designation is health 2, flammability
3, and reactivity 1. The primary uses are as a solvent for paints, varnishes, and
lacquers; in the extraction of uranium from fission products; and as a denaturant
for alcohol. The structure for methyl isobutyl ketone is shown in the following
illustration:

Methyl isobutyl ketone
$CH_3COiC_4H_9$

## Aldehydes

Aldehydes have a wide flammable range from 3 to 55% in air; they are toxic
and may polymerize. They have moderate boiling and flash points, and high
ignition temperatures. Fires involving aldehydes should be fought with polar-
solvent type foams because water may be ineffective. Aldehydes may also form
explosive peroxides as they age much the same way ethers do. Aldehydes are
composed of a carbon double-bonded to an oxygen with a hydrogen on the other
carbon connection. Aldehydes are carbonyls and, therefore, polar. The degree of
polarity is much the same as ketone and ester, and much less than alcohol and
organic acid. They are miscible in water and require the use of polar-solvent
foams to extinguish fires. Aldehydes have the general formula of **R–CHO.** There
is one radical attached to the carbon of the aldehyde functional group. Aldehydes
are one of the three derivatives in which the carbon in the functional group is
counted when naming the compound. The alternate terms for one- and two-carbon
radicals are also used with the aldehydes. A one-carbon aldehyde uses "form"
and a two-carbon uses "acet". The aldehydes are named by identifying the radical,
naming it, and ending with the word aldehyde. Aldehydes may also be named in
the same manner as the alternate names for alcohols; however, with the aldehydes,
the ending is "al" instead of "ol". For example, a one-carbon aldehyde is called
formaldehyde with the alternate name of methanal. In the following examples,
the structures, molecular formulas, and names are shown for one-, two-, and
three-carbon aldehydes.

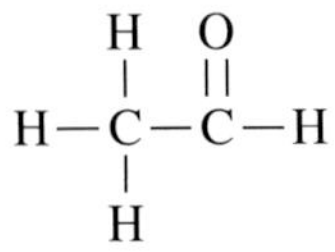

<pre>
    H   O                     H   H   O                     H   H   H   O
    |   ||                    |   |   ||                    |   |   |   ||
H — C — C — H          H — C — C — C — H          H — C — C — C — C — H
    |                        |   |                        |   |   |
    H                        H   H                        H   H   H
</pre>

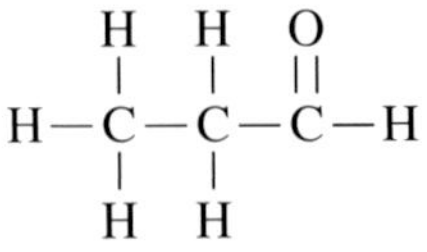

|  Acetaldehyde  |  Propionaldehyde  |  Butyraldehyde  |
| :---: | :---: | :---: |
|  Ethanal  |  Propanal  |  Butanal  |
|  $CH_3CHO$  |  $C_2H_5CHO$  |  $C_3H_7CHO$  |

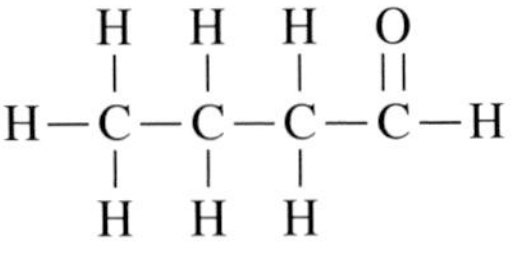

Aldehydes are carbonyls and thus polar. The polarity of the carbonyls is somewhat less than that of the alcohols and the organic acids. However, they are still miscible with water and require polar-solvent foam to extinguish fires.

**Acetaldehyde,** also known as ethanal, is a colorless liquid with a pungent, fruity odor. The odor is detectable at 0.07 to 0.21 ppm in air. It is highly flammable and a dangerous fire and explosion risk with a wide flammable range of 4 to 60% in air. The boiling point is 69°F, the flash point is –36°F, and the ignition temperature is 374°F. Acetaldehyde is miscible in water and the specific gravity is 0.78, which is lighter than water. The vapor density is 1.52, which is heavier than air. In addition to flammability, it is toxic (narcotic) and has a TLV of 100 ppm in air. Eye irritation occurs at 25 to 50 ppm in air. The four-digit UN identification number is 1089. The NFPA 704 designation is health 3, flammability 4, and reactivity 2. Its primary use is in the manufacture of other chemicals and artificial flavorings.

**Propionaldehyde,** also known as propanal and propylaldehyde, is a water-white liquid with a suffocating odor that is water soluble. It is a dangerous fire and explosion risk with a flammable range of 3 to 16% in air. The boiling point is 120°F, the flash point is 16°F, and the ignition temperature is 405°F. It is partially soluble in water and has a specific gravity of 0.81, which is lighter than water. The vapor density is 0.807, which is lighter than air. In addition to flammability, propionaldehyde is an irritant to the eyes, skin, and respiratory system. The four-digit UN identification number is 1275. The NFPA 704 designation is health 2, flammability 3, and reactivity 2. The primary uses are in the manufacture of other chemicals and plastics, as well as a preservative and disinfectant.

## Esters

The final flammable-liquid hydrocarbon derivatives are esters. In addition to being flammable, esters may polymerize. They have moderate boiling and flash points, narrow flammable ranges, and high ignition temperatures. Esters are made through a process referred to as esterification. An ester is formed when an alcohol is combined with an organic acid with water as a by-product. This process is illustrated in the following example by combining acrylic acid and methyl alcohol; the resulting ester compound is methyl acrylate.

Cone roof tank for flammable liquids.

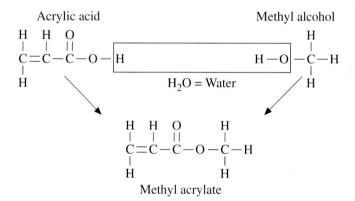

Esters have a carbon double bonded to one oxygen and a single bond with another oxygen. Esters are carbonyls and are polar. The degree of polarity is much less than organic acids and alcohols and is similar to ketones and aldehydes. Esters are miscible in water and require polar-solvent foams when fighting fires. In this book, only three esters will be discussed because of their common commercial use. The general formula for ester is **R–C–O–O–R,** or **R–C–O₂–R.** There are two radicals in the ester compounds. Because nothing is ever called ester, the word ester will not appear in the name of the compound. The radical that is attached to the carbon in the functional group determines which ester compound it will be. Esters are one of the functional groups in which all the carbons are counted, including the carbon in the functional group when naming the type of ester compound. In addition, esters use the alternate name for one- and two-carbon radicals that are attached to the carbon in the functional group only. The radical on the other side is named in the normal way. The ester is named with the radical on the right first, ending in the name of the type of ester on the left. Certain radicals attached to the carbon of the functional group will produce certain esters. For example, in a one-carbon ester the carbon in the functional group is used as the one carbon. A single hydrogen is attached to the carbon in the functional group to complete the bonding requirements. A one-carbon ester uses the alternate name for one carbon, which is "form"; esters end in "ate", so the name of a one-carbon ester is formate. Any radical can be attached to the oxygen in the functional group. If a methyl were added to the oxygen, the name of the ester compound is methyl formate. When a methyl radical is attached to the carbon in the functional group, it forms a two-carbon chain. The alternate name for two carbons, "acet", is used; the ending "ate" is used to indicate an ester, so a two-carbon ester is an acetate. The second radical added to the oxygen in the functional group determines what type of acetate the compound is. Theoretically, any radical can be used. If a vinyl radical were used, the compound would be vinyl acetate.

The last ester has a vinyl radical attached to the carbon in the functional group. This forms a three-carbon chain with one double bond in the chain. The name for a three-carbon radical with one double bond is "acryl". The ending for the ester is "ate"; so the ester is called acrylate. Any radical can be attached to the oxygen in the functional group. If a methyl is attached, the ester compound is called methyl acrylate. In the following illustrations, the structures and molecular formulas are shown for methyl formate, vinyl acetate, and methyl acrylate. There can be other radicals attached to the oxygen, which change the name of the compound.

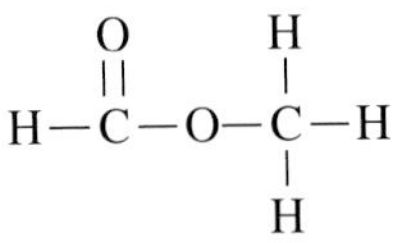

Methyl formate
HCOOCH$_3$

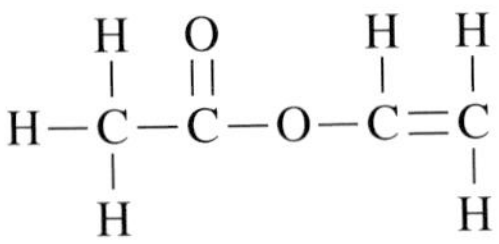

Vinyl acetate
CH$_3$COOC$_2$H$_3$

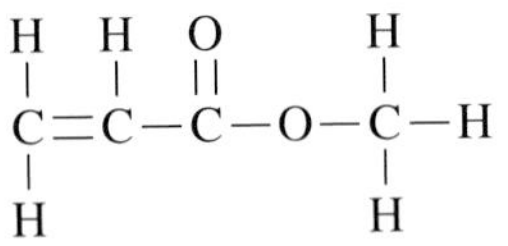

Methyl acrylate
C$_2$H$_3$COOCH$_3$

**Methyl formate** is a colorless liquid with an agreeable odor. It is a dangerous fire and explosion risk, with a flammable range of 5 to 23% in air. The boiling point is 89°F, the flash point is –2°F, and the ignition temperature is 853°F. In addition to being flammable and a polymerization hazard, it is also an irritant, with a TLV of 100 ppm in air. It is water soluble and has a specific gravity of 0.98, which is slightly lighter than water. The vapor density is 2.07, which is heavier than air. The four-digit UN identification number is 1243. The NFPA 704 designation is health 2, flammability 4, and reactivity 0. The primary uses of methyl formate are as a solvent, a fumigant, and a larvicide.

**Vinyl acetate** is included in the Top 50 industrial chemicals.

**Methyl acrylate** (inhibited) is a colorless, volatile liquid. It is a dangerous fire and explosion risk, with a flash point of 2.8 to 25% in air. The boiling point is 177°F, the flash point is 27°F, and the ignition temperature is 875°F. It is immiscible in water and has a specific gravity of 0.96, which is lighter than water. The vapor density is 0.957, which is slightly lighter than air. In addition to flammability and polymerization hazards, methyl acrylate is toxic by inhalation, ingestion, and skin absorption. It is an irritant to skin and eyes, with a TLV of 10 ppm in air. The four-digit UN identification number is 1919. The NFPA 704 designation is health 3, flammability 3, and reactivity 2. The primary uses of methyl acrylate are in polymers, in vitamin $B_1$, and as a chemical intermediate.

## TOP 50 INDUSTRIAL CHEMICALS

All of the chemicals mentioned in this section have a primary hazard of flammability; however, because they are different in chemical make-up, some require different firefighting tactics. As each of the different chemicals is presented, note the differences in physical and chemical characteristics. The secondary or hidden hazards of flammable liquids vary widely from one Top 50 chemical to another. Among the flammable liquids, many are also toxic, anesthetic, narcotic, and undergo polymerization. Many different hydrocarbon-derivative functional groups are represented in the flammable liquids. Earlier in this chapter, it was mentioned that almost 52% of all hazardous materials incidents involve flammable liquids. While many of those spills are hydrocarbon fuels, like gasoline and diesel fuel, they also involve industrial flammable liquids. Fifteen (30%) of the Top 50 industrial chemicals are flammable liquids.

Open floating roof tank used for hydrocarbon fuels.

**Ethylene dichloride** is the thirteenth most produced industrial chemical, with 17.26 billion lbs. in 1995. Ethylene dichloride is an alkyl halide hydrocarbon derivative. The primary hazard of the alkyl halides is toxicity. There are, however, some individual alkyl halides that are very flammable and classified as flammable liquids. Ethylene dichloride is a colorless, oily liquid with a chloroform-like odor and a sweet taste. It is a dangerous fire risk, with a flammable range of 6 to 16% in air. The boiling point is 183°F, the flash point is 56°F, and the ignition temperature is 775°F. Small fires involving ethylene dichloride should be fought with dry chemical and large fires with a hydrocarbon foam. Water may be ineffective and, if used, should be applied gently to the surface of the liquid. Water is generally ineffective against flammable liquid fires where the liquid has a flash point below 100°F. The farther below 100°F the liquid's flash point is, the less effective water will be. In addition to flammability, ethylene dichloride is toxic by ingestion, inhalation, and skin absorption; it is also a known carcinogen, with a TLV of 10 ppm in air. The vapor density is 3.4, which is heavier than air, so the vapors will stay close to the ground. The specific gravity is 1.3, which makes it heavier than water, so it will sink to the bottom. Alkyl halides are nonpolar and ethylene dichloride is only slightly miscible in water. The four-digit UN identification number is 1184. The NFPA 704 designation for ethyl dichloride is health 2, flammability 3, and reactivity 0. Ethylene dichloride is shipped in metal cans, drums, tank trucks, railcars, and barges. It is usually packaged under nitrogen gas, which is an inert material. Ethylene dichloride is used in the production of vinyl chloride and trichloroethane. It is also used in metal degreasing, as a paint remover, a solvent, and a fumigant.

$$Cl-\overset{\overset{\displaystyle H}{|}}{C}=\overset{\overset{\displaystyle H}{|}}{C}-Cl$$

Ethylene dichloride
$C_2H_2Cl_2$

**Benzene** is the sixteenth most heavily produced industrial chemical, with 15.97 billion lbs. in 1995. Benzene is the parent member of the aromatic hydrocarbon family. It is a colorless to light yellow liquid with a characteristic aromatic odor. Benzene is nonpolar and burns with incomplete combustion, producing a smoky fire. The flammable range is 1.5 to 8% in air. Benzene is also toxic, with a TLV of 10 ppm, and is a known carcinogen. Concentrations of 8000 ppm for 30 to 60 min are fatal. Its boiling point is 176°F, and the flash point is 12°F. The ignition temperature of benzene is 928°F. Its vapor density is 2.8, making it heavier than air. Small fires involving benzene should be fought with dry chemical and large fires with hydrocarbon foams. Water may be ineffective, and if used, should be applied gently to the liquid surface. Benzene is immiscible in water and has a specific gravity of 0.9. It is lighter than water and will float on the surface. The four-digit UN identification number for benzene is 1114. The NFPA

704 classification is health 2, flammability 3, and reactivity 0. Benzene reacts with oxidizing materials and should be stored away from them in fixed facilities. It is shipped in 55-gal drums, highway tank trucks, railcars, and barges. Benzene is used in the manufacture of many other chemicals and as a solvent. The structure and molecular formula for benzene are illustrated under the Aromatic Hydrocarbons in this chapter.

The twelveth most produced industrial chemical is **methyl *tert*-butyl ether,** with 17.62 billion lbs. produced in 1995. It is an ether hydrocarbon derivative and is highly flammable with a wide flammable range. Methyl *tert*-butyl ether is not considered toxic; however, it is mildly irritating to the eyes and skin. If inhaled, it may cause suffocation. It is a gasoline additive and there has been some controversy concerning potential health effects. Methyl *tert*-butyl ether is added to gasoline as an oxygenating compound that makes gasoline burn cleaner in the winter. The Centers for Disease Control conducted a study that was inconclusive: it did not vindicate the material, it just did not find enough evidence that the chemical is a health concern. Methyl *tert*-butyl ether is a colorless, nonpolar liquid with an anesthetic-like odor. It has a boiling point of 131°F and a flash point of –14°F. Extinguishing agents for ethers and other nonpolar, non-miscible, or slightly miscible liquids should be selected carefully. Small fires can be extinguished with dry chemical with some difficulty; remember that ether has an oxygen in the compound so excluding atmospheric oxygen may not be effective. Water may also be ineffective. Alcohol-type foams may be effective against materials that are slightly miscible. The higher molecular weight liquids will attack the alcohol-type foam and a hydrocarbon foam will be needed. The vapor density is 3, which makes it heavier than air. The specific gravity is 0.74, which is less than the weight of water; therefore, it will float on top of water. Methyl *tert*-butyl ether has a four-digit UN identification number of 2398. It is used primarily as an octane booster for unleaded gasoline. There is no NFPA 704 data available for methyl *tert*-butyl ether and the reference information on the material is sketchy. No ignition temperature or flammable range information is available in the common reference sources, including CAMEO (Computer-Aided Management of Emergency Operations). The structure and molecular formula are shown in the following illustration:

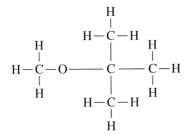

Methyl *tert*-butyl ether
$CH_3OtC_4H_9$

**Ethylbenzene** is the nineteenth most heavily produced industrial chemical, with 13.66 billion lbs. in 1995. It is a colorless aromatic hydrocarbon, with a characteristic odor. It is a dangerous fire risk, with a flammable range of 0.8 to 6.7% in air. The boiling point is 277°F, the flash point is 70°F, and the ignition temperature is 810°F. Small fires may be extinguished with dry chemical and large fires with hydrocarbon-type foam. Water may be ineffective and should be applied gently to the surface of the liquid if used. In addition to flammability, it is toxic by ingestion, inhalation, and skin absorption, with a TLV of 100 ppm in air. The vapor density is 3.7, which makes it heavier than air, and the vapors will tend to stay close to the ground. It is nonpolar, with a specific gravity of 0.9, which means it will float on water. Ethylbenzene is immiscible in water. The four-digit UN identification number is 1175. The NFPA 704 designation is health 2, flammability 3, and reactivity 0. It is shipped in cans, bottles, 55-gal drums, tank trucks, railcars, and barges. It should not be stored near oxidizing materials. The primary uses of ethylbenzene are as a solvent and an intermediate in the production of styrene. The structure and molecular formula are shown in the following example:

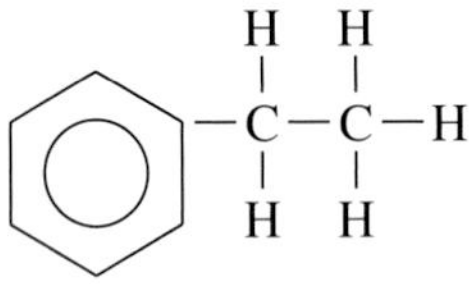

Ethylbenzene
$C_6H_5C_2H_5$

The twentieth most heavily produced industrial chemical is **styrene,** with 11.39 billions lbs. in 1995. Styrene is a colorless, oily, liquid aromatic hydrocarbon, with a characteristic odor. It is sometimes called vinylbenzene or phenylethylene. Styrene is a monomer and must be inhibited during transportation and storage to prevent polymerization. It is a moderate fire risk, with a flammable range of 1.1 to 6.1%. The boiling point is 295°F, with a flash point of 88°F, and an ignition temperature of 914°F. Small fires may be extinguished with dry chemical and large fires with hydrocarbon-type foam. Water may be ineffective and should be applied gently to the surface of the liquid if used. In addition to flammability, styrene is toxic by ingestion and inhalation, with a TLV of 50 ppm in air. The vapor density is 3.6, which is heavier than air. Styrene is nonpolar and it is immiscible in water, with a specific gravity of 0.9. The four-digit UN identification number for styrene monomer, inhibited, is 2055. The NFPA 704 designation is health 2, flammability 3, and reactivity 2. It is shipped in 55-gal drums, tank trucks, railcars, and barges. When stored, it should be kept away from oxidizers. The structure and molecular formula for styrene are shown in the following illustration:

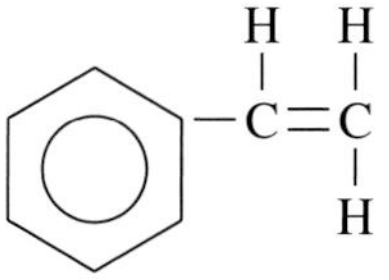

$$
\begin{array}{c}
\quad\quad H \quad H \\
\quad\quad | \quad\quad | \\
\bigcirc\!\!-\!C=C \\
\quad\quad\quad\quad | \\
\quad\quad\quad\quad H
\end{array}
$$

Styrene
$C_6H_5C_2H_3$

**Methanol,** also known as methyl alcohol and wood alcohol, is the twenty-first most produced industrial chemical, at 11.29 billion lbs. in 1995. Methyl alcohol is an alcohol hydrocarbon derivative. It is a clear, colorless liquid that is highly polar and miscible in water. Alcohol has hydrogen bonding and is the second most polar material of the hydrocarbon derivatives after organic acids. It is a dangerous fire risk, with a wide flammable range from 6 to 36.5% in air. Fighting fires will require the use of alcohol-type foam. The boiling point is 147°F, and the flash point is 52°F. Notice from the following structure that the molecular weight of methyl alcohol is 32. It has only one carbon and one oxygen, yet the boiling point and flash point are high. The ignition temperature of methyl alcohol is 867°F. The effects of polarity can readily be seen in this example. In addition to being flammable, methyl alcohol is toxic by ingestion and has a TLV of 200 ppm in air. The vapor density is 1.1, which is heavier than air. The specific gravity is 0.8, which makes it lighter than water. Methyl alcohol is miscible in water so it will mix rather than form defined layers. The UN 4 digit identification number is 1230. The NFPA 704 designation is health 1, flammability 3, and reactivity 0. Methyl alcohol is shipped in glass bottles, 55-gal drums, tank trucks, and railcars. The structure and molecular formula are shown in the following illustration:

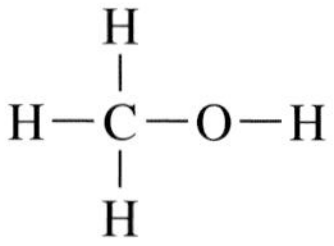

$$
\begin{array}{c}
\quad\quad H \\
\quad\quad | \\
H-C-O-H \\
\quad\quad | \\
\quad\quad H
\end{array}
$$

Methanol
Methyl alcohol
$CH_3OH$

The twenty-third most produced industrial chemical at 9.37 billion lbs. is **Xylene; para-Xylene** is the twenty-ninth most produced industrial chemical, at 6.34 billion lbs. produced in 1995. Xylene, sometimes referred to as dimethyl-benzene and xylol, is an aromatic hydrocarbon. The *Condensed Chemical Dictionary* refers to xylene as "a commercial mixture of the three isomers, "ortho", "meta", and "para". Xylene is a clear liquid that is nonpolar and immiscible in water. It is a moderate fire risk, with a flammable range of 0.9 to 7% in air. The boiling point is between 281 and 292°F, depending on the mixture. The flash point ranges from 81 to 90°F, and the ignition temperature ranges from 867 to 984°F. Small fires may be extinguished with dry chemical and large fires with hydrocarbon-type foam. Water may be ineffective and should be applied gently to the surface of the liquid if used. In addition to flammability, xylene is toxic by ingestion and inhalation, with a TLV of 100 ppm in air. The vapor density is 3.7; therefore it is heavier than air. The specific gravity is 0.9, so the xylenes will float on top of water. The four-digit UN identification number is 1307. The NFPA 704 designation for xylene is health 2, flammability 3, and reactivity 0. It is shipped in 55-gal drums, tank trucks, railcars, and barges. The structures and molecular formulas are shown in the Aromatic Hydrocarbon section of this chapter.

**Toluene** is the twenty-eighth most produced industrial chemical, at 6.73 billion lbs. in 1995. Toluene is an aromatic hydrocarbon. It is a colorless liquid with an aromatic odor. It is nonpolar and immiscible in water. Toluene is a dangerous fire risk, with a flammable range of 1.27 to 7% in air. The boiling point is 231°F, with a flash point of 40°F. The ignition temperature of toluene is 896°F. Small fires may be extinguished with dry chemical and large fires with hydrocarbon-type foam. Water may be ineffective and should be applied gently to the surface of the liquid if used. In addition to flammability, toluene is toxic by ingestion, inhalation, and skin absorption, with a TLV of 100 ppm in air. Its vapor density is 3.1, which is heavier than air. The specific gravity is 0.9, so it is lighter than water and will float on the surface. The four-digit UN identification number is 1294. The NFPA 704 designation is health 2, flammability 3, and reactivity 0. It is shipped in 55-gal drums, tank trucks, railcars, and barges. The structure and molecular formula are in the Aromatic Hydrocarbon section of this chapter.

Bulk containers above 119 gallons must be placarded on two sides.

Underflow dams used to stop the flow of flammable liquids that are lighter than water.

**Cumene,** isopropylbenzene, is the thirtieth most produced industrial chemical, with 5.63 billion lbs. in 1995. It is a colorless liquid that is insoluble in water. The specific gravity is 0.9, which is lighter than water. It is a moderate fire risk, with a flammable range of 0.9 to 6.5%. The boiling point is 306°F, the flash point is 96°F, and the ignition temperature is 795°F. Small fires may be extinguished with dry chemical, and large fires with hydrocarbon-type foam. Water may be ineffective and should be applied gently to the surface of the liquid if used. The vapor density is 4.1, which is heavier than air. In addition to flammability, cumene is toxic by ingestion, inhalation, and skin absorption; it is also a narcotic. The TLV is 50 ppm in air. The four-digit UN identification number is 1918. The NFPA 704 designation is health 2, flammability 3, and reactivity 0. The primary uses are in the production of phenol, acetone, and methylstryene solvents. The structure and molecular formula for cumene are shown in the following example:

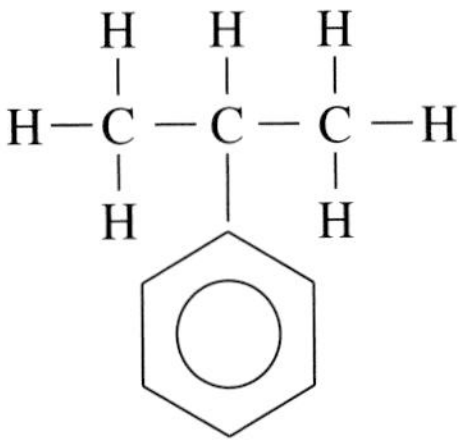

Cumene
Isopropylbenzene
$C_6H_5iC_3H_7$

**Propylene oxide** is the thirty-fifth most produced industrial chemical, at 4 billion lbs. in 1995. Propylene oxide is a cyclic ether hydrocarbon derivative, although it does not follow the trivial naming system for ethers. It has a cyclic structure between the oxygen and two carbons and is a colorless liquid with an ether-like odor. Propylene oxide is nonpolar, partially soluble in water, and is highly flammable, with a wide flammable range of 2 to 22% in air. The boiling point is 94°F, the flash point is 35°F, and the ignition temperature is 840°F. Small fires can be extinguished with dry chemical with some difficulty. Remember that ether has an oxygen in the compound, so excluding atmospheric oxygen may not be effective. Water may also be ineffective. Alcohol-type foams may be effective against materials that are slightly miscible. The higher molecular weight liquids will attack the alcohol-type foam and a hydrocarbon foam will need to be used. In addition to flammability, propylene oxide is an irritant with a TLV of 20 ppm in air and it is also corrosive. The vapor density is 2, so it is heavier than air. The specific gravity is 0.83, which is lighter than water, and it will float on the surface. The four-digit UN identification number is 1280. The NFPA 704 designation for propylene oxide is health 3, flammability 4, and reactivity 2. It is shipped in steel cylinders, tank trucks, railcars, and barges under the cover of nitrogen, which is an inert material. The structure and molecular formula for propylene oxide are shown in the following illustration:

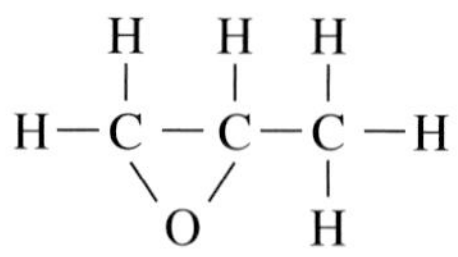

$$
\begin{array}{ccc}
\text{H} & \text{H} & \text{H} \\
| & | & | \\
\text{H}-\text{C}-\text{C}-\text{C}-\text{H} \\
\diagdown \diagup & | \\
\text{O} & \text{H}
\end{array}
$$

Propylene oxide
$CH_2OCHCH_3$

The fortieth most produced industrial chemical is **acrylonitrile,** at 3.21 billion lbs. in 1995. Acrylonitrile is sometimes referred to as vinyl cyanide. It is a colorless liquid with a mild odor, is nonpolar, and partially miscible in water. Acrylonitrile is a dangerous fire risk, with a flammable range of 3 to 17% in air. Small fires can be extinguished with dry chemical with some difficulty. Water may also be ineffective. Alcohol-type foams may be effective against materials that are slightly miscible. The higher molecular weight liquids will attack the alcohol-type foam and a hydrocarbon foam will need to be used. The boiling point is 171°F, the flash point is 32°F, and the ignition temperature is 898°F. In addition to flammability, acrylonitrile is toxic by inhalation and skin absorption. Acrylonitrile may polymerize because of the double bond in the vinyl and the triple bond between the carbon and nitrogen; it is always shipped with an inhibitor. The TLV is 2 ppm in air and it is considered a human carcinogen. The vapor density is 1.8, which is heavier than air. The specific gravity is 0.8, which is lighter than water, so it will float on the surface. The four-digit UN identification number is 1093. The NFPA 704 designation is health 4, flammability 3, and reactivity 2. It is shipped in 55-gal drums, tank trucks, railcars, and barges. It should not be stored or shipped uninhibited. The structure and molecular formula are illustrated in the following example.

$$\begin{array}{c} \overset{\displaystyle H}{\underset{\displaystyle |}{}} \quad \overset{\displaystyle H}{\underset{\displaystyle |}{}} \\ C = C - C \equiv N \\ \underset{\displaystyle |}{} \\ H \end{array}$$

Acrylonitrile
$CH_2CHCN$

**Vinyl acetate** is the forty-first most produced industrial chemical, at 2.89 billion lbs. in 1995. Vinyl acetate is an ester hydrocarbon derivative. It is a colorless liquid that has been stabilized with an inhibitor. Although it is a polar compound because of the carbonyl structure, it is only slightly miscible in water. Vinyl acetate is a highly flammable liquid, with a flammable range of 2.6 to 13.4% in air, and it may polymerize without the inhibitor or when exposed to heat or an oxidizer during an accident. The boiling point is 161°F, the flash point is 18°F, and the ignition temperature is 756°F. Fighting fires will require the use of an alcohol-type foam. In addition to flammability, vinyl acetate is toxic by inhalation and ingestion, with a TLV of 10 ppm in air. The vapor density is 3, so it is heavier than air. The specific gravity is 0.9, which means it will float on water. The four-digit UN identification number is 1301. The NFPA 704 designation is health 2, flammability 3, reactivity 2. It is shipped in 55-gal drums, tank trucks, railcars, and barges. It should be stored separately from oxidizing materials. The structure for vinyl acetate is shown in the Esters portion of the Hydrocarbon Derivatives section of this chapter.

The forty-third most produced industrial chemical is **acetone,** with 2.76 billion lbs. in 1995. Acetone, also known as dimethyl ketone, is a ketone hydrocarbon derivative. It is a colorless, volatile liquid with a sweetish odor. Acetone is a carbonyl, is polar, and is miscible with water, which means it will require a polar-solvent foam for fire extinguishment. This compound is highly flammable, with a range of 2.6 to 12.8% in air. The boiling point is 133°F, the flash point is –4°F, and the ignition temperature is 869°F. Fighting fires will require the use of an alcohol-type foam. In addition to flammability, acetone is a narcotic at high concentrations, and is moderately toxic by ingestion and inhalation, with a TLV of 750 ppm in air. The vapor density is 2, which is heavier than air. The specific gravity is 0.8, so it is lighter than water, but is very miscible and will mix with the water rather than form layers. The four-digit UN identification number is 1090. The NFPA 704 designation is health 1, flammability 3, and reactivity 0. It is shipped in pails, drums, tank trucks, railcars, and barges. Acetone is the solvent used to dissolve acetylene in cylinders to keep it stable. The structure and molec-

Geodesic dome retrofitted as closed floating roof tank.

ular formula for acetone are shown in the Ketones portion of the Hydrocarbon Derivatives section of this chapter.

**Butyraldehyde,** also known as butanal, is the forty-fifth most produced industrial chemical, with 2.68 billion bounds in 1995. It is a water-white liquid with a pungent aldehyde odor. Butyraldehyde is a dangerous fire risk, with a flammable range of 2.5 to 12.5% in air. The boiling point is 168°F, the flash point is 10°F, and the ignition temperature is 446°F. It is slightly soluble in water with a specific gravity of 0.8, which is lighter than water. The vapor density is 0.804, which is lighter than air. In addition to flammability, butyraldehyde is corrosive and causes severe eye and skin burns. It may be harmful if inhaled. The four-digit UN identification number is 1129. The NFPA 704 designation is health 3, flammability 3, and reactivity 2. The primary uses of butyraldehyde are in plastics and rubber, and as a solvent. The structure and formula are shown in the Hydrocarbon Derivatives section of this chapter.

**Cyclohexane** is the forty-seventh most produced industrial chemical, at 2.13 billion lbs. in 1995. Cyclohexane is a cyclic alkane hydrocarbon with all single bonds between the carbons. The liquid is colorless, nonpolar, and immiscible with water. Cyclohexane is highly flammable, with a flammable range of 1.3 to 8% in air. The boiling point is 179°F, the flash point is –4°F, and the ignition temperature is 473°F. Small fires may be extinguished with dry chemical and large fires with a hydrocarbon-type foam. Water may be ineffective and should be applied gently to the surface of the liquid if used. In addition to being flammable, cyclohexane is toxic by inhalation, with a TLV of 300 ppm in air. The vapor density is 29, so it is significantly heavier than air. The specific gravity is 0.8, which is lighter than water so it will float on the surface. The four-digit

UN identification number is 1145. The NFPA 704 designation for cyclohexane is health 1, flammability 3, and reactivity 0. It is shipped in 55-gal drums, tank trucks, railcars, and barges. The structure and molecular formula are shown in the following illustration:

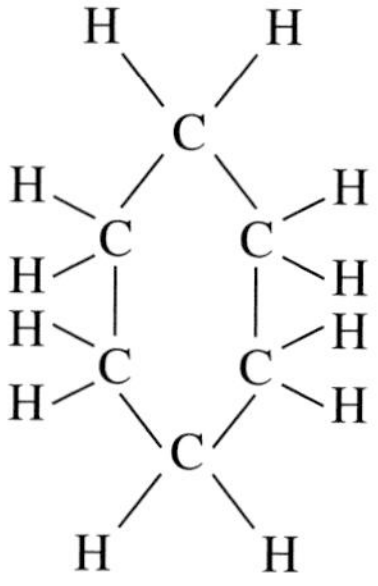

Cyclohexane

$c\text{-}C_6H_{12}$

**n-Butyl alcohol** was the forty-ninth most produced industrial chemical at 1.45 billion lbs. in 1994; however, it fell from the Top 50 in 1995. Butyl alcohol is an alcohol hydrocarbon derivative, and is a colorless liquid with a wine-like odor. Alcohols are highly polar and are miscible in water. Butyl alcohol is a moderate fire risk, with a flammable range of 1.4 to 11.2% in air. The boiling point is 243°F, the flash point is 98°F, the ignition temperature is 650°F. Fires should be fought with alcohol-type foams. In addition to being flammable, butyl alcohol is toxic when inhaled for long periods, irritating to the eyes, and absorbed through the skin. The TLV is 50 ppm in air. The vapor density of butyl alcohol is 2.6, so it is heavier than air. The specific gravity is 0.8, which is lighter than water; however, it is miscible in water and will mix rather than form layers. The four-digit UN identification number is 1120. The NFPA 704 designation for butyl alcohol is health 1, flammability 3, and reactivity 0. It is shipped in glass bottles, pails, 55-gal drums, tank trucks, railcars, and barges. The structure and molecular formula are shown in the Alcohols portion of the Hydrocarbon Derivatives section of this chapter.

The final flammable liquid in the Top 50 is **isopropyl alcohol,** which was number 50 in 1994, with 1.39 billion pounds produced; in 1995, isopropy alcohol also fell from the Top 50. Isopropyl alcohol is an alcohol hydrocarbon derivative. The liquid is colorless with a pleasant odor, highly polar, and miscible in water. Firefighting will require polar-solvent foam for extinguishment. Isopropyl alcohol is highly flammable, with a flammable range of 2 to 12% in air. The boiling point is 181°F, the flash point is 53°F, and the ignition temperature is 750°F. In addition to flammability, isopropyl alcohol is toxic by ingestion and inhalation, with a TLV of 400 ppm in air. The vapor density is 2.1, which is heavier than air. The specific gravity is 0.8, which is lighter than water; however, it is miscible in water and will mix rather than form layers. The four-digit UN identification number is

1219. The NFPA 704 designation is health 1, flammability 3, and reactivity 0. It is shipped in glass bottles, pails, 55-gal drums, tank trucks, railcars, and barges. The structure and molecular formula for isopropyl alcohol are shown in the following example:

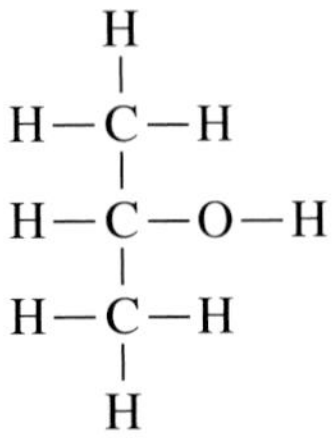

$$
\begin{array}{c}
\text{H} \\
| \\
\text{H}-\text{C}-\text{H} \\
| \\
\text{H}-\text{C}-\text{O}-\text{H} \\
| \\
\text{H}-\text{C}-\text{H} \\
| \\
\text{H}
\end{array}
$$

Isopropyl alcohol
$C_3H_7OH$

## OTHER FLAMMABLE LIQUIDS

**Carbon disulfide** is a nonmetal compound that is a clear, colorless, or faintly yellow liquid, and almost odorless. It is highly flammable, with a flammable range of 1.3 to 50% in air, and can be ignited by friction. The boiling point is 115°F, the flash point is –22°F, and the ignition temperature is very low at 194°F. Contact with a steampipe or a lightbulb could ignite carbon disulfide. It is slightly water soluble and has a specific gravity of 1.26, which is heavier than water. In addition to being highly flammable, carbon disulfide is also a poison and is toxic by skin absorption, with a TLV of 10 ppm in air. The four-digit identification number is 1131. The NFPA 704 designation is health 3, flammability 4, and reactivity 0. The primary uses of carbon disulfide are as a solvent and in the manufacture of rayon, cellophane, and carbon tetrachloride.

## CHEMICAL RELEASE STATISTICS

There are two flammable liquids in the EPA Top 15 Chemical Releases from fixed facilities during 1988 through 1992: benzene and toluene. The Top 15 does not include the spills of hydrocarbon fuels, which are flammable liquids and are the most frequently released hazardous material. In the statistical period, there were 1048 incidents involving benzene and 579 involving toluene. These releases resulted in 63 injuries and over 12 million lbs. of material released into the environment.

The DOT reports that during 1995 flammable liquids were the second most released hazardous material in incidents with 1139. This figure does not take into account fuels spilled from vehicles other than commercial transportation. The largest number of fatalities and serious injuries occurred from flammable liquid incidents.

The Association of American Railroads reported that 26% of all non-derailment leaks of hazardous materials involved flammable liquids. Of all tank car shipments of hazardous materials on the rails 28% involve flammable liquids. Of these, an average of approximately 20% are of gasoline, 10% of fuel oil flammable, 17% of fuel oil combustible, 8% of methyl alcohol, and 17% of denatured alcohol. Fuel oil is the tenth highest volume commodity shipped by rail, methyl alcohol is nineth, denatured alcohol fourteenth, and gasoline twentieth. In 1994, 279 tank cars were found leaking flammable liquids.

## INCIDENTS

Most of the more commonly encountered flammable liquids are fuels: gasoline, diesel fuel, heating oil, and jet fuel. Other spills include materials such as alcohols, ketones, aldehydes, paint thinners, pesticides, benzene, toluene, and xylene, along with other industrial solvents. These flammable liquids can present responders with special problems as well as hidden hazards. Because they are liquids, they can flow away from the scene, following the terrain into storm drains, sanitary sewers, waterways, and other low-lying areas. In addition to the flammability hazard, responders will need to stop the flow of the product in some situations, if properly trained and equipped to do so. Incidents range from leaks in vehicle fuel tanks, to transportation accidents involving tank trucks resulting in leaks and fires, to large bulk-storage tank fires.

In Kansas City, KS, a fire occurred at a gasoline station that also was associated with a bulk-fuel storage and loading area. This incident occurred prior to the use of underground storage tanks for gasoline stations, and was one of the primary forces that led to the requirements for the use of underground tanks. The fire started at a loading rack on top of a tanker being loaded with fuel. Eventually the fire spread to horizontal storage tanks used to supply the fuel dispensers at the gasoline station. The elevated horizontal tanks had unprotected steel supports. These supports quickly collapsed from the heat of the fire. When one of the tanks hit the ground it opened up, sending burning fuel into the street where firefighters were directing large streams of water onto the fire. Six firefighters lost their lives when they were overrun by the burning fuel. This incident resulted in a number of code changes involving the storage of flammable liquids.

A fire in Norfolk, VA occurred involving an MC 306 gasoline tanker truck. When firefighters arrived, they found the tanker fully involved, with leaking, burning fuel flowing down the street into storm sewers leading to a retention pond. They had fires on multiple fronts, including the tanker, parked cars in an adjacent parking lot, on the street, in the sewer, and on the retention pond. Exposures included a senior-citizen apartment building next to the fire in the retention pond. The fire was extinguished without incident. During mop-up operations, a firefighter was injured when cut by a jagged edge on the burned-out shell of the tanker as he fell into a portion of the tank.

MC/DOT 306/406 tanker used to transport flammable fuels, such as gasoline and diesel fuel.

## REVIEW QUESTIONS

1. The boiling point of a liquid is defined as the point in which _____ equals _____.

2. The flash point of a liquid has to do with the _____ temperature.

3. Liquids that have low boiling points also have _____ and _____.

4. Liquids that have high flash points have _____ vapor pressure.

5. There are three factors that can affect boiling point: _____, _____, and _____.

6. Which of the following compounds are polar and which are nonpolar?
   _____ Benzene             A. Polar
   _____ Propyl alcohol      B. Nonpolar
   _____ Ethyl ether         C. Super polar
   _____ Methyl ethyl ketone D. Super-duper polar
   _____ Formic acid
   _____ Pentane

7. Polymers are long-chained molecules that are made up of _____.

8. Identify the following formulas as alkanes, alkynes, alkenes, or aromatics.

   $C_6H_{14}$     $C_7H_8$     $C_7H_{14}$     $C_6H_6$     $C_8H_{16}$

9. Provide structures, names, and hazards for the following hydrocarbon derivative compounds.

   HCHO     $C_2H_3COOCH_3$     HCOOH     $CH_3CHO$     $CH_3COOH$

10. Provide the names, formulas, and hazards for the following structures.

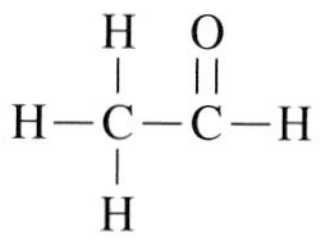

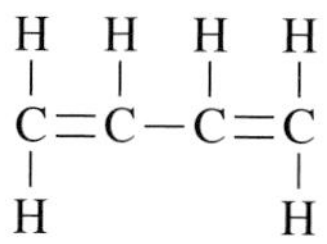

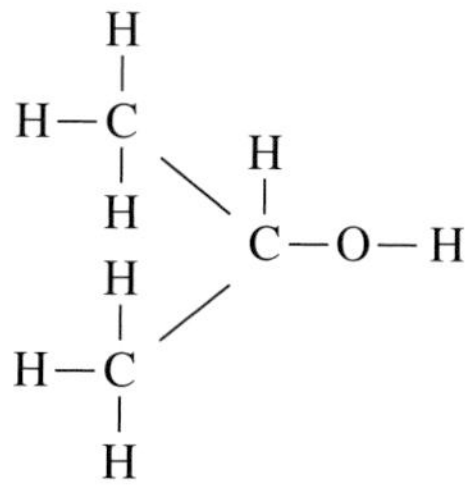

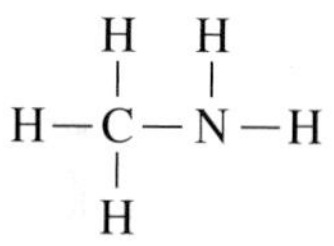

# 5                 FLAMMABLE SOLIDS

Hazard Class 4 contains materials that are solids. Additionally, some pyrophoric and water-reactive liquids are included under the 4.2 spontaneously combustible and 4.3 dangerous when wet divisions. The primary hazard of these solids is flammability. Flammable solids can be categorized into five groups according to the hazards of the materials: flammable metals, spontaneous combustibles, intensely burning or difficult-to-extinguish flash point solids, and water reactives. The solid material may take on many physical forms, including fine powder, filings, chips, and various sized solid chunks. The smaller the solid, the more dangerous it becomes in terms of flammability and potential explosiveness.

Some of the solid materials, such as white phosphorus, may be pyrophoric; that is, they spontaneously ignite on exposure to air. Calcium carbide must be wet before it becomes a hazard by releasing the flammable gas acetylene. When wet, some solids release flammable hydrogen gas. Many materials release poisonous gases, such as phosphine, nitrous oxide, or chlorine. There are solid materials that release oxygen, such as sodium peroxide, when they are wet. This can present added danger to firefighters, because the primary extinguishing agent they use is water. Water can cause the release of oxygen from water-reactive materials. If fire is present, the oxygen will accelerate the combustion process and make the fire more difficult to extinguish.

Solid materials may spontaneously combust when exposed to water; the reaction is exothermic or heat producing. If combustible materials are present, the heat of the reaction can cause combustion to occur. Some materials like phosphorus are shipped under water, whereas others such as picric acid are shipped with 10 to 50% water in the container.

Solids, such as picric acid, are considered wetted explosives. If allowed to dry out, they are classified as Class 1 Explosives. However, they present only a limited hazard as long as the water is present and, therefore, they are considered flammable solid materials.

The flammable solid hazard class is divided into three divisions by the DOT: 4.1 flammable solids, 4.2 spontaneous combustibles, and 4.3 dangerous when wet materials. Each division presents its own particular hazards when encountered in a hazardous materials incident. It is important that emergency responders have

Dry-bulk truck for transporting solid materials.

a thorough understanding of each of the divisions, the types of materials that make them up, and the hazards posed in a release.

## CLASS 4.1 FLAMMABLE SOLIDS

Class 4.1 materials are flammable solids. The Department of Transportation (DOT) defines 4.1 materials as: "(1) Wetted explosives that, when dry, are Class 1 Explosives. (2) Self-reactive materials that are liable to undergo a strongly exothermic decomposition at normal or elevated temperatures caused by excessively high transport temperatures or by contamination. (3) Readily combustible solids that may cause a fire through friction, such as matches, show a burning rate faster than 0.087 inches per second, or any metal powders that can be ignited and react over the whole length of a sample in 10 minutes or less."

Flammable solids can be elements, metals in various physical consistencies, or salts. The alkali metals in column one on the Periodic Table are considered flammable solids, particularly lithium, sodium, and potassium. Alkali metals are incompatible with carbon tetrachloride, carbon dioxide, water, and the halogens. The alkaline earth metals in column two are also flammable to some degree, magnesium and calcium more so than the others. The alkaline earth metals are incompatible with carbon tetrachloride, chlorinated hydrocarbons, halogens, and carbon dioxide. With both families, the degree and intensity of combustion depends on the physical form of the solid material. For example, solid magnesium is very difficult to ignite; however, if the magnesium is in the form of filings, shavings, or powder, it will burn explosively. Other metals, such as aluminum and titanium, can also be dangerously flammable and explosive as filings, shavings, or powders.

## FLASH-POINT SOLIDS/SUBLIMATION

There is a small group of flammable solid materials that go through a process called sublimation at normal temperatures. Sublimation is a process by which solid materials go directly from a solid to a vapor without becoming a liquid. Even though these materials do not become liquids, they still have a flash point, hence the term flash-point solid is given to the group. The flash points are generally above 100°F. Once ignition occurs, the solid material melts and flows like a flammable liquid. In addition to being flammable, flash-point solids may also be narcotic and toxic. Two common flash-point solids are camphor and paradichlorobenzene, also known as mothballs or flakes. Mothballs are placed in areas where clothing is stored to prevent moths from doing damage to the clothing. The fact that the mothballs are flash-point solids allows them to pass from a solid to a vapor without becoming a liquid. Because of this feature, the vapor from the mothballs repels the moths without harming the clothing.

**Camphor,** also known as gum camphor and 2-camphanone, is a naturally occurring ketone that comes from the wood of the camphor tree. Camphor is composed of colorless or white crystals, granules, or easily broken masses. It has a penetrating aromatic odor that sublimes slowly at room temperature. The flash point is 150°F and the autoignition temperature is 871°F. Flammable and explosive vapors are evolved when heated. While searching old newspapers for articles on fires, the author discovered an article from the late 1800s that told of a bottle of camphor placed on a wood stove. The bottle exploded from the increased vapor pressure from the heat and severely burned the resident. Camphor is slightly water soluble, undergoes sublimation, and is a flash-point solid. The molecular formula and structure of camphor are illustrated in the following example:

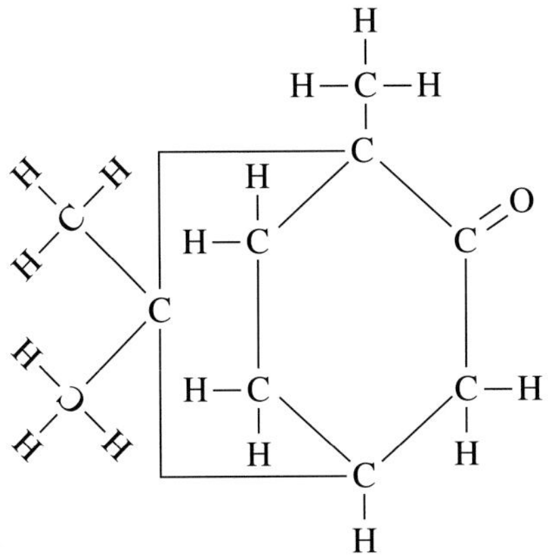

Camphor
$C_{10}H_{16}O$

**Paradichlorobenzene, PDB,** also commonly known as mothballs, are white, volatile crystals, with a penetrating odor. The boiling point is 345°F, the flash point is 150°F, and the melting point is 127°F. Paradichlorobenzene is insoluble in water and has a specific gravity of 1.5, which is heavier than water. The vapor density is 5.1, which is heavier than air. In addition to being flammable, it is also toxic by ingestion, and an irritant to the eyes, with a TLV of 75 ppm in air. The four-digit UN identification number is 1592. The NFPA 704 designation for paradichlorobenzene is health 2, flammability 2, and reactivity 0. The primary uses are as a moth repellent, a general insecticide, a soil fumigant, and in dyes. The structure and molecular formula are shown in the following example:

Paradichlorobenzene
$C_6H_4Cl_2$

The following are some examples of 4.1 flammable solid materials, including wetted explosives and flammable metals.

**Barium azide, Ba(N₃)₂,** is a crystalline solid with not less than 50% water by mass that explodes when shocked or heated. Barium azide decomposes and gives off nitrogen at 240°F and is soluble in water. The four-digit UN identification number for barium azide is 1571. Its primary use is in high explosives.

**Fusees** are flares used on the highway or rail as warning devices and are considered flammable solids that are ignited by friction. They produce a temperature of 1200°F and a 70-candle flame visible for over one quarter mile. Fusees are primarily composed of strontium nitrate (72%), which produces the characteristic red color and provides oxygen; potassium perchlorate (8%), which is an oxidizing agent; and sulfur (10%), which is an oxidizer and combustion controller. Oil, wax, and sawdust act as binding agents that aid in the control of burning. Flares absorb water and should be stored in a dry place. They are said to have an indefinite shelf life.

**Magnesium, Mg,** is a metallic element of the alkaline earth metal family. It is found as a silvery soft metal, a powder, as pellets, turnings, or ribbons. Magnesium is flammable and a dangerous fire hazard, which has an ignition temperature of about 1200°F. It is insoluble in water; however, when burning, it reacts violently with water. In the form of a powder, a pellet, or as turnings, it can explode in contact with water. Dry sand or talc are used to extinguish small fires involving magnesium. The four-digit UN identification number for magnesium powder or magnesium alloys with more than 50% magnesium powder is 1418. The number for magnesium pellets, turnings, ribbons, or magnesium alloys with more than 50% magnesium turnings, ribbons, or pellets is 1869. The NFPA 704 designation is health 0, flammability 1, and reactivity 1. The primary uses of magnesium are in diecast auto parts, missiles, space vehicles, powder for pyrotechnics, flash photography, and dry and wet batteries.

Flares used by police officers are shipped as flammable solid materials.

**Trinitrophenol,** also known as picric acid, is composed of yellow crystals and is a nitro hydrocarbon derivative. It is shipped with not less than 10% water as a wetted explosive. There is a severe explosion risk when shocked or heated to 572°F, and it reacts with metals or metallic salts. In addition to being flammable and explosive, it is toxic by skin absorption. Picric acid has caused disposal problems in school and other Chemistry laboratories where the moisture has evaporated from the container as the material ages. When the picric acid dries out, it becomes a high explosive closely related to TNT. Picric acid has been found in various amounts in school labs across the country. In a dry condition, picric acid is very dangerous and should be handled by the bomb squad. The structure and molecular formula for picric acid are illustrated in the following example:

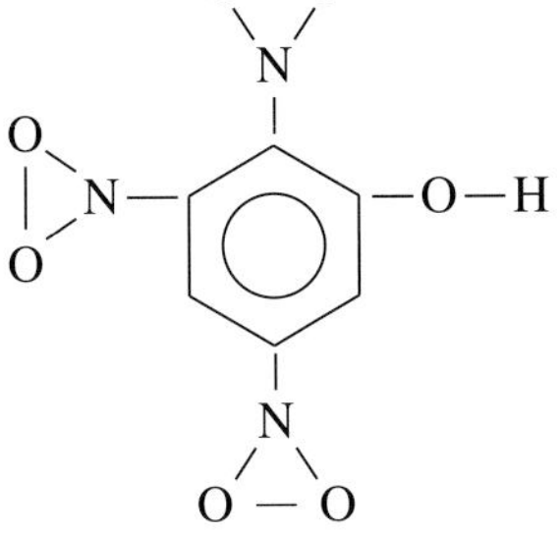

Trinitrophenol
Picric acid
$C_6H_2(NO_2)_3OH$

**Ammonium picrate** is a nitro hydrocarbon derivative. It is composed of yellow crystals with not less than 10% water by mass. Ammonium picrate is highly explosive when dry and a flammable solid when wet and is slightly soluble in water. The four-digit UN identification number for ammonium picrate with not less than 10% water is 1310. The primary uses are in pyrotechnics and explosives. The structure and molecular formula are shown in the following example:

$$NO_2$$

Ammonium picrate
$C_6H_2(NO_2)_3ONH_4$

**Matches** are shipped as flammable solids and will ignite by friction. The "safety" match was invented in 1855. The primary composition is mostly potassium chlorate (an oxysalt), which is a strong oxidizer, antimony III sulfide (a binary salt), and a glue on the match head. The striking surface is composed of a mixture of red phosphorus, antimony III sulfide, a little iron III oxide (a binary oxide), and powdered glass held in place by glue. The match functions when a strike produces heat that converts a tiny trace of red phosphorus to white phosphorus, which instantly ignites. The heat ignites the chemicals in the match head and their short blaze ignites the wood or paper of the match stick.

The "strike-anywhere" match was invented in 1898. It is primarily composed of potassium chlorate, tetraphosphorus trisulfide, ground glass, and the oxides of zinc and iron. This match functions when the strike gives enough heat to initiate a violent reaction between the $KClO_3$ and the $P_4S_3$; the heat from the reaction ignites the matchstick. The four-digit UN identification numbers for matches are as follows: fusee matches 2254, safety matches 1944, strike-anywhere matches 1331, and wax (Vesta) matches 1945.

**Sulfur, S,** is a nonmetallic element composed of pale yellow crystals. It is a dangerous fire and explosion risk in a finely divided form. In the molten form, it has a flash point of 405°F, a boiling point of 832°F, and an ignition temperature of 450°F. Sulfur is insoluble in water and its specific gravity is 1.8, which is heavier than water. The four-digit UN identification number as a dry solid is 1350. The NFPA 704 designation is health 2, flammability 1, and reactivity 0. The primary uses of sulfur are in the manufacture of sulfuric acid, carbon disulfide, and petroleum refining.

**Titanium, Ti,** is a metallic element that is a silvery solid or dark gray amorphous powder. Titanium is shipped and stored in a number of physical forms. These include powder, sheets, bars, tubes, wire, rods, sponges, and single crystals. Titanium is a dangerous fire and explosion risk and will burn in a nitrogen atmosphere. It has an ignition temperature of 2192°F when suspended in air. Water and carbon dioxide are ineffective extinguishing agents for fires involving titanium. The four-digit UN identification number for dry titanium powder when dry is 2546. The number for titanium powder with not less than 20% water is 1352. Titanium sponge, granules, or powder have a number of 2878. The primary uses of titanium are in the manufacture of structural material in aircraft, jet engines, missiles, marine equipment, surgical instruments, and orthopedic appliances.

**Urea nitrate, $CO(NH_2)_2HNO_3$,** is a colorless crystal shipped with not less than 20% water by mass. It is a dangerous fire and explosion risk, decomposes at 152°C, and is slightly soluble in water. The four-digit UN identification number is 1357. It is used in the manufacture of explosives and urethane.

**Zirconium, Zr,** is a metallic element with a grayish, crystalline scale or gray amorphous powder form. It is flammable or explosive in the form of a powder or dust and as borings and shavings. The powder should be kept wet in storage. Zirconium is a suspected carcinogen, with a TLV of 5 mg/m$^3$ of air and it is insoluble in water. The four-digit UN identification numbers depend on the form and amount of water present. The number for zirconium dry, as wire, sheeting, or in the form of strips is 2009. The number for zirconium dry, as a wire, sheeting, or as strips that are thinner than 254 μm, but not thinner than 18 μm, is 2858. Dry zirconium metal powder is 2008. Wet zirconium powder is 1358. Zirconium metal in a liquid suspension is 1308. The primary uses are as a coating on nuclear fuel rods, photo flash bulbs, pyrotechnics, explosive primers, and laboratory crucibles.

**White phosphorus, P,** also known as yellow phosphorus, is a nonmetallic element that is found in the form of crystals or a wax-like transparent solid. It ignites spontaneously in air at 86°F which is also its ignition temperature. White phosphorus should be stored under water and away from heat. It is a dangerous fire risk, with a boiling point of 536°F and a melting point of 111°F. White phosphorus is toxic by inhalation and ingestion and contact with skin produces burns. The TLV is 0.1 mg/m$^3$ of air and it is insoluble in water, with a specific gravity of 1.82, which is heavier than air. White phosphorus is shipped and stored under water to keep it from contacting air. The four-digit UN identification number is 2447. The NFPA 704 designation is health 4, flammability 4, and reactivity 2. The primary uses are in rodenticides, smoke screens, and analytical chemistry.

## INCIDENTS

An incident occurred in Gettysburg, PA involving phosphorus being shipped under water in 55-gal drums because phosphorus is air reactive. One of the drums developed a leak and the water drained off. This allowed the phosphorus to be exposed to air, which caused it to spontaneously ignite. The fire spread to the other containers and eventually consumed the entire truck. The ensuing fire was fought with large volumes of water and in the final stages covered with wet sand.

Rail tank car of air-reactive phosphorus shipped under water.

Clean-up created problems because as the phosphorus and sand mixture was shoveled into over-pack drums, the phosphorus was again exposed to air and reignited small fires. When phosphorus burns, it also gives off toxic vapors.

A train derailment in Brownson, NE, resulted in a tank car of phosphorus overturning and the phosphorus igniting upon contact with air. Phosphorus is shipped under water so there was water inside the tank car. Chemtrec was called and responders were told correctly that the phosphorous would not explode. However, the water inside the tank car was turned to steam from the heat of the phosphorus fire. The pressure from the steam caused a boiler-type of explosion that had nothing to do with the phosphorus! Just another example of the hidden hazards that responders must be aware of in dealing with hazardous materials. Not only do the hazardous materials have to be considered, but also the container, and any "inert" materials that may be involved with the product.

## COMBUSTIBLE DUSTS

There are solid materials that are not listed in a DOT hazard class, but may become flammable solids because of their physical state. These materials are combustible dusts, which are very finely divided particles of some other ordinary Class A combustible material. This group may include such ordinary materials as sawdust, grain dusts, flour, and coal dust. These materials are not considered a hazard class by the DOT. They may, however, be shipped in dry-bulk transportation containers. The combustible dusts become a problem when they are suspended in air in the presence of an ignition source. If this happens in a fixed facility, such as a grain elevator or flour mill, an explosion may occur. These materials may also be suspended in air in a transportation accident, and in the presence of an ignition source, may create an explosion. Even though combustible dusts are not considered in any of the DOT hazard classes, they present a significant fire and explosion hazard under certain conditions and responders should be aware of this hazard.

For several years Nebraska had the unwelcome distinction of having more deaths as a result of grain-elevator explosions than any other state. The State Fire Marshal implemented an intensive grain-elevator inspection program. The inspections focused on housekeeping, maintenance, and compliance with state and NFPA codes and regulations. The initial inspection of all grain elevators throughout the state took approximately 2 years to complete. Since that time, it is the author's understanding that there have not been any deaths from grain-elevator explosions in the state. These types of explosions are also discussed in Chapter 2.

## CLASS 4.2 SPONTANEOUSLY COMBUSTIBLE

Class 4.2 materials are spontaneously combustible. The DOT defines them as pyrophoric materials. Even though this Hazard Class is flammable solids, these materials may be found as solids or liquids. They can ignite without an external ignition source within 5 min after coming into contact with air. There are other 4.2 materials that may be self-heating, i.e., in contact with air and without an energy supply (ignition source), they are liable to self-heat, which can result in a fire involving the material or other combustible materials nearby.

Containers of spontaneously combustible materials in a warehouse.

**EXAMPLES OF MATERIALS SUBJECT TO SPONTANEOUS HEATING**

| | | |
|---|---|---|
| Alfalfa Meal | Hides | Peanut Oil |
| Used Burlap Bags | Castor Oil | Charcoal |
| Coal | Powdered Eggs | Lanolin |
| Coconut Oil | Lard Oil | Linseed Oil |
| Cottonseed Oil | Manure | Soybean Oil |
| Fertilizers | Metal Powders | Fish Meal |
| Fish Oil | Olive Oil | Whale Oil |

**Figure 5.1**

## SPONTANEOUS IGNITION

**Activated carbon** is flammable and the dust is toxic by inhalation. It is incompatible with calcium hypochlorite and all oxidizing agents. When some carbon-based materials, such as activated carbon or charcoal briquettes, are in contact with water, an oxidation reaction occurs between the carbon material, the water, and pockets of trapped air. The reaction is exothermic, which means heat is produced in the reaction and slowly builds up until ignition occurs spontaneously. Materials subject to spontaneous heating are listed in Figure 5.1.

**Carbon-based animal or vegetable oils,** such as linseed oil, cooking oil, and cottonseed oil can also undergo spontaneous combustion when in rags or other combustible materials. This type of spontaneous heating *cannot* occur in the case of petroleum oils. Petroleum oils do not have double bonds in the compounds. The oxidation reaction that occurs with animal and vegetable oils is different from the reaction with the carbon-based materials. The oxygen from the air trapped in the mass reacts with the double bonds present in the animal and vegetable oils. The breaking of the double bonds creates heat, which ignites the materials. Petroleum oils do not contain these double bonds and, therefore, cannot undergo this type of spontaneous heating to cause a fire. Fires started by this spontaneous heating process can be difficult to extinguish because they usually involve deep-seated fires. In order for enough heat to be sustained to cause combustion, there must be insulation. This insulation can be the material itself or may be in the form of some other combustible material, such as rags.

The high-rise fire at 1 Meridian Plaza in Philadelphia started in rags soaked with linseed oil during construction operations. These rags were improperly discarded and underwent slow heating, which eventually led to spontaneous ignition.

**Charcoal briquettes** are a dangerous fire risk. They may undergo spontaneous ignition when they become wet. This is a slow process and the heat generated must be confined as it builds up and ignites.

## PYROPHORIC SOLIDS AND LIQUIDS

**Diethyl zinc** is an organo-metal compound and is a dangerous fire hazard. It spontaneously ignites in air and reacts violently with water, releasing flammable

Bag of charcoal, an ORM-D material, that may undergo spontaneous heating in contact with water. If heat is confined, combustion may occur.

vapors and heat. It is a colorless pyrophoric liquid with a specific gravity of 1.2, which is heavier than water, so it will sink to the bottom. It decomposes explosively at 248°F. It has a boiling point of 243°F, a flash point of –20°F, and a melting point of –18°F. The four-digit UN identification number is 1366. The NFPA 704 designation is health 3, flammability 4, and reactivity 3. The white space at the bottom of the diamond has a W with a slash through it to indicate water reactivity. The primary uses of diethyl zinc are in the polymerization of olefins, high-energy aircraft and missile fuel, and the production of ethyl mercuric chloride. The molecular formula and structure are illustrated in the following example:

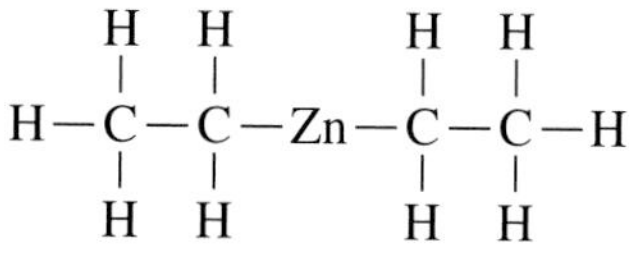

$$
\begin{array}{ccccccc}
 & H & H & & H & H & \\
 & | & | & & | & | & \\
H- & C- & C- & Zn- & C- & C- & H \\
 & | & | & & | & | & \\
 & H & H & & H & H &
\end{array}
$$

Diethyl zinc
$Zn(C_2H_5)_2$

**Pentaborane** is a nonmetallic, colorless liquid with a pungent odor. It decomposes at 300°F if it has not already ignited and will ignite spontaneously in air if impure. It is a dangerous fire and explosion risk, with a flammable range of 0.46 to 98% in air. The boiling point is 145°F, the flash point is 86°F, and the ignition temperature is 95°F, which is extremely low. Any object that is 95°F or above can be an ignition source. Ignition sources can be ordinary objects on a hot day in the summer, such as the pavement, metal on vehicles, and even the air. In addition to extreme flammability, it is also toxic by ingestion or inhalation and is a strong irritant. The TLV is 0.005 ppm in air and it is immiscible in water. The four-digit UN identification number is 1380. The NFPA 704 designation for pentaborane is health 4, flammability 4, and reactivity 2. The primary uses are as fuel for air-breathing engines and as a propellant. The structure and molecular formula are shown in the following example:

$$
\begin{array}{ccccccc}
\text{H} & \text{H} & \text{H} & \text{H} & \text{H} \\
| & | & | & | & | \\
\text{H}-\text{B} = \text{B} = \text{B} = \text{B} = \text{B}-\text{H} \\
| & & & & | \\
\text{H} & & & & \text{H}
\end{array}
$$

Pentaborane

$B_5H_9$

**Aluminum alkyls** are colorless liquids or solids. They are pyrophoric and may ignite spontaneously in air. They are often in solution with hydrocarbon solvents. Aluminum alkyls are pyrophoric materials in a flammable solvent. The vapors are heavier than air, water reactive, and corrosive. Decomposition begins at 350°F. The four-digit UN identification number is 3051. The NFPA 704 designation is health 3, flammability 4, and reactivity 3. The white space at the bottom of the diamond has a W with a slash through it, indicating water reactivity. They are used as catalysts in polymerization reactions.

**Aluminum phosphide, AlP,** is a binary salt, one of the "North Carolina Highway Patrol" (NCHP) (Chapter 1). These salts have the specific hazard of giving off poisonous and pyrophoric phosphine gas when in contact with moist air, water, or steam. They will also ignite spontaneously in contact with air. This compound is composed of gray or dark yellow crystals and is a dangerous fire risk. Aluminum phosphide decomposes on contact with water and has a specific gravity of 2.85, which is heavier than water. The four-digit UN identification number is 1397. The NFPA 704 designation is health 4, flammability 4, and reactivity 2. The white section at the bottom of the diamond has a W with a slash through it, indicating water reactivity. Aluminum phosphide is used in insecticides, fumigants, and semiconductor technology.

**Potassium sulfide, $K_2S$,** is a binary salt. It is a red or yellow-red crystalline mass or fused solid. It is deliquescent in air, which means it absorbs water from the air, and it is also soluble in water. Potassium sulfide is a dangerous fire risk and may ignite spontaneously. It is explosive in the form of dust and powder. It decomposes at 1562°F and melts at 1674°F. The specific gravity is 1.74, which is heavier than air. The four-digit UN identification number is 1382. The NFPA 704 designation is health 3, flammability 1, and reactivity 0. Potassium sulfide is used primarily in analytical chemistry and medicine.

**Sodium hydride, NaH,** is a binary salt that has a specific hazard of releasing hydrogen in contact with water. It is an odorless powder that is violently water reactive. The four-digit UN identification number is 1427. The NFPA 704 designation is health 3, flammability 3, and reactivity 2. The white space at the bottom of the diamond has a W with a slash through it, indicating water reactivity.

## INCIDENTS

A series of fires have occurred in laundries around the country since 1989. One in six commercial, industrial, or institutional laundries reports a fire each year, which results in over 3000 fires. The primary cause is thought to be spontaneous combustion. Chemicals, including animal and vegetable oils, may be left behind in fabrics after laundering. The heat from drying may cause the initiation of the chemical reaction that causes spontaneous ignition. Spontaneous combustion, according to the *Handbook of Fire Prevention Engineering,* "is a runaway temperature rise in a body of combustible material, that results from heat being generated by some process taking place within the body." On June 16, 1992, a fire in a nursing home laundry in Litchfield, IL, caused $1.5 million in damage. The cause was determined to be spontaneous ignition of residual chemicals in the laundered fabric reacting to heat from the dryer. In Findlay, OH, on July 2, 1994, a fire destroyed a commercial laundry causing over $5 million in damage. Traces of linseed oil were found in a pile of clean, warm garments piled in a cart waiting to be folded.

The 1 Meridian Plaza fire in Philadelphia was caused by linseed oil-soaked rags that spontaneously ignited. The rags were used to apply linseed oil to restore and clean wood paneling on the twenty-second floor of the high-rise building. Three firefighters lost their lives fighting that fire. Fires in restaurants have also occurred involving residual animal or vegetable oils in cleaning rags. The oils are never completely removed by laundering. When placed in the dryer, the rags are heated. When put away on the storage shelf this heat can become trapped, along with the oil remaining on rags, when confined. The spontaneous combustion process begins very slowly and the heat of the reaction increases until combustion occurs.

## CLASS 4.3 DANGEROUS WHEN WET

Class 4.3 materials are dangerous when wet. The DOT definition is "a material that, by contact with water, is liable to become flammable or to give off flammable or toxic gas at a rate greater than 1 liter per kilogram of the material per hour." Examples of water-reactive materials include zinc powder, trichlorosilane, sodium phosphide, sodium aluminum hydride, and the metallic elements potassium, sodium, and lithium. Potassium, lithium, and sodium come from family one on the Periodic Table, known as the alkali metals. They are in the first column on the table and, as with other families on the chart, they have similar chemical characteristics. They are silvery, soft metals that are reactive with air and violently reactive with water. Contact with water causes spattering, the release of free hydrogen gas, and the production of heat. The heat can be so great that it ignites the hydrogen gas.

Metallic elements, such as **magnesium** and **calcium,** are from family two on the Periodic Table. These materials are known as the alkaline earth metals. Unlike the alkali metals, magnesium must be burning before it reacts with water or it must be in a finely divided form, such as filings and powder. The filings, flakes, dusts, and powders can ignite explosively in contact with water to evolve flammable hydrogen gas and heat. The heat may be great enough to ignite the hydrogen gas. Magnesium is insoluble in water. The ignition temperature of magnesium is about 1200°F and the melting point is approximately the same. When magnesium ignites, the temperatures can reach 7200°F. In contact with burning magnesium, water produces a violent explosion. Water in contact with magnesium fillings or powder can produce a spontaneous explosion. Talc, dry sand, Met-L-X, foundry flux, and G-1 powder should be used to extinguish small magnesium fires. Large fires should be fought with flooding volumes of water from unmanned monitors and aerial devices. Calcium can be expected to behave in a similar fashion.

**Calcium carbide, $CaC_2$,** is a binary salt. It is a grayish-black, hard solid that reacts with water to produce acetylene gas, a solid corrosive that is calcium hydroxide, and release heat. Acetylene gas is manufactured by reacting calcium carbide with water. Because acetylene is so unstable, it is not shipped in bulk quantities. Calcium carbide is shipped to acetylene-generating plants where it is reacted with water in a controlled reaction. After the reaction process, the acetylene gas is placed into specially designed containers for shipment and use. Calcium carbide has a specific gravity of 2.22, which is heavier than water. The four-digit UN identification number for calcium carbide is 1402. The NFPA 704 designation is health 3, flammability 3, and reactivity 2. The white section at the bottom of the diamond contains a W with a slash through it, indicating water reactivity. It is shipped in metal cans, drums, and specially designed covered bins on rail cars and trucks. When shipped and stored, it should be kept in a cool dry place. The primary uses are in the generation of acetylene gas for welding, vinyl acetate monomer, and as a reducing agent.

**Phosphorus pentasulfide, $P_4S_{10}$,** is a nonmetallic inorganic compound. It is a yellow to greenish-yellow crystalline mass with an odor similar to hydrogen sulfide. It is a dangerous fire risk and ignites by friction or in contact with water. The boiling point is 995°F and the ignition temperature is 287°F. It decomposes on contact with water or moist air, liberating toxic and flammable hydrogen sulfide gas. The specific gravity is 2.09, so it is heavier than water. It is toxic by inhalation with a TLV of 1 mg/m$^3$ of air. The four-digit UN identification number is 1340. The NFPA 704 designation is health 2, flammability 1, and reactivity 2. The primary uses are in insecticides, safety matches, ignition compounds, and sulfonation.

**Methyl dichlorosilane, $CH_3SiHCl_2$,** is a colorless liquid with a sharp, irritating odor. It is a dangerous fire risk, corrosive, and water reactive. The flammable range is wide, from 6% on the lower end to 55% on the upper end. The boiling point is 107°F, the flash point is 15°F, and the ignition temperature is more than 600°F. The specific gravity is 1.11, which is heavier than water. Vapors are heavier than air and will travel to ignition sources. It is immiscible in water and decomposes on contact to release hydrogen chloride gas. Methyl dichlorosilane is very toxic by inhalation and skin absorption; it is irritating to the skin, eyes, and respiratory system. Contact with the material may cause burns to the eyes and skin. The four-digit UN identification number is 1242. The NFPA 704 designation

Closed containers of water-reactive calcium carbide, which produces acetylene gas upon contact with water.

is health 3, flammability 3, and reactivity 2. The white space at the bottom of the diamond has a W with a slash through it, indicating water reactivity. The primary use is in the manufacture of siloxanes, which are straight-chained compounds similar to paraffin hydrocarbons.

**Potassium, K,** is a metallic element, also known as kalium. It is an alkali metal that is soft, silvery, and rapidly oxidized in moist air. Potassium is a combustible solid that may ignite spontaneously on contact with moist air. It is a dangerous fire risk and it reacts violently with water and moisture in the air to release hydrogen gas and form potassium hydroxide, which is a corrosive liquid. The boiling point is 1410°F and the melting point is 146°F. The specific gravity is 0.86, which is lighter than water. The reaction with water is also exothermic, and the heat produced is enough to ignite the hydrogen gas that is released. Potassium metal is usually stored under kerosene to keep it from reaching the air. As it ages it also can form explosive peroxides, much like ethers do. When these peroxides are present, it may explode violently if handled or cut. Potassium coated with peroxides should be destroyed by burning. The four-digit UN identification number is 2257. The NFPA 704 designation is health 3, flammability 3, and reactivity 2. The white section at the bottom of the diamond has a W with a slash through it, indicating water reactivity.

## TOP 50 INDUSTRIAL CHEMICALS

There is one flammable solid in the Top 50 industrial chemicals. **Carbon black** is the thirty-seventh most produced industrial chemical, with 3.32 billion

Open- and closed-leg grain elevators are a primary source of dust explosions.

lbs. in 1995; it is a finely divided form of carbon. It may ignite explosively if suspended in air in the presence of an ignition source or slowly undergo spontaneous combustion in contact with water. In addition, it is toxic by inhalation, with a TLV of 3.5 mg/m$^3$ in air. The primary uses are in the manufacture of tires, belt covers, plastics, carbon paper, colorant for printing inks, and as a solar-energy absorber.

## INCIDENTS

A fire in a warehouse in Chicago involved barrels of oil-soaked magnesium shavings and filings. Fighting the fire with water produced violent explosions, resulting in whiplash injuries to firefighters who were on aerial apparatus over the burning magnesium. The facility was in a residential neighborhood, so firefighters had to try to control the fire with water even though the material is water reactive when burning.

In New York City three firefighters died and several others were injured in a fire involving sodium metal. The firefighters were extinguishing a fire in a 55-gal drum of molten sodium, when a small amount of water on a shovel came in contact with the sodium. This triggered a chemical reaction and explosion producing temperatures in excess of 2000°F, splattering the molten sodium on the firefighters. The sodium burned through their turnouts, station uniforms, and underwear. Contact with moisture on the skin caused burning of the tissue below. Water in contact with water-reactive and molten metals can produce violent reactions. Had the firefighters tried to extinguish the fire in the drum with a hose line, many more may have died.

## FIRE-EXTINGUISHING AGENTS

Class 4.3 materials are water reactive. When large amounts of these materials are involved in fire, water is the only extinguishing agent available in quantities large enough to extinguish the fires; just understand that when water is used, there may be violent reactions and explosions. Preparations need to be made for the safety of personnel based upon the hazards of the materials. Small fires of water reactive materials, especially metallic-based materials, can be extinguished with a dry-powder extinguishing agent. For other flammable solid materials, water is also the agent of choice in most cases. It is important, however, to make sure there is a positive identification of the product as with all hazardous materials. Once the product is identified, the proper extinguishing agent for any individual material can be identified through reference materials such as the *North American Emergency Response Guide Book,* the CAMEO (Computer-Aided Management of Emergency Operations) computer data base, CHEMTREC (Chemical Transportation Emergency Center), or some other reference source.

## REVIEW QUESTIONS

1. There are three subclasses in hazard Class 4, _____, _____, and _____.

2. Sublimation is a process by which a solid changes from a _____ to a _____, without going through the _____.

3. Phosphorus is a 4.1 flammable solid that will _____ when exposed to _____.

4. Which of the following may provide combustible dusts?
   A. Flour
   B. Grain
   C. Wood
   D. Coal
   E. All of the above

5. Which of the following compounds may undergo spontaneous ignition?
   A. Fish oils
   B. Motor oil
   C. Linseed oil
   D. Cottonseed oil
   E. Gasoline
   F. Methyl ethyl ketone
   G. Isopropyl ether

6. Which of the following extinguishing agents would be appropriate for large fires involving water-reactive materials?
   A. Carbon dioxide
   B. Dry chemical

    C. Dry powder
    D. Water
    E. None of the above

7. Which families from the Periodic Table are water reactive in their elemental state?

8. When calcium carbide contacts water, what is produced?

# 6          OXIDIZERS

Hazard Class 5 (oxidizers) is separated into two divisions: 5.1 and 5.2. Class 5.1 materials are solids and liquids that, according to the Department of Transportation (DOT), "by yielding oxygen can cause or enhance the combustion of other materials." Although that is not a very technical definition from a chemistry standpoint, it gets right to the point for emergency response. While oxidizers themselves do not burn, if present in a fire situation, they will make the fire burn faster and become more difficult to extinguish.

The National Fire Protection Association (NFPA) classifies oxidizers into four groups. These groups are identified in Figure 6.1. Common groups of oxidizers include oxysalts, inorganic peroxides (salt peroxides), certain acids, elements, and organic peroxides. Examples of oxidizers from each of the NFPA classes are illustrated in Figure 6.2.

Oxygen is probably the most recognized oxidizer. Even though oxygen is essential for life to exist, it can be a very dangerous material. Oxygen is found in transportation and storage as a compressed gas and as a cryogenic liquid with a temperature of −183°F. Oxygen does not burn, but in contact with organic materials it can become very explosive. Oxygen-enriched atmospheres can be deadly to emergency responders if there is a fire or heat source nearby. Many times, the enriched atmosphere is not visible or detectable to responders without the use of monitoring instruments. While the oxidizers in this chapter are solids and liquids, through physical and chemical reactions they can release oxygen gas, which can cause some of the same problems as compressed oxygen.

Class 5.2 materials are organic peroxides. Unlike peroxide salts, which contain metals, these are organic compounds that contain carbon in their formula. These materials contain oxygen in the bivalent –O–O– structure and may be considered a derivative of hydrogen peroxide. The structure and molecular formula for hydrogen peroxide are illustrated in the following example:

$$H - O - O - H$$

Hydrogen peroxide
$$H_2O_2$$

## NFPA CLASSES OF OXIDIZERS

**Class 1** - Solid or liquid that readily yields oxygen or oxidizing gas or that readily reacts to oxidizer combustible materials.

**Class 2** - Oxidizing material that can cause spontaneous ignition when in contact with combustible materials.

**Class 3** - Oxidizing material that can undergo vigorous self-sustained decomposition when catalyzed or exposed to heat.

**Class 4** - Oxidizing material that can undergo an explosive reaction when catalyzed or exposed to heat, shock, or friction.

**Figure 6.1**

In organic peroxide compounds, both of the hydrogen atoms in hydrogen peroxide have been replaced by organic radicals. One of the major hazards of 5.2 organic peroxides is the instability of the compounds. The oxygen-to-oxygen single bond is an unstable bond. It is this same bond that is responsible for the explosiveness of the nitro compounds discussed in Chapter 2. Oxidizers, especially the organic peroxides, should be treated with a great deal of respect. They can be just as dangerous and explosive as Class 1 compounds.

## CLASS 5.1 OXIDIZERS

Oxidizers may be elements, acids, or salts classified into families with specific hazards associated with each family. There are elements found on the Periodic Table that are oxidizers in their elemental state. These include oxygen, chlorine, fluorine, and bromine.

**Oxygen, $O_2$,** can be encountered as a gas, cryogenic liquid, and liquid or solid in compound with other materials. Though nontoxic, it is very reactive with hydrocarbon-based materials. Oxygen is a strong oxidizer. Liquid oxygen in contact with an asphalt surface, such as a parking lot or highway, can create a

## EXAMPLES OF OXIDIZERS IN THE NFPA CLASSES

| Class 1 | Class 2 |
|---|---|
| Aluminum Nitrate | Calcium Hypochlorite |
| Calcium Peroxide | Nitric Acid (above 70%) |
| Potassium Persulfate | Sodium Peroxide |
| Sodium Nitrite | Potassium Permanganate |

| Class 3 | Class 4 |
|---|---|
| Ammonium Dichromate | Perchloric Acid (60% - 72%) |
| Calcium Hypochlorite (over 50%) | Hydrogen Peroxide (>90%) |
| Hydrogen Peroxide (52% - 91%) | Potassium Superoxide |

**Figure 6.2**

contact explosive. Dropping an object on the area, driving over, or even walking on the area can cause an explosion to occur. Oxygen is a nonmetallic gaseous element. Oxygen makes up approximately 21% of the atmosphere. The boiling point of oxygen is –297°F. It is nonflammable, but supports combustion. Liquid oxygen can explode when exposed to heat or organic materials.

**Chlorine, fluorine,** and **bromine** in their elemental form are all strong oxidizers even though they are placarded and labeled as poisons. Two terms commonly associated with oxidizers are oxidation (or oxidation reaction) and reduction. **Oxidation** is the loss of electrons by one reactant, and **reduction** is the gaining of electrons by another. Metals usually lose electrons, and nonmetals usually gain electrons. The elements in the upper right corner of the Periodic Table are very electronegative, or electron drawing. Fluorine is the most electron-drawing element known. Chlorine and oxygen are also very electron drawing. Oxidation and reduction always occur together. No substance is ever oxidized unless something else is reduced. For example, when sodium and chlorine combine ionicly, the electron of sodium is given to chlorine; sodium has been oxidized. Chlorine receives the electron of sodium; chlorine has been reduced. The substance that accepts the electrons is known as the oxidizing agent. Therefore, in the reaction between sodium and chlorine, chlorine is the oxidizing agent. Chlorine is an oxidizer. Sodium is the reducing agent because sodium helps chlorine to be reduced. Chlorine is reduced to the chloride ion in the reaction with sodium. In summary, the substance that is oxidized is the reducing agent (gives up its electrons). The substance that is reduced is the oxidizing agent (receives electrons).

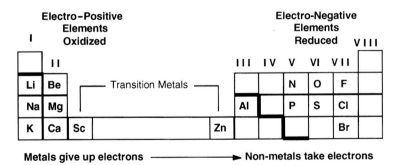

Chlorine is a dense, greenish-yellow gas. Although it may be a gas or a liquefied gas, it can also be released from solid compounds that are oxidizers. Chlorine is not combustible; however, it will support combustion just like oxygen. Chlorine does not occur freely in nature. It is found in compound within the minerals halite (rock salt), sylvite, and carnallite, and as the chloride ion in sea water.

**Fluorine, F,** is the most powerful oxidizing agent known. Like its relative chlorine, it is classified as a 2.3 poison gas by the DOT and is toxic by inhalation, with a TLV of 1 ppm. However, it is also a powerful and dangerous oxidizing agent. Liquid fluorine is such a strong oxidizer that it can cause concrete to burn. It is a pale yellow gas or cryogenic liquid with a pungent odor. It reacts violently

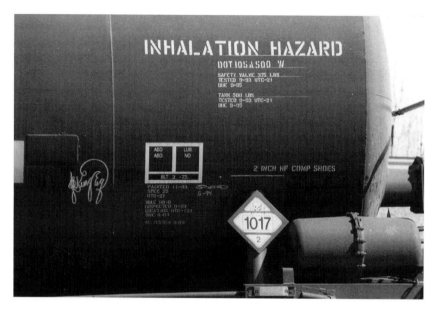

Chlorine, though placarded as a poison, is also a strong oxidizer and will support combustion just like oxygen.

with a wide range of organic and inorganic compounds and is a dangerous fire and explosion risk when in contact with these materials.

**Bromine** is a dark, reddish-brown liquid with irritating fumes. It attacks most metals and reacts vigorously with aluminum and explosively with potassium. It is a strong oxidizing agent and may ignite combustibles on contact.

## OXYSALTS

Oxysalts are combinations of metals and covalently bonded, nonmetal oxyradicals. They end in "ate" or "ite" and may have the prefixes "per" or "hypo". Nine oxysalt radicals will be presented with this group (Figure 6.3). The first six all have −1 charges: $FO_3$ (fluorate), $ClO_3$ (chlorate), $BrO_3$ (bromate), $IO_3$ (iodate), $NO_3$ (nitrate), and $MnO_3$ (manganate). The next two have −2 charges: $CO_3$ (carbonate) and $SO_4$ (sulfate). The last oxyradical is $PO_4$ (phosphate), which has a −3 charge (see Figure 1.8). All of the radicals listed above are considered to be in their base state, the "normal" number of oxygen atoms present in that oxyradical. When a metal is added to any oxyradical in the base state, the compound ends in "ate", such as sodium phosphate. Oxyradicals may be found with varying numbers of oxygen atoms. There may be more or less oxygen atoms in a compound than the base state. Regardless of the number of oxygen atoms on the oxyradical, the charge of the radical **does not** change. When naming the compounds with one additional oxygen atom, the prefix "per" is used; to indicate excess oxygen over the base state, the ending is still "ate". When the number of oxygen atoms is one less than the base state of an oxyrad-

## OXYSALTS

| | |
|---|---|
| Fluorate | $FO_3$ |
| Chlorate | $ClO_3$ |
| Bromate | $BrO_3$ |
| Iodate | $IO_3$ |
| Nitrate | $NO_3$ |
| Manganate | $MnO_3$ |
| Carbonate | $CO_3$ |
| Sulfate | $SO_4$ |
| Phosphate | $PO_4$ |

**Figure 6.3**

ical, the ending of the oxyradical name is "ite". An example is magnesium sulf*ite*. Finally, an oxyradical can have two less oxygen atoms than the base state; the oxyradical name will now have a prefix "hypo" and will end in "ite". An example is aluminum *hypo*phosph*ite*. All oxysalt compounds are salts and have the hazard of being oxidizers. They contain, among other things, fluorine, chlorine, and oxygen, which are all strong oxidizers. Most oxysalts do not react with water, but are soluble in water. In the process of mixing with water, they may liberate oxygen, fluorine, or chlorine. Oxysalts have varying numbers of oxygens in their compounds. The common oxysalt compound has three or four oxygen atoms, which is known as the base state. Some oxysalts are loaded with oxygen, such as the "per–ate" compounds, which will have four or five oxygen atoms. The compounds that have two oxygen atoms will end in "ite". The "hypo–ite" compounds will have one oxygen atom. The fact that the "hypo–ites" have only one oxygen atom does not mean that they are not dangerous oxidizers. Some of the "hypo–ite" compounds, in addition to oxygen, also have chlorine atoms, which are oxidizers.

Perchlorates and other oxyradicals in the "per–state" contain one more oxygen atom than the base state chlorates. They are loaded with oxygen and want to give it up readily. Perchlorates can form explosive mixtures with organic, combustible, or oxidizable materials. Contact with acids, such as sulfuric acid, can form explosive mixtures.

**Lithium perchlorate, $LiClO_4$,** is an oxysalt that is a colorless, deliquescent crystal. Oxysalt "per–ate" compounds are loaded with excess oxygen and will readily give it up in a reaction. Lithium perchlorate is a powerful oxidizing agent. It has more available oxygen than liquid oxygen on a volume basis. Lithium perchlorate has a specific gravity of 2.429, which is heavier than water and is water soluble. Lithium perchlorate is a dangerous fire and explosion risk in contact with organic materials. It is an irritant to skin and mucous membranes. The primary use of lithium perchlorate is as a solid rocket propellant.

Chlorates are strong oxidizing agents. When heated, they give up oxygen readily. Contact with organic or other combustible materials may cause spontaneous combustion or explosion. They are incompatible with ammonium salts, acids, metal powders, sulfur, and finely divided organic or combustible substances.

**Potassium chlorate, $KClO_3$,** is a transparent, colorless crystal or white powder. It is soluble in boiling water and decomposes at approximately 750°F, giving off oxygen gas. Potassium chlorate is a strong oxidizer and forms explosive mixtures with combustible materials, such as sugar, sulfur, and others. Potassium chlorate is incompatible with sulfuric acid, other acids, and organic material. The four-digit UN identification number is 1485. Its primary uses are as an oxidizing agent in the manufacture of explosives and matches; in pyrotechnics; and as a source of oxygen. Sodium and potassium chlorates have similar properties.

Chlorites are powerful oxidizing agents. They have one less oxygen than the base state oxysalts. They form explosive mixtures with combustible materials. In contact with strong acids they can release explosive chlorine dioxide gas.

**Calcium chlorite** is an oxysalt with a molecular formula of $Ca(ClO_2)_2$. It is a white crystalline material that is soluble in water. It is a strong oxidizer and a fire risk in contact with organic materials. The four-digit UN identification number is 1453.

Hypochlorites have two less oxygen atoms than the base-state compounds. They can cause combustion in high concentrations when in contact with organic materials. When heated or in contact with water, they can give off oxygen gas. At ordinary temperatures, they can give off chlorine and oxygen in contact with moisture and acids. They are commonly used as bleaches and swimming pool disinfectants.

**Calcium hypochlorite, $Ca(ClO)_2$,** is an oxysalt; it is a crystalline solid and an oxidizer that decomposes at 212°F. Calcium hypochlorite is a dangerous fire risk in contact with organic materials. It is also a common swimming pool chlorinator and decomposes in contact with water, releasing chlorine into the water. If a container of calcium hypochlorite becomes wet in storage, the result can be an exothermic reaction. If combustible materials are present, a fire may occur. The chlorine in the compound will be released by contact with the water and will then accelerate the combustion process. The four-digit UN identification number for dry mixtures with not less that 39% available chlorine (8.8% oxygen) is 1748; hydrated with not less than 5.5% and not more than 10% water, the number is 2880; mixtures that are dry, with not less than 10% but not more than 39% available chlorine, are numbered as 2208. The NFPA 704 designation for calcium hypochlorite is health 3, flammability 0, and reactivity 1. The white section at the bottom of the diamond has the prefix "oxy", indicating an oxidizer. The primary uses are as a bleaching agent, a swimming pool disinfectant, a fungicide, in potable water purification, and as a deodorant.

Metal nitrates are oxysalts and as a group have a wide range of hazards. Common to many of them, however, is the fact they are oxidizers and are heat and shock sensitive. When heated they will melt, releasing oxygen, which will increase the combustion process. Molten nitrates react with organic materials very violently. When solid streams of water are used for fire suppression, steam explosions may occur upon contact with the molten materials. Nitrates can be very dangerous oxidizers and will explode if contaminated, heated, or shocked. Most nitrates have similar properties.

**Aluminum nitrate, $Al(NO_3)_3$,** is a white crystal material that is soluble in cold water. It is a powerful oxidizing agent that decomposes at approximately 300°F. Aluminum nitrate should not be stored near combustible materials. The four-digit UN identification number is 1438. The primary uses are in textiles, leather tanning, as an anticorrosion agent, and as an antiperspirant.

**Sodium nitrate,** also known as Chile saltpeter and soda niter, has a molecular formula of $NaNO_3$. Sodium nitrate is a colorless, odorless, transparent crystal. It oxidizes when exposed to air and is soluble in water. This material explodes at 1000°F, certainly much lower than temperatures encountered in many fires. Sodium nitrate is toxic by ingestion and has caused cancer in test animals. When used in the curing of fish and meat products, it is restricted to 100 ppm. Sodium nitrate is incompatible with ammonium nitrate and other ammonium salts. The four-digit UN identification number is 1498. Sodium nitrate is used as an antidote for cyanide poisoning and in the curing of fish and meat.

**Potassium nitrate** (saltpeter) has a molecular formula of $KNO_3$. It is found as a transparent or colorless to white crystalline powder and as crystals. Potassium nitrate is water soluble and is a dangerous fire and explosion risk when heated or shocked or in contact with organic materials. It is a strong oxidizing agent with a four-digit UN identification number of 1486. Potassium nitrate is used in the manufacture of pyrotechnics, explosives, and matches. It is often used in the illegal manufacture of home-made pyrotechnics and explosives.

Persulfates are strong oxidizers and may cause explosions during fires. Oxygen may be released by the heat of the fire and cause explosive rupture of the containers. Explosions may also occur when persulfates are in contact with organic materials.

**Potassium persulfate, $K_2S_2O_8$,** is composed of white crystals that are soluble in water and decomposes below 212°F. Potassium persulfate is a dangerous fire risk in contact with organic materials. It is a strong oxidizing agent and an irritant with a four-digit UN identification number of 1492. The primary uses are in bleaching, as an oxidizing agent, as an antiseptic, as a polymerization promoter, and in the manufacture of pharmaceuticals.

Permanganates mixed with combustible materials may ignite from friction or spontaneously in the presence of inorganic acids. Explosions may occur with either solutions or dry mixtures of permanganates.

**Potassium permanganate, $KMnO_4$,** is composed of dark purple, odorless, crystals with a blue metallic sheen. It is soluble in water, decomposes at 465°F, and is a powerful oxidizing material. Potassium permanganate is a dangerous fire and explosion risk in contact with organic materials. Potassium permanganate is incompatible with sulfuric acid, glycerine, and ethylene glycol. The four-digit UN identification number is 1490. The primary uses of potassium permanganate are as an oxidizer, a bleach, a dye, during radioactive decontamination of the skin, and in the manufacture of organic chemicals.

Some ammonium compounds are oxysalts. Although ammonia is not a metal, in the case of the ammonium ion it acts like a metal when attached to the oxyradicals. When ammonia gas is added to water, it readily dissolves and remains as $NH_3$. One of the hydrogens leaves water but leaves its electrons behind. The protons of the hydrogen then attach to the unbonded electrons on nitrogen to complete its duet. This hydrogen is loosely held to the nitrogen and comes off easily. The hydrogen ions in the water are attracted to the negative side of the ammonia molecule where that unbonded pair of electrons is located (you can think of the ammonia as a slightly polar molecule). The ammonium ion is positive because the hydrogen ion contributes no electrons to the ion. It is not important to understand why this happens, but rather that the hazards of the compounds will be similar to the rest of the oxysalts, i.e., they are oxidizers. This is a complex covalent sharing arrangement and is one of those chemistry concepts that should be accepted rather than explained for the purpose of emergency response. The ammonium ion is shown in the following illustration:

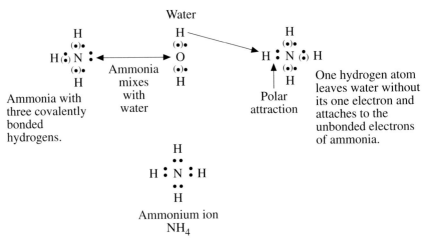

The ammonium ion is formed.

**Ammonium iodate, $NH_4IO_3$,** is an oxidizing agent and a dangerous fire and explosion risk in contact with organic materials. It is a white odorless powder.

**Ammonium chlorate** is an ammonium compound with a molecular formula of **$NH_4ClO_3$**. It is a colorless or white crystal that is soluble in water. Ammonium chlorate is a strong oxidizer and when contaminated with combustible materials can spontaneously ignite. It is shock sensitive and can detonate when exposed to heat or vibration. One of its primary uses is in the manufacture of explosives.

## PEROXIDE SALTS

Peroxide salts should not be confused with organic peroxides. Peroxide salts contain a metal and a peroxide radical. Organic peroxides are made up of non-metal hydrocarbon radicals and the peroxide functional group. Peroxide salts are water reactive and give off oxygen, evolve heat, and produce a corrosive liquid in contact with water. The corrosive liquid is the hydroxide of the metal in the compound. The heat produced may be sufficient to ignite nearby combustible materials. Metal peroxides may also decompose when exposed to heat, with similar results to their reaction with water. The hazards and physical, and chemical characteristics of metal peroxides are similiar. Sodium peroxide and barium peroxide are very common metal peroxides and are detailed below. Keep in mind that they are not the only metal peroxides.

**Sodium peroxide** has a molecular formula of $Na_2O_2$ and is an inorganic peroxide salt. It is a yellowish-white powder that turns yellow when heated. Sodium peroxide absorbs water and carbon dioxide from the air and is soluble in cold water. It is a strong oxidizing agent. It is corrosive and can cause burns to the eyes and skin and is also toxic by ingestion and inhalation. It is water reactive and a dangerous fire and explosion risk in contact with water, alcohol, or acids. Sodium peroxide forms self-igniting mixtures with powdered metals and organic materials. It is incompatible with ethyl or methyl alcohol, glacial acetic acid, carbon disulfide, glycerine, ethylene glycol, and ethyl acetate. The four-digit UN identification number is 1504. The NFPA 704 designation is health 3, flammability 0, and reactivity 1. The 704 diamond has the prefix *oxy* in the white space at the bottom. It is used as a bleach and an oxygen-generating material for diving bells and submarines.

**Barium peroxide, $BaO_2$,** is a grayish-white powder that is slightly soluble in water. Barium peroxide is a dangerous fire and explosion risk in contact with organic materials and decomposes around 1450°F. It is also toxic by ingestion, is a skin irritant, and should be kept cool and dry in storage. The four-digit UN identification number is 1449. The primary uses of barium peroxide are in bleaching, in thermal welding of aluminum, as an oxidizing agent, and in dyeing of textiles.


## INORGANIC ACID OXIDIZERS

At higher concentrations some acids can be strong oxidizers and cause combustion in contact with organic materials. Nitric acid is a very dangerous oxidizing acid; it is also corrosive and its vapors are toxic.

**Nitric acid, $HNO_3$,** is considered an oxidizer above 40% concentration. It is shipped in bulk quantities in MC/DOT 312 tanker trucks. The MC 312 tanker is a small-diameter tank with reinforcing rings around the outside. The corrosive materials carried in these tanks are very heavy. By recognizing the shape of the MC 312, you will realize that there is a corrosive material in the tank even though it may be placarded with the oxidizer placard. Nitric acid will be discussed further in Chapter 9.

**Chromic acid, $CrO_3$,** is composed of dark, purplish-red, odorless crystals that are soluble in water. The specific gravity is 2.7, which is heavier than water. It is a powerful oxidizing agent and may explode in contact with organic materials.

Chromic acid is a poison, corrosive to the skin, and has a TLV of 0.05 mg/m³ of air. Chromic acid is a known human carcinogen. The four-digit UN identification number is 1463. The NFPA 704 designation is health 3, flammability 0, and reactivity 1. The white section at the bottom of the 704 diamond has an "oxy" prefix, indicating it is an oxidizer.

**Perchloric acid** has a molecular formula of $HClO_4$. At concentrations of more than 50% but less than 72% by volume it is placarded as an oxidizer. Concentrations above 72% are forbidden in transportation. It is a colorless, fuming liquid, and is unstable in its concentrated form. It is a strong oxidizing agent, is corrosive, and is placarded as a Class 8 corrosive in concentrations less than 50%. Perchloric acid will ignite vigorously in contact with organic materials or detonate by shock or heat. It is toxic by inhalation and ingestion and is a strong irritant. Perchloric acid is water soluble and has a specific gravity of 1.77, which means it is heavier than water. However, it is water soluble and will mix rather than form layers. Upon contact with water, heat is produced. The boiling point is 66°F and the vapor density is 3.46, which is heavier than air. Perchloric acid is incompatible with acetic anhydride, bismuth and its alloys, alcohols, paper, wood, and other organic materials. The four-digit UN identification number for concentrations greater than 50% is 1873. Concentrations less than 50% have a four-digit identification number of 1802. The NFPA 704 designation is health 3, flammability 0, and reactivity 3. There is an "oxy" prefix in the white section of the 704 diamond, indicating an oxidizer.

Oxidizers, like many of the other hazard classes, can have more than one hazard. They can be corrosive, poisonous, and explosive under certain conditions. Many are water reactive and some may react violently with other chemicals, particularly organic materials.

Oxygen is one of the components of the original fire triangle and the more recent fire tetrahedron. The combustion process involves heat, oxygen, fuel, and a chemical chain reaction. Combustion is simply a rapid oxidation reaction that is accompanied by the emission of energy in the form of heat and light. When we think about combustion, we usually think about atmospheric oxygen allowing combustion to occur. In the combustion process, the fuel source is heated; the molecules start to vibrate rapidly, and if the vibrations are strong enough, the molecules break into small fragments with incomplete bonds. These "free radical" fragments are unstable and cannot remain as fragments, so they want to bond with some other element to complete their electron requirements. As the heat increases, the radical fragments rise and encounter oxygen from the air. The oxygen is very electronegative (electron drawing) and quickly attracts the electrons from the radical fragments. Bonding occurs between the radical fragments and the oxygen, forming a chemical bond. This bonding process is exothermic, and the heat energy is fed back into the combustion process, allowing the combustion to continue.

Without oxygen or chemical oxidizers, most combustion cannot occur. Materials that contain oxygen in their compound can support combustion even in the absence of atmospheric oxygen. When extra oxygen is added, over and above atmospheric oxygen, the process of combustion is accelerated. The more oxygen that is present, the more bonds that can take place between the radical fragments and the oxygen. The higher the oxygen concentration, the more accelerated the combustion process.

Oxidizers can undergo spontaneous combustion by three different means. First, there is slow oxidization. This occurs when an oxidizer comes in contact with a material that has double bonds (also known as Pi bonds), such as animal or vegetable oils and alkenes. Animal and vegetable oils are actually large esters, with the exception of turpentine which is a pure hydrocarbon compound. The problem is the same, however; the double bonds in turpentine can be attacked by oxygen from the air and can undergo spontaneous ignition. Shown below is the structure and molecular formula for turpentine; notice the double bonds:

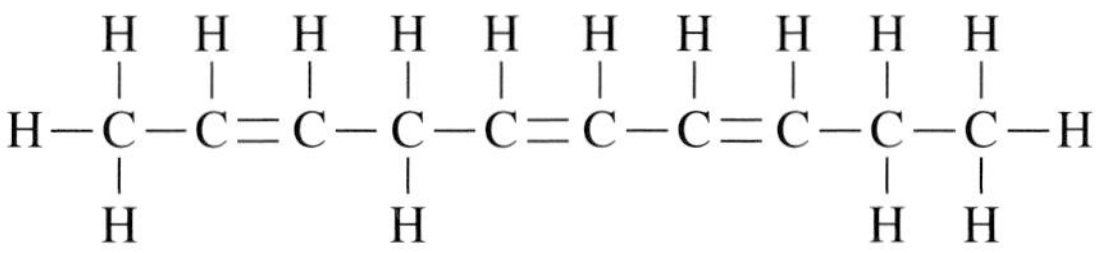

$$\begin{array}{c} \quad\; H \;\; H \;\; H \;\; H \;\; H \;\; H \;\; H \;\; H \;\; H \;\; H \\ \quad\; | \;\;\; | \;\;\; | \;\;\; | \;\;\; | \;\;\; | \;\;\; | \;\;\; | \;\;\; | \;\;\; | \\ H-C-C=C-C-C=C-C=C-C-C-H \\ \quad\; | \qquad\quad | \qquad\qquad\quad | \;\; | \\ \quad\; H \qquad\quad H \qquad\qquad\quad H \;\; H \end{array}$$

Turpentine
$C_{10}H_{16}$

As the oxidizer (which can be oxygen in the air) breaks the double bonds, heat is produced. If there is enough insulating material present, the heat can build up to the point of spontaneous ignition.

Second, the reaction of an inorganic oxidizer with water can be exothermic, i.e., it gives off heat. They also generate oxygen gas in contact with water. The heat and oxygen represent two sides of the fire triangle or tetrahedron. All that is needed is fuel. If combustible materials are present, combustion can occur spontaneously. When that occurs, there is also excess oxygen present that accelerates the rate of combustion. Third, the corrosive action of a strong acid, such as nitric acid, can generate heat. The heat can be sufficient to ignite combustible materials to which the corrosive is exposed.

Some oxidizers mix with water rather than react. The liquid mixture can impregnate combustible and noncombustible materials. When the water evaporates from the material, the oxidizer is left behind. The materials will now burn with great intensity if exposed to heat or fire because of the oxidizers present. Firefighter turnouts can become impregnated with oxidizer materials during firefighting operations. While firefighter turnouts are resistant to combustion, the oxidizer in the fabric can cause the turnouts to burn with great intensity if exposed to heat or fire. The turnouts must be decontaminated prior to being used again or the firefighters will be in unnecessary danger when exposed to heat or fire.

Some compounds can undergo spontaneous combustion when they contact water. This is a slow process, as in the case of charcoal. The wetted charcoal is oxidized by the oxygen in the air. If the heat produced is accumulated in the material because of the water reaction, combustion may occur. This air oxidation can also occur with animal and vegetable oils when they are present in combustible materials, such as rags. The double bond (also referred to as a Pi bond) in the animal and vegetable oils is oxidized by the oxygen in the air. This action on the double bond creates heat. If the heat is confined, spontaneous combustion may occur.

The corrosive action of strong acids can generate heat. The heat may be enough to ignite combustible materials. When the corrosive material is also an oxidizer, the oxidizer will contribute to the acceleration of the combustion process.

An explosion is really nothing more than a very rapid combustion. A chemical explosion, as discussed in Chapter 2, requires a chemical oxidizer to be present. Without the chemical oxidizer, the combustion does not accelerate fast enough to allow an explosion to occur. Oxidizers that allow for explosions can themselves explode when heated or shocked.

In Henderson, NV, a fire occurred in the Pepcon chemical plant. The fire heated ammonium perchlorate to the point that a detonation occurred. Ammonium perchlorate is an oxidizer used in the manufacture of solid rocket fuel, but it is not an explosive. However, this oxidizer detonated, creating a shock wave that blew the windshields out of responding fire apparatus, injuring several firefighters.

**Ammonium perchlorate, $NH_4ClO_4$,** is a white crystalline material that is soluble in water. It is a strong oxidizing agent and a skin, eye, and respiratory irritant. Ammonium perchlorate is shock-sensitive and may explode or detonate when exposed to heat-reducing agents or by spontaneous chemical reaction. (A reducing agent is a material that removes the oxygen from the compound.) Oxidizers can be dangerously explosive materials even though they are not classified as explosives. Closed containers can rupture violently when heated. Ammonium perchlorate decomposes at 464°F and produces oxides of nitrogen, hydrogen chloride, and ammonia. It is incompatible with acids, alkalis, powdered metals, and organic materials. It has a four-digit UN identification number of 1442. The NFPA 704 classification is health 1, flammability 0, and reactivity 4. The prefix "oxy" is placed in the white section at the bottom of the 704 diamond. Ammonium perchlorate is usually shipped in fiber drums, bags, steel drums, and tote bins. It is primarily used in the manufacture of explosives.

## TOP 50 INDUSTRIAL CHEMICALS

**Sodium carbonate, $Na_2CO_3$,** also known as soda ash, is an oxysalt that is the eleventh most produced industrial chemical, with 22.28 billion lbs. in 1995. It can be found naturally or can be synthetic. It is a grayish-white powder or lumps containing up to 99% sodium carbonate. Sodium carbonate is soluble in water. It is not a particularly hazardous material and is not regulated in transportation by the DOT. The primary uses are in the manufacture of other chemicals and products, including glass, paper, soaps, cleaning compounds, petroleum refining, and as a catalyst in coal liquefaction.

**Ammonium nitrate, $NH_4NO_3$,** an ammonium compound, is the fifteenth most produced industrial chemical, with 15.99 billion lbs. in 1995. The 1994 production level was a 6.7% increase over the 1993 level; however, the production

## AMMONIUM NITRATE 4-DIGIT IDENTIFICATION  NUMBERS

More than 0.2% combustible material - 0222
Not more than 0.2% combustible material - 1942
With organic coating - 1942
Ammonium Nitrate Fertilizer - 2067
Fertilizer that is liable to explode - 0223
Fertilizer with ammonium sulfate - 2069
Fertilizer with calcium carbonate - 2068
Fertilizer with not more than 0.4% combustible material - 2071
Fertilizer with phosphate or potash - 2070
Fertilizers - 2071, 2072
Fertilizers N.O.S. - 2072
Ammonium nitrate fuel oil mixture - 0331
Mixed fertilizers - 2069

**Figure 6.4**

fell in 1995 by about 1%. Ammonium nitrate is a colorless crystal that is soluble in water. It is a strong oxidizer. The specific gravity is 1.72, which is heavier than water. Ammonium nitrate is soluble in water and decomposes at 410°F, evolving nitrous oxide gas. It may explode under confinement and at high temperatures. Large amounts of water should be applied using unmanned appliances to fight fires, with all personnel evacuated to a safe distance. Ammonium nitrate is incompatible with acids, flammable liquids, metal powders, sulfur, chlorates, and any finely divided organic or combustible substance. The four-digit UN identification numbers are listed in Figure 6.4. The NFPA 704 designation for ammonium nitrate is health 0, flammability 0, and reactivity 3. In the white space at the bottom of the diamond, the prefix "oxy" indicates it is an oxidizer. The primary

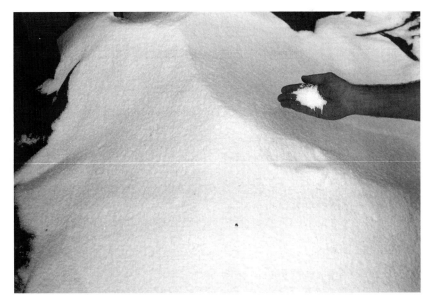

Commercial grade ammonium nitrate, an oxidizer used to make blasting agents.

**RELATIVE STRENGTHS OF SOME OXIDIZING AGENTS**

| | |
|---|---|
| 1 - Fluorine | 7 - Sulfuric Acid (concentrated) |
| 2 - Ozone | 8 - Oxygen |
| 3 - Hydrogen Peroxide | 9 - Bromine |
| 4 - Metallic Chlorates | 10 - Iron III compounds |
| 5 - Nitric Acid | 11 - Iodine |
| 6 - Chlorine | 12 - Sulfur |

**Figure 6.5**

uses of ammonium nitrate are in fertilizers, explosives, pyrotechnics, herbicides, and insecticides. It is also used as an oxidizer in solid rocket fuel.

**Ammonium sulfate, $(NH_4)_2SO_4$,** an ammonium compound, is the thirty-first most produced industrial chemical, with 5.24 billion lbs. in 1995. It is a brownish-gray to white crystal. Ammonium sulfate is soluble in water and is nonflammable. This compound is an oxidizer with a specific gravity of 1.77, which is heavier than water. The primary uses are in fertilizers, water treatment, fermentation, fireproofing compositions, and as a food additive.

There are a number of other chemicals in the Top 50 that are oxidizers and may be found as a gas or a liquid. However, they are not classified as oxidizers because under the DOT hazard system they have a more severe hazard or are compressed gases. Oxygen is the third most produced chemical, with 53.48 billion lbs. annually; chlorine is number 10, with 25.09 billion lbs. produced annually; nitric acid is number 14, with 17.24 billion lbs. annually. Figure 6.5 lists some examples of other common oxidizing materials in order of relative strength.

## INCIDENTS

Ammonium nitrate fertilizer, mixed with a hydrocarbon fuel, was used in the explosives that rocked the Federal Building in Oklahoma City in April 1995, killing 166 people including 19 children, and injuring 450 others. The damage to the building was so extensive that it had to be demolished. In addition, several other buildings in the downtown area were damaged by the explosion. Ammonium nitrate was also used in the bombing of the World Trade Center in New York City.

Commercial-grade ammonium nitrate was involved in the explosion in Kansas City, MO in 1988 that killed six firefighters. The firefighters were responding to a construction site where explosives were in a box trailer being used for storage. The storage trailer was on fire; the firefighters may have been unaware the explosives were stored there and fought the fire. The resulting explosion totally destroyed one fire engine and damaged another beyond repair, in addition to the deaths of the six firefighters.

As a result of that tragic explosion in Kansas City, OSHA has issued a new regulation involving the use of DOT placards and labels in fixed storage. All hazardous materials that require DOT placarding and labeling in transportation must continue to be placarded and labeled in fixed storage. The placards and

Hopper truck hauling ammonium nitrate.

labels must remain on the containers until the materials are used up and the containers have been purged or properly discarded.

Ammonium nitrate fertilizer can be made resistant to flame and detonation by an exclusive process involving the addition of 5 to 10% ammonium phosphate.

## CLASS 5.2 ORGANIC PEROXIDES

**Class 5.2** organic peroxides are assigned to seven generic types by the DOT (Figure 6.6) and are classified by the extent to which they will detonate or deflagrate. Organic peroxides are a hydrocarbon derivative and the major hazard is explosion. They are highly dangerous materials used as initiators and catalysts for polymerization reactions. Organic peroxides are highly reactive because of the presence of the oxidizer and the fuel within the formula. They can start their own decomposition process when contaminated, heated, or shocked. Organic peroxides are nonpolar and immiscible in water. Some examples of organic peroxide compounds and compounds subject to peroxide formation are listed in Figure 6.7. The organic peroxide functional group is composed of two oxygens single-bonded to each other. There is a hydrocarbon radical on each side of the single-bonded oxygens. Their general formula is **R–O–O–R.** Peroxides are named much the same way as ethers and ketones. There must be two radicals, which are named smallest to largest with the word "peroxide" at the end of the name. If the radicals are the same, the prefix "di" is used to indicate two of the same radicals. For example, if the radicals "methyl" and "ethyl" are attached to the peroxide, the compound is named methyl ethyl peroxide. Shown in the following illustration are the names, molecular formulas, and structures of two common organic peroxides. Note that benzoyl peroxide does not follow the trivial naming system for peroxides; however the "peroxide" in the name provides the family information that will help in determining the hazard, which is explosive.

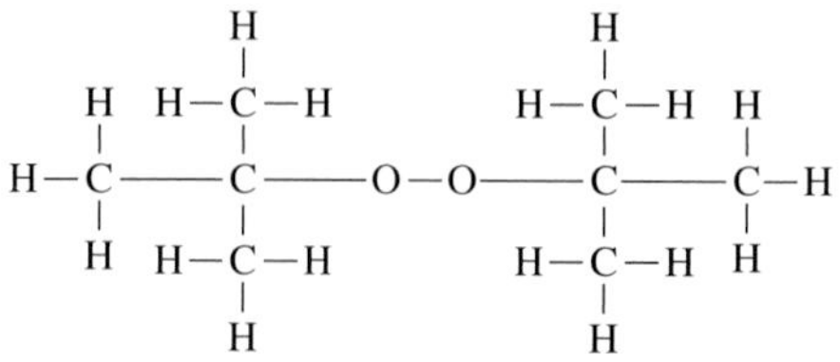

Di-*tert*-butyl peroxide
t-C$_4$H$_9$OOt-C$_4$H$_9$

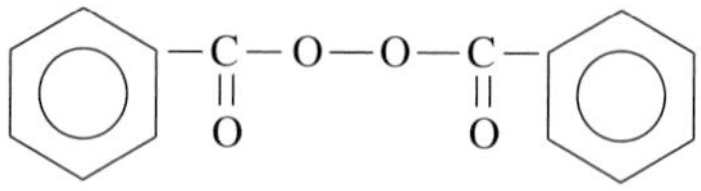

Benzoyl peroxide
(C$_6$H$_5$CO)$_2$O$_2$

Organic peroxides are widely used in the plastics industry as polymerization reaction initiators. All organic peroxides are combustible. Many can be decomposed by heat, shock, or friction; some, such as methyl ethyl ketone peroxide,

GENERIC TYPES OF ORGANIC PEROXIDES

TYPE A          Organic peroxide that can detonate or deflagrate rapidly as packaged for transport.  Transportation of type A organic peroxides is forbidden.

TYPE B          Organic peroxide as packaged for transport, neither detonates nor deflagrates rapidly, but can undergo a thermal explosion.

TYPE C          Organic peroxide as packaged for transport, neither detonates nor deflagrates rapidly and cannot undergo a thermal explosion.

TYPE D          Organic peroxide which:

(I)    Detonates only partially, but does not deflagrate rapidly and is not affected by heat when confined;
(II)   Does not detonate, deflagrates slowly, and shows no violent effect if heated when confined; or
(III) Does not detonate or deflagrate, and shows a medium effect when heated under confinement.

TYPE E          Organic peroxide which neither detonates nor deflagrates and shows low, or no, effect when heated under confinement.

TYPE F          Organic peroxide which will not detonate in a cavitated state, does not deflagrate, shows only a low, or no, effect if heated when confined, and has low, or no, explosive power.

TYPE G          Organic peroxide that will not detonate in a cavitated state, will not deflagrate, shows no effect when heated under confinement, has no explosive power, and is thermally stable.

**Figure 6.6**

## ORGANIC PEROXIDES AND PEROXIDIZABLE COMPOUNDS

| | |
|---|---|
| Methyl Ethyl Ketone Peroxide | Dioxane |
| Benzoyl Peroxide | Furan |
| Ether Peroxides | Butadiene |
| Peracetic Acid | Vinyl Chloride |
| Potassium Metal | Styrene |
| Vinylidene | Cyclohexane |
| Methyl Acetylene | Tetrahydrofuran |
| Cyclopentane | Cumene |

**Figure 6.7**

can detonate. Organic peroxides can be liquids or solids and are usually dissolved in a flammable or combustible solvent. Organic peroxides can be dangerously explosive materials. Organic peroxides have a self-accelerating decomposition temperature (SADT), and they are shipped and stored under refrigeration to keep them cool. The SADT temperatures range from 0° to 50°F and higher. SADT is the temperature at which the compounds will start to decompose. This decomposition reaction may result in a violent detonation that cannot be stopped by anything responders might try to do. The best bet is to make sure that the materials **do not** reach their SADT. Some organic peroxides are so unstable that they are forbidden in transportation. Organic peroxides exhibit the following hazards to emergency responders: they are unstable, flammable, highly reactive, may explode in a fire, are corrosive, may be toxic, and are oxidizers. They all have SADTs, and once this reaction starts there is little responders can do to stop it. Responders should withdraw and treat the material as Class 1 explosives.

**Methyl ethyl ketone peroxide** is an organic peroxide, even though there is also a ketone functional group in the compound. It is a colorless liquid and a strong oxidizing agent. It is a fire risk in contact with organic materials. Methyl ethyl ketone peroxide is a strong irritant to the skin and tissues. The TLV ceiling is 0.2 ppm in air. The four-digit UN identification number is 2550. The primary uses are in the production of acrylic resins and as a hardening agent for fiberglass-reinforced plastics. The molecular formula and structure are shown in the following example:

Methyl ethyl ketone peroxide
MEK peroxide
$C_8H_{16}O_4$

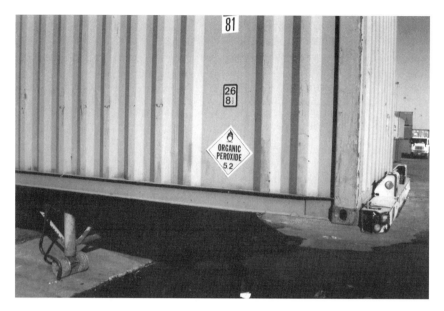

Organic peroxides have self-accelerating decomposition temperatures and may decompose explosively on exposure to heat.

**Di-tertiary butyl peroxide** is a clear, water-white liquid. It has a specific gravity of 0.79, which is lighter than water and it will float on the surface. It is nonpolar and insoluble in water. Di-tertiary butyl peroxide is a strong oxidizer and may ignite organic materials or explode if shocked or in contact with reducing agents. In addition to being an oxidizer, di-tertiary butyl peroxide is highly flammable. It has a boiling point of 231°F and a flash point of 65°F. The NFPA 704 designation is health 3, flammability 2, and reactivity 4. The prefix "oxy" for oxidizer is placed in the white section at the bottom of the 704 diamond. The molecular formula and structure are shown at the beginning of the Organic Peroxides section of this chapter.

**Hydrogen peroxide** is a colorless liquid organic peroxide that is a powerful oxidizer. It is also a dangerous fire and explosion risk and is toxic in high concentrations, with a TLV of 1 ppm in air. It is soluble in water, with a boiling point of 258°F. Hydrogen peroxide is also corrosive and a strong irritant. It is shipped and stored in 40 to 60% and in greater than 60% concentrations. Common commercial strengths are 27.5, 35, 50, and 70%. Hydrogen peroxide is incompatible with most metals and their salts, alcohols, organic substances, and any flammable substances. The four-digit UN identification number for the 40 to 60% concentrations is 2014. The NFPA 704 designation is health 2, flammability 0, and reactivity 1. The white section at the bottom of the 704 diamond has an "oxy" for oxidizer.

The four-digit UN identification number for concentrations greater than 60% is 2015. The NFPA 704 designation is health 2, flammability 0, and reactivity 3. The white section at the bottom of the 704 diamond has an "oxy" for oxidizer.

Intermodal tank of hydrogen peroxide, a strong oxidizer.

The structure and molecular formula for hydrogen peroxide are shown at the beginning of this chapter.

In its pure form, **benzoyl peroxide** is a white, granular crystalline solid that ignites very easily, burns with great intensity, and may explode. It has a faint odor of benzaldehyde and is tasteless. Benzoyl peroxide is slightly water soluble. It may explode spontaneously when dry with less than 1% water. It decomposes explosively above 105°C. Its autoignition temperature is 176°F. It should not be mixed unless at least 33% water is present. Benzoyl peroxide is highly toxic by inhalation, with a TLV of 5 mg/m$^3$ of air. The burning characteristics are very similar to black powder. Benzoyl peroxide decomposes very rapidly when heated, and if the material is confined detonation will occur. The molecular formula and structure for benzoyl peroxide are shown at the beginning of the organic peroxide section of this chapter.

Most ethers, when stored for more than 6 months, will form explosive **ether peroxides** in their containers. The primary ethers to be concerned about are ethyl ether, ethyl tertiary butyl ether, ethyl tertiary amyl ether, and isopropyl ether. Isopropyl ether is considered the worst hazard in storage. Ether peroxides that form inside containers are organic peroxides and are very sensitive to shock and heat. When these peroxides are concentrated or heated, they may detonate. Peroxide formation can be detected as early as one month in storage. Ethers are usually stored in amber glass bottles. Light and heat are important contributors to peroxide formation, although light seems to have more effect than heat. There is no effective means of inhibiting peroxide formation in ether containers. If aging containers of ethers are encountered, they should be treated like bombs. The bomb squad should be called for handling and disposal. There have been cases where employees and response personnel have been severely injured when an aging can of ether has exploded in their hands. Ethers may be found in high schools and college laboratories and throughout industry. Ether is also used in the processing of illegal drugs and may be encountered in clandestine drug lab operations.

The dangers presented by all oxidizers are similar. They are dangerous in contact with organic materials. They accelerate combustion. They can be a serious fire and explosion hazard. Even though some are water reactive, water is the extinguishing agent of choice for fires involving oxidizers. Extinguishing agents that work by excluding atmospheric oxygen will not always work with oxidizers. Oxidizers have their own oxygen supply within the compound and do not need atmospheric oxygen to support combustion. Emergency responders should treat oxidizers with the same respect as incidents involving explosives because they may be just as dangerous.

## CHEMICAL RELEASE STATISTICS

The DOT lists one oxidizer among the Top 50 incidents involving hazardous materials: hydrogen peroxide was involved in 89 incidents during 1995. There are no organic peroxides in the Top 50 industrial chemicals, the EPA Top 15 released chemicals, or the railroad Top 25 commodities. This does not mean, however, that they will not be encountered in storage or transportation. These are dangerous materials and should be handled very carefully by emergency responders. The same precautions taken for explosives should be followed for organic peroxides.

## INCIDENT

Two workers were killed and 13 people injured, including three firefighters, in an explosion and fire at a chemical plant. The facility manufactured organic peroxides, including methyl ethyl ketone (MEK) peroxide, which was involved in the fire. Also involved in the fire were bunkers for storing the MEK peroxide and benzoyl peroxide. Arriving companies were ordered to stage until the explosions subsided and an aggressive attack could be mounted safely.

## REVIEW QUESTIONS

1. Name four elements from the Periodic Table that are oxidizers: _____, _____, _____, and _____.

2. Which families of compounds are considered oxidizers?

3. Oxidizers may be shipped under which of the following placards?
   A. Dangerous
   B. Oxidizer

C. Poison or poison gas

D. All of the above

4. Name the following oxysalt compounds and balance the formulas if needed. (Any compounds with transition metals are balanced.)

$AlSO_5$      $LiClO_2$      $NaFO_3$      $MgPO_2$      $CuClO_3$

5. Provide formulas and structures for the following organic peroxide hydrocarbon derivative compounds.

Methyl ethyl peroxide        Di-vinyl peroxide

Iso-propyl butyl peroxide

# 7

Hazard Class 6 materials are poisons that are solids and liquids. Some of the liquids also produce vapors that are inhalation hazards. If Class 6 poisons do present an inhalation hazard, the transport vehicle must be placarded regardless of the quantity, just as with the Class 2.3 poisons. Class 6 is divided into two subclasses: 6.1 and 6.2. The Department of Transportation (DOT) defines a Class 6.1 poison as a material, other than a gas, known to be so toxic to humans as to afford a hazard to health during transportation or which, in the absence of adequate data, is presumed to be toxic to humans because it falls within any one of the following categories when tested on laboratory animals.

**Oral toxicity.** A liquid with a lethal dose (LD) of 50 or not more than 500 milligrams per kilogram or a solid with an $LD_{50}$ of not more than 200 mg/kg of body weight of the animal. ($LD_{50}$ is the single dose that will cause the death of 50 percent of a group of test animals exposed to it by any route other than inhalation.)

**Dermal toxicity.** A material with an $LD_{50}$ for acute dermal toxicity of not more than 1000 mg/kg.

**Inhalation toxicity.** A dust or mist with a lethal concentration (LC) of 50 for acute toxicity on inhalation of not more than 10 mg/kg. ($LC_{50}$ is the concentration of a material in air that, on the basis of laboratory tests through inhalation, is expected to kill 50 percent of a group of test animals when administered in a specific period.)

Simply stated, the DOT definition really says that a very small amount of a poison is dangerous to life and that the material should be considered very dangerous. A medical definition for a poison material is "the ability of a small amount of a material to produce injury by a chemical action." A chemical definition is "the ability of a chemical to produce injury when it comes in contact with a susceptible tissue." Perhaps a better definition of a poison would be that "a poison is a chemical that, in relatively small amounts, has the ability to produce injury by chemical action when it comes in contact with a susceptible tissue." Corrosives are not usually thought of as poisonous materials. However, if the combined chemical definition is applied to corrosives, in fact they are poisonous to the tissue that they contact.

Allyl alcohol (DOT/UN identification number 1098) is toxic by absorption, inhalation, and ingestion. However, like many other hazardous materials, it also has multiple hazards. Allyl alcohol is placarded as a 6.1 poison primary hazard, but it is also a flammable liquid.

Materials in Subclass 6.2 are infectious substances, meaning "they are viable microorganisms or their toxins which may cause diseases in humans or animals." This section also includes regulated medical waste.

Poisons are among the most dangerous materials for emergency responders. Many of the effects of poisons do not present themselves right away; in fact, the toxic effect may not appear for days, months, or years. Because the effects may not present themselves right away, responders may be led to believe there is no danger. One of the main reasons decontamination is done for hazardous materials incidents is to prevent the spread of toxic materials away from the "hot zone".

**Toxicology** is the science of the study of poisons and their effects on the human body. It is also the study of detection in the body systems and of antidotes to counteract poisonous effects. The science of toxicology is relatively new. It evolved out of the concern for worker health and safety that became a concern in the early twentieth century. Some of the first concerns for worker health and safety came from the trade unions. Eventually, the Occupational Health and Safety Act was passed by Congress in the 1970s creating the Occupational Safety and Health Administration (OSHA). These events led to the eventual formation of the science of industrial hygiene. This new science involved the protection of workers in the workplace. Industrial hygiene is the application of industrial hygiene concepts to the work environment. Many of the toxicological measuring terms found in reference sources come from the recommendations of safe workplace exposures (Figure 7.1). When responding to poison incidents, it is important to remember that, first of all, the emergency scene is the workplace for emergency responders. Second, the concentrations that emergency responders will encounter at a spill will often be much higher than the "normal" or acceptable workplace measurements.

## OSHA PERMISSIBLE EXPOSURE LIMITS (PELS)

| Chemical | 8 hr. PEL | Odor Threshold |
|---|---|---|
| Ozone | 0.1 ppm | 0.1 ppm |
| Chlorine | 0.5 ppm | 0.3 ppm |
| Benzene | 1 ppm | 5-12 ppm |
| Hydrogen Sulfide | 10 ppm | Fatigues Nose |
| Carbon Monoxide | 35 ppm | Odorless |
| Trichloroethylene | 50 ppm | 20 ppm |
| Toluene | 100 ppm | 2.0 ppm |
| Gasoline(TLV) | 300 ppm | 10 ppm |
| Freon 113 | 1000 ppm | 350 ppm |
| Methane | Simple Asphyxiant | Odorless |

**Figure 7.1**

## TYPES OF EXPOSURE

Types of exposure include:

**Acute:** A one-time, short-duration exposure. Depending on the concentration and duration of exposure, there may or may not be toxic effects. A one-time exposure can cause illness or death; however, it cannot be cumulative. In order for cumulative effects to occur, there must be multiple exposures. If multiple exposures occur, it is considered chronic rather than acute.

**Subacute:** Involves multiple exposures with a period of time between exposures. The effect is actually less than an acute exposure. The theory is that as long as there are periods of time between exposures, there will be no ill effects. There are no cumulative effects of subacute exposure because of the time between exposures. This concept is similiar to the time factor when dealing with radioactive materials. Personnel can be exposed to certain levels of radioactivity for short periods of time without any ill effects.

**Chronic:** As with subacute, chronic exposures are multiple exposures; however, with chronic exposure, there can be cumulative effects. Cumulative effects are simply a build-up of poison in the body. After the first exposure, some or all of the toxic material stays in the body. The first exposure may not cause any illness or damage. As additional exposures occur and the poison builds up in the body, it can reach toxic levels where illness, damage, or death can occur.

All of these exposures are usually considered workplace events. The emergency responders' workplace is the incident scene. An emergency responder can have an acute exposure to a toxic material at the scene of a HazMat incident or other type of emergency response. A single exposure may not produce any symptoms or illness, but multiple exposures may cause damage. It is important to monitor exposures to personnel to determine whether they experience any illness several days after an incident. Sometimes it takes several days for a toxic material to reach the susceptible target organ. The tendency is to assume that when a person is exposed to a poison at an incident scene the ill effects, if there are going to be any, will occur right away. This is not always the case with many poisons. When an illness occurs following a hazardous materials incident, the personnel should be checked out just in case.

Once a poison has entered the body, it can behave in a number of ways. First, the effect on the body may be localized, i.e., it only affects the tissue that it has directly contacted. Second, the effect may be systemic or a whole body effect. In this instance, the effect on the contact tissue is little if any. The poison enters the bloodstream and travels throughout the body until a target organ is reached or the material is secreted through the body's waste removal system. Last, the effects may be a combination of localized and systemic.

## ROUTES OF EXPOSURE

In order for a person to be affected by a poison, the poison must directly contact the body or enter into the body. There are four routes by which toxic

**ROUTES OF EXPOSURE**

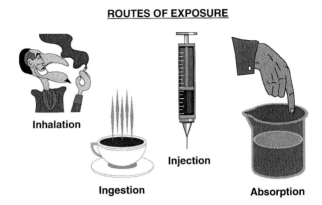

Inhalation

Injection

Ingestion

Absorption

**Figure 7.2**

materials can enter the body and cause damage (Figure 7.2): inhalation, absorption, ingestion, and injection.

**Inhalation** involves poison gas or vapor entering the body through the respiratory system, where most of the damage usually occurs. Once in the lungs, the poison can injure the respiratory tissues or enter the bloodstream or both. Examples of toxicants that produce disease of the respiratory tract are shown in Figure 7.3.

With **absorption,** poison solids, liquids, or gases enter the body through the skin, eyes, or some other tissue (Figure 7.4). Damage may occur at the point of contact or the material may travel to the susceptible target organ and cause harm there. Effects may be local, such as irritation and death of body tissues, which occurs through direct contact. When naphthalene contacts the eyes, it can cause cataracts and retina damage. Phenothiazine (insecticide) damages the retina.

**TOXICANTS THAT PRODUCE DISEASE OF THE RESPIRATORY TRACT**

| Toxicant | Acute Effect | Chronic Effect |
|---|---|---|
| Ammonia | Irritation, edema | Bronchitis |
| Arsenic | Bronchitis, irritation, pharyngitis | Cancer |
| Chlorine | Cough, irritation, asphyxiation | |
| Phosgene | Edema | Bronchitis, fibrosis, pneumonia |
| Toluene | Bronchitis, edema, bronchospasm | |

**Figure 7.3**

**EXAMPLES OF CHEMICALS TOXIC BY SKIN ABSORPTION**

| | |
|---|---|
| Acetaldehyde | Acetone |
| Acrolein | Ammonia |
| Aniline | Arsenic |
| Benzene | Barium |
| Camphor | Carbon Disulfide |
| Carbon Tetrachloride | Chlordane |
| Butyric Acid | Chlorine |
| Cumene | Bromine |

**Figure 7.4**

Thallium causes cataracts and optic nerve damage. Methanol causes optic nerve damage.

**Ingestion** occurs when the solid, liquid, or gaseous poison enters the body through the mouth and is swallowed. Damage may occur to the tissues contacted or the poison may enter the bloodstream. Absorption can also occur after ingestion. The poisonous material may be absorbed through the tissue in the mouth, stomach, intestines, or some other tissue it contacts after ingestion occurs.

**Injection** involves a jagged or sharp object that has been contaminated with a toxic material creating or entering through an open wound in the skin. The poison enters the bloodstream once injected into the skin.

The exposure method may greatly impact the severity of the damage produced. A chemical that is extremely poisonous by one method of exposure may have little if any effect by other methods. For example, carbon monoxide is very toxic by inhalation. Just a 1% concentration in air, if inhaled, is fatal in one minute. However, you could stay in a 100% concentration indefinitely provided there was an outside air supply, such as self-contained breathing apparatus (SCBA). Carbon monoxide is not absorbed through the skin. Rattlesnake venom is poisonous if it gets in the bloodstream; it damages the cells. It must be injected to cause damage. If ingested, it may cause nausea, but it will not enter the bloodstream.

## EFFECTS OF EXPOSURE

Toxicology relates to the physiological effect, source, symptoms, and corrective measures for toxic materials. Poisons can be divided into several general categories: asphyxiants, corrosives, sensitizers, carcinogens, mutagens, teratogens, and irritants.

### Short-Term Effects

When a poison contacts a tissue and a chemical action occurs, the poison will produce an injury to that tissue. The effects of an exposure to a poison, however, will differ based on the poison involved, the type of exposure, and the method of exposure. There are three types of effects that can occur as the result of an exposure to any given poison: immediate, long-term, and etiologic.

Immediate effects depend on the dosage received by the person exposed to the poison. Dosages can be large or small, short-term or long-term, and can be acute, subacute, or chronic. The effects from those types of exposure can vary from none, to slight discomfort, to illness and even death. The effect may also depend on the person exposed. Not all people are affected by the same poison, the same dosage, and the same exposure in the same way. The differences are based upon individual body chemistry, age, health, sex, and size. Immediate effects can include asphyxiation, corrosive damage, sensitizing, and irritation.

Airborne **asphyxiants** act upon the body by displacing the oxygen in the air. These materials, such as nitrogen, hydrogen, helium, and methane, are known as simple asphyxiants, they dilute the oxygen in the air. In the case of Class 6 poisons, however, asphyxiants act by interfering with the blood's ability to convert or carry oxygen in the bloodstream. These are known as chemical asphyxiants. Death also results from a lack of oxygen, but it is not a result of a lack of oxygen in the air. Examples of chemical asphyxiants include hydrogen cyanide, benzene, toluene, and aniline.

**Corrosives** are acids or bases that can in small amounts cause damage to tissue. They damage tissue in much the same way as a thermal burn; however, they are much more damaging. The type of damage is the same whether exposed to acids or bases. Examples are nitric acid, sulfuric acid, phosphoric acid, sodium hydroxide, and potassium hydroxide.

**Sensitizers,** on first exposure, cause little or no harm in humans or test animals, but on repeated exposure, they may cause a marked response not necessarily limited to the contact site. This response is similar to the process that occurs in allergies that humans develop. It is a physiological reaction to a sensitizing material. For example, a person who moves into an area that has high pollen counts and other airborne allergens may not experience any effects at first, but the longer the exposure occurs, the more the symptoms that develop. Examples of sensitizers are isocyanates and epoxy resins.

**Irritants** are materials that cause irritation to the respiratory system, other body organs, or the surface of the skin. The irritation may be a corrosive action, localized irritation, inflammation, pulmonary edema, or a combination. Effects may include minor discoloration of the skin, rashes, and tissue damage. Examples of irritants are tear gas and some disinfectants. Several factors influence toxic effects, one of which is the concentration of a material (Figure 7.5). Concentrations may be expressed in terms of percentage, parts per million or per billion

**EXPRESSIONS OF CONCENTRATION**

• **Percentages**
• **Parts per million (ppm) or billion (ppb)**
• **Milligrams per cubic meter,**
        **foot, kilogram, and cubic liter**

**Figure 7.5**

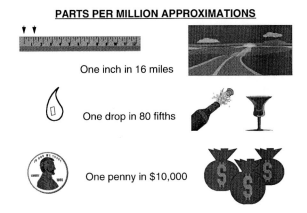

**PARTS PER MILLION APPROXIMATIONS**

One inch in 16 miles

One drop in 80 fifths

One penny in $10,000

**Figure 7.6**

per milligram per kilogram, the higher the concentration, the more serious the effects. Figure 7.6 illustrates some parts per million approximations. There are also materials that can have an anesthetic effect. This may range from loss of feeling and sensation to unconsciousness. Anesthetics include nitrous oxide, ethers, and hydrocarbons.

## Long-Term Effects

Long-term effects can be divided into three types: carcinogenic, teratogenic, and mutagenic. All three types of effect take from the time of exposure up to 15 years or even longer to develop symptoms. The effects can also be aggravated by exposures to other materials or other unrelated health problems.

**Carcinogens** are materials that cause cancer. This is a chronic toxic effect. Information about cancer-causing materials has been obtained by studying populations exposed to materials for long periods, usually in the workplace. Data has also been obtained through tests using laboratory animals. This does not mean that a material will actually cause cancer in humans, but it may be a good indication. Examples of known carcinogens include benzene, asbestos, arsenic, arsenic compounds, vinyl chloride, and mustard gas.

**Mutagens** cause mutations or alterations in genetic material, altering the DNA. The changes may be chromosomal breaks, rearrangement of chromosome pieces, gain or loss of entire chromosomes, or damage within a gene. Effect may involve the current generation that was exposed or future generations. Examples of mutagens include arsenic, chromium, dioxin, mercury, ionizing radiation from X-rays or other radioactive material, caffeine, LSD, marijuana, and nitrous oxide. Additionally, ethylene oxide, ethyleneimine, hydrogen peroxide, benzene, and hydrazine are also mutagens.

**Teratogens** cause one-time birth defects in offspring resulting from maternal or paternal exposure to toxic materials. The word "teratology" is derived from the Latin meaning the study of monsters. It is actually the study of congenital malformations, which started with the study of the correlation of German measles to birth defects. Later, there was an industrial link to teratogens discovered involving the chemical methyl mercury. The result of exposure to teratogenic chemicals or living organisms is the alteration of developing cells, leading to improper functioning of these cells. This may result in the death of the embryo or fetus. Specific types of birth defects can be caused by specific types of chemicals. These chemicals do not permanently damage the reproductive system and normal children can be produced as long as repeated exposure does not occur. Examples of teratogens include thalidomide, ethyl alcohol, and *o*-benzoic sulfimide (the artificial sweetener saccharin). Thalidomide is a drug that was used in the 1960s as a treatment for morning sickness in pregnant mothers. However, it was discovered too late for 10,000 babies who were born with various malformations. In Japan, where fish is a staple food, mothers ate fish contaminated with mercury compounds, causing children to be born with cerebral palsy. Ethyl alcohol is found in alcoholic beverages and is a known teratogenic material, which is the reason doctors tell pregnant women not to drink alcohol. Ethyl alcohol causes growth failure and impaired brain development. Additional teratogens include heavy metals, methyl mercury, mercury salts, lead, thallium, selenium, penicillin, tetracyclines, excess Vitamin A, and carbon dioxide.

## ETIOLOGIC EFFECTS

Class 6.2 infectious substances is the last type of toxic effect that will be presented here. These materials are in a class all their own. The DOT defines an etiologic agent as "a viable (live) micro-organism, or its toxin that is capable of producing disease in humans." The main reason this group is included with poisons is that the living organisms produce toxins (poisons). These toxins are biological wastes produced by the microorganisms. In small doses, the body's defense systems can handle the toxins. However, as the volume of toxin increases, it overwhelms the body's ability to defend against it. Toxins travel through the body until they reach a susceptible tissue and cause damage. Etiological toxins are among the most poisonous materials known. For example, the bacterium *Clostridium botulinum*, the cause of botulism, is a single cell that can release a toxin so potent that four-hundred thousandths of an ounce is enough to kill 1,000,000 laboratory guinea pigs.

## VARIABLES OF TOXIC EFFECTS

Not all people are affected by toxic materials in the same way. Variables include age, genetic make-up, sex (in terms of the genetic difference between males and females, most of which are related to reproduction and nurturing functions), weight, general health, body chemistry (the chemicals produced in each individual are unique, not only by form, but by quantity), and physical condition. These variables can all affect the way individuals respond to any toxic

Infectious medical waste is a DOT Class 6.2 infectious substance.

material. Infants and children are often more sensitive to toxic materials than younger adults. Elderly persons have diminished physiological capabilities to deal with toxic materials. This age group may be more susceptible to toxic effects at relatively lower doses. Chemicals may be more toxic to one sex than the other. Males may be affected by chemicals that do not affect females; some chemicals can affect both. Chemicals may affect the reproductive system of either the male or female. Females who are pregnant may be affected by a toxic material that will cause damage to the fetus, whereas a male may not be affected at all. Damage to the male reproductive system may include sterility, infertility, abnormal sperm, low sperm count, or reduced hormonal activity. These types of chemicals include lead; mercury; PCBs; 2,4,D; paraquat; ethanol; vinyl chloride; and DDT. Females are affected by DDT, parathion, PCBs, cadmium, methyl mercury, and anesthetic gases.

## DOSE/RESPONSE

Dose can be expressed in three ways: the amount of the substance actually in the body, the amount of material entering the body, and the concentration in the environment. Response is the health damage resulting from a specified dose. All chemicals, no matter what their makeup, are toxic if taken in a large enough dose. Even water can be toxic if too much is ingested at one time. Paracelsus (1493–1541) observed that "all substances are poisons; there is none which is not a poison. The right dose differentiates a poison and a remedy." There is a no-adverse-effects level for almost all materials. The dose is more than just the amount of a toxic material that has caused the exposure. More correctly, the dose is related to the weight of the individual exposed. This is the basis for many of the toxicological terms that have been developed based upon tests on laboratory animals. For example, a 1-lb. rat given 1 oz. of a toxic material should have the same response as a 2000-lb. elephant given 2000 oz. of the same toxic material. In both cases, the amount per weight is the same: 1 oz. for each pound of animal weight. This amount per weight is known as the dose.

Response of an individual plant or animal species to a hazardous material is based on concentration; length, type, and route of exposure; and susceptible target organ. Additionally, other health-related variables include age, sex, physical condition, and size (mass). This relationship is referred to as dose/response, or in other words, the amount of the exposure and the resulting biological effect. Each plant or animal species has its own individual response to a given chemical. A substance administered at a dose large enough to be lethal to rabbits may have a lesser effect on rats or dogs.

## SUSCEPTIBLE TARGET ORGANS

When a poison enters the body through one of the four routes of exposure, it will cause damage to some bodily function. This is referred to as the susceptible "target organ" (Figure 7.7). It may take a toxic material from several hours to several days to reach a susceptible target organ and cause damage. The target organ should not be confused with routes of exposure, which are the manner in which the poison enters the body. The susceptible target organ is the organ or system to which the poison does its damage once it enters the body (Figure 7.8). For example, a pesticide may enter the body through inhalation, but may have an effect on the central nervous system. In this instance, the route of exposure is inhalation through the lungs; the target organ is the central nervous system. Target organs in the body include the respiratory system, liver, kidneys, central nervous system, blood, bone marrow, skin, cardiovascular system, and other body tissues.

The skin is the largest single organ of the body. It provides a barrier between the environment and other organs, except for the lungs and eyes, and is a defense

### TARGET ORGANS AND CHEMICALS THAT AFFECT THEM

| | |
|---|---|
| Lungs | Halogens, Hydrogen sulfide |
| Liver | Vinyl Chloride, Aromatics, Chlorinated HC |
| Kidneys | Mercury, Calcium, Carbon Tetrachloride |
| Blood | Carbon Monoxide, Chlorinated HC |
| Neurologic | Organophosphates, Carbon Monoxide |
| Skeletal | Fluorides, Selenium |
| Skin | Arsenic, Chromium, Beryllium |

**Figure 7.7**

## SUSCEPTIBLE TARGET ORGANS

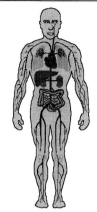

Nervous System
Skeleton & Bone Marrow
Thyroid
Lungs & Respiratory System
Cardiovascular System
Blood
Liver
Kidneys
Intestines
Skin

**Figure 7.8**

against many chemicals. Studies show that 97% of chemicals to which the body is exposed are deposited on the skin.

## RATES OF EXPOSURE

Exposure rate involves measurement of workplace exposure of hazardous materials based upon tests conducted on laboratory animals. These values are translated to humans based on the weight ratio between an animal and a human. These are only estimations and should be used with caution. Toxicology information is expressed in parts per million (ppm) or milligrams per cubic meter ($mg/m^3$), which are terms indicating concentration of the toxic material. These terms are not related but they do have one thing in common: the smaller the numerical value, the more toxic the material being measured. The emergency responder's workplace is the incident scene and the concentrations of toxic materials are much higher than the values indicated through the following terms:

**TLV-TWA:** Threshold limit value, time-weighted average, is the average concentration for a normal 8-hour workday and a 40-hour workweek to which nearly all workers can be exposed repeatedly, day after day, without adverse effect and without protective equipment. Some examples are shown in Figure 7.9.

**TLV-STEL:** Threshold limit value, short-term exposure limit is the maximum concentration averaged over a 15-min period to which healthy adults can be exposed safely. Exposures should not occur more than four times a day, and there should be at least 60 min between exposures.

**TLV-C** or **TLV-ceiling:** The concentration that should not be exceeded during any part of the work day. **This is the only reliable measurement that should be used by emergency responders on incident scenes.**

## TOXICITY COMPARISONS OF SOME HAZMAT

| Common Name | TLV | LC$_{50}$ |
|---|---|---|
| Hydrogen Cyanide | 10 ppm | 300 ppm |
| Hydrogen Sulfide | 10 ppm | 600 ppm |
| Sulfur Dioxide | 5 ppm | 1000 ppm |
| Chlorine | 1 ppm | 1000 ppm |
| Carbon Monoxide | 50 ppm | 1000 ppm |
| Ammonia | 50 ppm | 10,000 ppm |
| Carbon Dioxide | 5,000 ppm | 10% |
| Methane | 90,000 ppm | Simple Asphyxiant |

**Figure 7.9**

**PEL:** The permissible exposure limit is the maximum concentration averaged over 8 hours to which 95% of healthy adults can be repeatedly exposed for 8 hours per day, 40 hours per week.

**IDLH:** Immediately dangerous to life and health, the maximum amount of a toxic material that a healthy adult can be exposed to for up to 15 min (NIOSH) and escape without irreversible health effects. There are also IDLH values established for oxygen deficient atmospheres, and explosive or near-explosive atmospheres (above, at, or near the lower explosive limits). At the scene of a hazardous materials incident, it should be assumed that the concentrations present are above the IDLH, and only responders wearing SCBA and proper chemical protective clothing should be allowed near the incident scene.

**LD$_{50}$** is the lethal dose by ingestion or absorption for 50% of the laboratory animals exposed. (There is a wide variation among species. The LD$_{50}$ for one type of animal could be thousands of times less than for another type.) Examples of LD$_{50}$ values for common chemicals are shown in Figure 7.10.

**LC$_{50}$** is the lethal concentration by inhalation for 50% of the laboratory animals exposed. Some examples are shown in Figure 7.9.

**NOAEL** is no-observable-adverse-effect level.

**GRAS** is generally recognized as safe.

## LD$_{50}$ VALUES FOR COMMON CHEMICALS

| | |
|---|---|
| Sucrose (table sugar) | 29,000 mg/kg |
| Ethyl Alcohol | 14,000 |
| Sodium Chloride | 3,000 |
| Vitamin A | 2,000 |
| Vanillin | 1,580 |
| Aspirin | 1,000 |
| Chloroform | 800 |
| Copper Sulfate | 300 |
| Caffeine | 192 |
| Phenobarbital, Sodium Salt | 162 |
| DDT | 113 |
| Sodium Nitrate | 85 |
| Nicotine | 53 |
| Sodium Cyanide | 6.4 |
| Strychnine | 2.5 |

**Figure 7.10**

Toxicity measures are based on dose and size or weight of the test animal. This is then projected to humans, based on the weight ratio of the dose, the weight of the animal, and the weight and health of the human. All of the values should be considered nothing more than an educated guess. Many factors can influence the way any given individual will react if exposed to a toxic material. Some factors may be allergies and previous illness or operation. Individuals that have had a splenectomy are much more susceptible to poisons than those who have not. Individuals with such a history should be aware that they are at additional risk whenever they are exposed to toxic chemicals. Consider all toxicity data as an estimate; do not stake your life on toxicity data!

## DEFENSE MECHANISMS FOR TOXIC MATERIALS

There are three types of defense against toxic materials: internal, antidotal, and external; three things can happen once a chemical is taken into the body: metabolism, storage, and excretion. Internal defenses are the ability of the body to get rid of a toxic material, sometimes referred to as metabolism. The body normally excretes waste materials through the feces or urine. Additionally, women can also excrete through the ova and breast milk. In these instances, the excretions from the mother represent exposure to the offspring. Two main organs in the body filter materials and remove them through excretion: the liver and the kidneys. All of the blood in the body passes through these two organs and is filtered. The filtered material is passed from the liver to the intestine through the gallbladder or to the kidney through the blood, after some degree of chemical break-down (metabolizing). If the liver is unable to break down the poison, it may store the toxic material within its own tissue. Toxic materials, such as lead, can be stored in the bone tissue. Some may be bound by blood proteins and stored in the blood. It is this storage of the toxic material that causes most of the long-term damage. The kidneys also filter poisons from the blood and may incur damage in the process. The materials filtered are then passed on to the urine, which is held in the bladder. This can lead to bladder cancer through chronic exposure to the toxic materials. Toxic materials that enter the body cannot always be excreted and are stored in the fatty tissues. Examples include the organochlorines, DDT, PCBs, and chlordane. When a person loses weight, for whatever reason, toxic materials that have been stored are released back into the body as the levels of fat decrease. When this occurs, the body is reexposed to the chemicals and illness can occur. Chemicals that cannot be excreted from the body may be bound by blood proteins and stored in the blood. Lead is an example of a material that cannot be excreted from the body and is stored in the bones.

Polarity has an effect on the ability of the body to excrete toxic materials. If a poison is polar, it is usually soluble in water, because water is also polar and is more easily removed from the body. The body also has a system of converting nonpolar compounds so that they can also be removed; however, the body cannot convert all nonpolar compounds. Therefore, some poisons are difficult to convert and may stay in the body for long periods of time. For example, table salt (sodium chloride) is polar and is easily excreted. DDT, a now banned pesticide, is nonpolar and is not easily excreted, so it stays in the body for long periods. This is one of the main reasons DDT is no longer allowed to be used on food crops. The characteristic of polarity or nonpolarity can have a very crucial impact on the effect a toxic material has on the body.

Two other mechanisms by which the body defends against toxic materials is through breathing and sweating. The lungs are also able to remove materials from the blood. This can be noted by the odor of alcohol on the breath of a person who has been drinking. Odor detected from a person who is sweating is toxic material being removed from the body. While all three methods contribute to the removal of toxic materials from the body, the last two are of only minor significance. Internal defenses of the body against toxic materials do a good job against many types of chemicals. However, not all toxic materials can be removed by the body's systems. It is best not to rely on the body to remove toxic materials but rather to take precautions to ensure that the toxic materials do not enter the body in the first place.

Antidotes are taken into the body to counteract the effects of some toxic materials. The definition of an antidote is "any substance that nulls the effects of a poison on the spot, and prevents its absorption or blocks its destructive action once absorbed." The problem with antidotes is there are only about 20 in existence and they do not work on all types of poisons (Figure 7.11). Another problem with antidotes is availability, the antidote must be administered immediately after exposure to a poison or the victim will die anyway. Most EMS units do not carry antidotes for toxic exposures; by the time a person is taken to a medical center and the antidote is administered, it will be too late. So do not depend on an antidote to save your life!

External defenses against toxic materials are by far more effective than the internal defenses of the body or antidotes. What it amounts to in simple terms is: do not let toxic materials into the body to begin with; the idea is to place a barrier between responders and the poison. These barriers include chemical protective clothing and SCBA. Chemical protective clothing provides protection

### ANTIDOTES FOR POISONS

| | |
|---|---|
| Cyanide | Amyl Nitrate, Sodium Nitrate Sodium Thiosulfate |
| Organophosphate pesticides | Atropine, Pralidoximine |
| Methanol or Ethylene Glycol | IV Ethanol, Hemodialysis |
| Nitrites | Oxygen, Methylene Blue |
| Hydrocarbons | Oxygen |

**Figure 7.11**

against absorption and contact tissue damage. SCBA prevents inhalation and ingestion of toxic materials. Firefighter turnouts do not provide any type of chemical protection from toxic materials. Turnouts may prevent some types of injection; however, the toxic material will contaminate the turnouts. There are also some preventive measures that can be taken to prevent ingestion of toxic materials. Do not eat, drink, smoke, or place anything in your mouth until decontamination has been completed.

## TOXIC ELEMENTS

There are a number of elements whose main hazard is toxicity. These include arsenic; mercury; heavy metals, such as lead and cadmium; and the halogens: fluorine, chlorine, and bromine. Fluorine and chlorine are covered in Chapter 3 in the Poison Gas section. It is important to note the uses of elements, because they give an indication where these materials may be found in storage and manufacturing.

**Arsenic, As,** is a nonmetallic element that is a silver-gray, brittle, crystalline solid that darkens in moist air. Arsenic is a carcinogen and a mutagen. It is insoluble in water and reacts with nitric acid. The primary routes of exposure are through inhalation absorption, skin or eye contact, and ingestion. The exposure limit is 0.002 mg/m$^3$. The IDLH is 100 mg/m$^3$ in air. The target organs are the liver, kidneys, skin, lungs, and lymphatic system. Arsenical compounds are incompatible with any reducing agents. The four-digit UN identification number is 1558. The primary uses are alloying additives for metals, especially lead and copper; in boiler tubes, high-purity semiconductors, special solders, and medicines.

**Mercury, Hg,** is a liquid metallic element that is silvery in color and very heavy. It is insoluble in water and has a specific gravity of 13.59, which is heavier than water. Mercury is highly toxic by skin absorption and inhalation of fumes or vapor. The TLV is 0.05 mg/m$^3$ of air and the IDLH is 10 mg/m$^3$. All inorganic compounds of mercury are toxic by ingestion, inhalation, and absorption. Most organic compounds of mercury are also highly toxic. The target organs affected are the central nervous system, kidneys, skin, and eyes. Mercury is incompatible with acetylene and ammonia. The four-digit UN identification number is 2809. The uses of mercury are in electrical appliances, instruments, mercury vapor lamps, mirror coating, and as a neutron absorber in nuclear power plants.

**Bromine, Br$_2$,** is a nonmetallic, fuming liquid element of the halogen family on the Periodic Table. It is dark reddish-brown in color with irritating fumes. Bromine is slightly soluble in water and attacks most metals. The boiling point is 138° F and the specific gravity is 3.12, which is heavier than water. Bromine is toxic by ingestion and inhalation and is a severe skin irritant. The TLV is 0.1 ppm in air and the IDLH is 10 ppm. The target organs are the respiratory system, the eyes, and the central nervous system. It is also a strong oxidizing agent and may ignite combustible materials on contact. It is listed by the DOT as a Class 8 corrosive; however, it carries the corrosive and poison label. The four-digit UN identification number for bromine is 1744. The NFPA 704 designation for bromine and bromine solutions is health 3, flammability 0, and reactivity 0. The white section at the bottom of the 704 diamond has the prefix "oxy", indicating an oxidizer. The primary uses are in anti-knock compounds for gasoline, bleaching, water purification, as a solvent, and in pharmaceuticals.

## TOXIC SALTS

Binary salts have varying hazards, one of which is toxicity. Some of the binary salts are highly toxic, such as sodium fluoride, calcium phosphide, and mercuric chloride. Cyanide salts are also highly toxic, such as sodium cyanide and potassium cyanide. The remaining salts, binary oxides, peroxide, hydroxides, and oxysalts are generally not considered toxic.

**Calcium phosphide, $Ca_3P_2$,** is a binary salt made up of red-brown crystals or gray granular masses. It is water reactive and evolves phosphine gas in contact with water, which is highly toxic and flammable. The four-digit UN identification number is 1360. The primary uses are as signal fires, torpedoes, pyrotechnics, and rodenticides.

**Sodium fluoride, NaF,** is a binary salt that is a clear, lustrous crystal or white powder. The insecticide grade is frequently dyed blue. It is soluble in water and has a specific gravity of 2.558, which is heavier than water. Sodium fluoride is highly toxic by ingestion and inhalation, and is also strongly irritating to tissue. The TLV is 2.5 mg/m³ of air. The four-digit UN identification number is 1690. The primary uses are fluoridation of municipal water at 1 ppm, as an insecticide, rodenticide, and fungicide, and in toothpastes and disinfectants.

**Mercuric chloride (mercury II chloride), $HgCl_2$,** is a binary salt composed of white crystals or powder. It is odorless and soluble in water. It is highly toxic by ingestion, inhalation, and skin absorption. The TLV is 0.05 mg/m³ of air. The four-digit UN identification number is 1624. The primary uses of mercuric chloride are in embalming fluids, insecticides, fungicides, wood preservatives, photography, textile printing, and dry batteries.

**Sodium cyanide, NaCN,** is a cyanide salt that is a white, deliquescent, crystalline powder and is soluble in water. The specific gravity is 1.6, which is heavier than water. Sodium cyanide is toxic by inhalation and ingestion, with a TLV of 4.7 ppm and 5 mg/m³ of air. The target organs are the cardiovascular system, central nervous system, kidneys, liver, and skin. Reactions with acids can release flammable and toxic hydrogen cyanide gas. Cyanides are incompatible with all acids. The four-digit UN identification number is 1689. The NFPA 704 designation is health 3, flammability 0, and reactivity 0. The primary uses are in gold and silver extraction from ore, electroplating, fumigation, and insecticides.

**Potassium cyanide, KCN,** is a cyanide salt that is found as a white, amorphous, deliquescent lump or crystalline mass with a faint odor of bitter almonds. It is soluble in water and has a specific gravity of 1.52. It is a poison that is

A 55-gallon drum of parathion, a highly toxic pesticide.

absorbed through the skin. Target organs are the same as for sodium cyanide. Reaction with acids releases flammable and toxic hydrogen cyanide gas. The four-digit UN identification number is 1680. The NFPA 704 designation is health 3, flammability 0, and reactivity 0. The primary uses are in gold and silver ore extraction, insecticides, fumigants, and electroplating.

## HYDROCARBONS

Most of the alkane, alkene, and alkyne hydrocarbon compounds are considered to be flammable as their major hazard and the toxicity is considered as moderate to low. The vapors are more likely to be asphyxiant than toxic. TLVs range from 50 ppm for hexane to 300 ppm for octane. Decane is listed as having a narcotic effect. Many of these hydrocarbons are found in mixtures and it will be necessary to look at the Material Safety Data Sheets (MSDS) to obtain toxicity information on particular mixtures.

Benzene, toluene, and xylene are aromatic hydrocarbons. They are considered highly toxic and human carcinogens. Benzene has a TLV of 0.1 ppm in air, according to the *NIOSH Guide 1994 Addition,* and a STEL of 1 ppm. The OSHA STEL is 5 ppm and a PEL of 1 ppm. Toluene is toxic by ingestion, inhalation, and skin absorption. The TLV for toluene is 100 ppm in air. Xylenes are toxic by inhalation and ingestion, with a TLV of 100 ppm. The target organs are the blood, skin, bone marrow, eyes, central nervous system, and respiratory system.

## HYDROCARBON DERIVATIVES

Several hydrocarbon derivatives are toxic as a primary hazard. While some compounds in each of the groups are toxic, not all compounds are toxic. It is important, however, to consider them all toxic within a group until the specific material can be looked up in a reference source and the exact hazards verified. Families with toxicity as a primary hazard are the alkyl halides and amines. Other groups that are toxic, although they may not be the primary hazard, include the alcohols, aldehydes, and organic acids. The ethers are considered anesthetic; however, there are some ethers that are listed as Class 6.1 poisons by the DOT. They are compounds that have chlorine added to the ether. The toxicity comes from the chlorine. For example, 2,2-dichlorodiethyl ether and dichlorodimethyl ether have NFPA 704 health designations of 3 and 4, respectively. Epichlorohydrin, also known as chloropropylene oxide, is a Class 6.1 poison and is an ether with chlorine added to the compound. Methychloromethyl ether is another compound with chlorine added. The NFPA 704 designation for health is 3. The ketones as a group are considered narcotic. The primary hazard of the esters is polymerization. Many of them are flammable liquids and polymers. The DOT does list some ester compounds as Class 6.1 poisons and, again, these compounds have chlorine added, which accounts for their toxicity. For example, ethyl chloroformate is an ester that has an NFPA health designation of 4. Remember that too much of any chemical can be toxic, so too much of an anesthetic or a narcotic can be toxic.

### Alkyl Halides

The alkyl halide functional group is composed of a hydrocarbon radical and some combination of halogens from family seven on the Periodic Table. The halogens are all toxic and, therefore, it is not difficult to see that the alkyl halides are also going to be toxic. The general formula for alkyl halide is **R–X.** The "R" represents one or more hydrocarbon radical(s) and the "X" represents one or more halogen(s). The "X" can be replaced by fluorine (F), chlorine (Cl), bromine (Br), and iodine (I), or combinations of two or more. It is important to remember that hydrocarbon derivatives started out as hydrocarbons before hydrogens were removed and other elements were added. Many of the hydrocarbon names are still used in the naming process with the alkyl halides. There are three ways alkyl halides can be named. They are all correct naming conventions and the compounds may be listed under any one of the possibilities. When looking up the compounds in reference books, you may have to look under the alternate names to find information on the compound. The first naming convention is one in which the radical is named first, the "ine" is dropped from the halogen, and an "ide" ending is added. For example, if the compound has one carbon, the radical for one carbon is methyl. If there is chlorine attached to the methyl radical, the alkyl halide compound is named methyl chloride. The structure, molecular formula, and names for the compound are shown in the following illustration:

$$
\begin{array}{c}
H \\
| \\
H-C-Cl \\
| \\
H
\end{array}
$$

Chloromethane
Methyl chloride
$CH_3Cl$

If fluorine is attached to a one-carbon radical, the name is methyl fluoride, and so on.

The second convention would be to name the halogen first and then the hydrocarbon radical. In this case, the "ine" ending is dropped from the halogen, and an "o" is added to the abbreviated name for the halogen. In the case of chlorine, it is "chloro." The radical is on the end of the name. When the radical is on the end, the name reverts back to the hydrocarbon that was used to form the radical. Methyl is a radical of the one-carbon alkane, methane. So if the halogen chlorine is added to methane, and the halogen is named first, the name is chloromethane. If bromine is the halogen, the name is bromomethane. If the radical is a two-carbon radical and the halogen is fluorine, the name of the compound is fluoroethane, and so on. Illustrated in the following example are the names and structural formulas for some one-, two-, and three-carbon alkyl halides:

$$
\begin{array}{c}
H \\
| \\
H-C-Br \\
| \\
H
\end{array}
\qquad
\begin{array}{c}
H \quad H \\
| \quad | \\
H-C-C-F \\
| \quad | \\
H \quad H
\end{array}
\qquad
\begin{array}{c}
H \quad H \quad H \\
| \quad | \quad | \\
H-C-C-C-Cl \\
| \quad | \quad | \\
H \quad H \quad H
\end{array}
$$

Methyl bromide   Ethyl fluoride   Propyl chloride
Bromomethane   Fluoroethane   Chloropropane
$CH_3Br$     $C_2H_5F$     $C_3H_7Cl$

It is possible to use more than one halogen to form alkyl halide compounds. If the multiple halogens are the same type, the prefix "di" is used for two, "tri" for three, and "tetra" for four. Some chemicals, as has been previously mentioned, have trade names. There are also trade names for some of the alkyl halides. For example, a one-carbon radical with three chlorine atoms attached is called trichloro methane, or methyl trichloride; however, the trade name for the compound is chloroform. A methyl radical with four chlorines attached is named tetrachloro methane or methyl tetrachloride. The trade name for the compound is carbon tetrachloride, a material that was used as a fire-extinguishing agent. It is no longer approved as an extinguishing agent because when it contacts a hot surface it decomposes to phosgene gas. Shown in the following examples are the names, molecular formulas, and structures for some alkyl halides with multiple numbers and combinations of halogens in the compounds:

$$\begin{array}{ccc} \text{Cl} & \text{H} & \text{Cl} \\ | & | & | \\ \text{Cl}-\text{C}-\text{Cl} & \text{Br}-\text{C}-\text{Br} & \text{Cl}-\text{C}-\text{Cl} \\ | & | & | \\ \text{H} & \text{H} & \text{Cl} \end{array}$$

| Trichloro methane | Dibromo methane | Tetrachloro methane |
|:---:|:---:|:---:|
| Methyl trichloride | Methyl dibromide | Methyl tetrachloride |
| Chloroform | $CH_2Br_2$ | Carbon tetrachloride |
| $CHCl_3$ | | $CCl_4$ |

Some alkyl halide compounds also have double bonds: dichloroethylene, dichloropropene, dichlorobutene, and trichloroethylene.

**Trichloroethylene, CHClCCl$_2$,** is a stable, low-boiling, colorless liquid with a chloroform-like odor. It is not corrosive to the common metals even in the presence of moisture. It is slightly soluble in water and is nonflammable. It is toxic by inhalation, with a TLV of 50 ppm and an IDLH of 1000 ppm in air. The FDA has prohibited its use in foods, drugs, and cosmetics. The four-digit UN identification number is 1710. The NFPA 704 designation is health 2, flammability 1, and reactivity 0. Its primary use is in metal degreasing, dry cleaning, as a refrigerant and fumigant, and for drying electronic parts.

## Amines

The next toxic hydrocarbon derivative functional group is the amines. Amines are toxic irritants, in addition to being flammable. The amines are considered slightly polar when compared to nonpolar materials. The amine functional group is represented by a single nitrogen surrounded by two or less hydrogens. The general formulas for the amines are **R–NH$_2$, R$_2$NH,** and **R$_3$N.** It is the nitrogen that identifies the amine group, not the number of hydrogens attached to the nitrogen. The amines are covered in detail in Chapter 4 under the Hydrocarbon Derivatives section.

The degree of toxicity of amines varies from compound to compound. Many of them are strong irritants. TLV values range in the low double digits from 5 to 10 ppm. Diethylamine is toxic by ingestion and is a strong irritant. It has a TLV of 10 ppm in air. Butyl amine is a skin irritant with a TLV of 5 ppm in air. It is important to obtain further information when dealing with amines. Look the materials up in reference books and MSDS sheets to determine the toxic characteristics of a given amine compound.

**Aniline, C$_6$H$_5$NH$_2$,** also known as phenylamine, is a colorless, oily liquid with a characteristic amine odor and taste. It rapidly turns brown when exposed to air. It is soluble in water, with a specific gravity of 1.02, which is very slightly heavier than water. Aniline is an allergen and is toxic if absorbed through the skin. The TLV is 2 ppm in air and the IDLH is 100 ppm. The target organs are the blood, cardiovascular system, liver, and kidneys. Aniline is incompatible with nitric acid and hydrogen peroxide. The four-digit UN identification number is 1547. The NFPA 704 designation is health 3, flammability 0, and reactivity 1. The primary uses are in dyes, photographic chemicals, isocyanates for urethane foams, explosives, herbicides, and petroleum refining. The structure and molecular formula for aniline are shown in the following illustration:

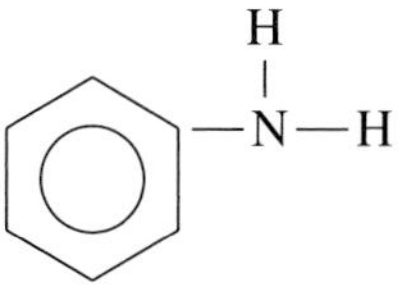

Aniline
Phenylamine
$C_6H_5NH_2$

## Alcohols

All alcohols are toxic to some degree. Ethyl alcohol is drinking alcohol and, when consumed in moderation, has limited toxic effects. However, ethyl alcohol, if taken in excess, can have toxic effects. In the short-term, the effects of ethyl alcohol are sedative and depressant; in the long run, cancer may occur and/or liver damage. Methyl alcohol is toxic by ingestion and can cause blindness. It has a TLV of 200 ppm. Propyl alcohol is toxic by skin absorption. It has a TLV of 200 ppm. However, isopropyl alcohol is used as rubbing alcohol and is applied to the skin. Isopropyl alcohol is toxic by ingestion and inhalation with a TLV of 400 ppm. Butyl alcohol is toxic by prolonged inhalation, is an eye irritant, and is absorbed through the skin. The TLV ceiling is 50 ppm in air. There are also some alcohols that have chlorine added to the compound, which increases their toxicity. For example, ethylene chlorohydrin is an alcohol that has an NFPA 704 designation for health of 4. As you can see, the hazards and routes of entry vary widely among the alcohols. It should be assumed the alcohol in a spill is the worst-case scenario until the exact hazard of the compound can be researched. The alcohols are covered in detail in Chapter 4 under Hydrocarbon Derivatives.

**Cresols, (o-,m-,p-,), CH₃C₆H₄OH,** are alcohol hydrocarbon derivatives. They are colorless to yellowish or pinkish liquids. They are found in the "ortho", "meta", and "para" isomers, like the xylenes in Chapter 4. Cresol has a characteristic phenolic odor and is soluble in water with a specific gravity of 1.05, which is slightly heavier than water. It is an irritant, corrosive to the skin and mucous membranes, and is absorbed into the skin. The TLV is 5 ppm in air and the IDLH is 250 ppm. The target organs are central nervous system, respiratory system, liver, kidneys, skin, and eyes. The four-digit UN identification number is 2076. The NFPA 704 designation is health 3, flammability 2, and reactivity 0. The primary uses are as a textile scouring agent, herbicide, phenolic resins, and ore flotation; the para isomer is used in synthetic food flavors. Shown in the following example are the structures and molecular formulas for the isomers of cresol:

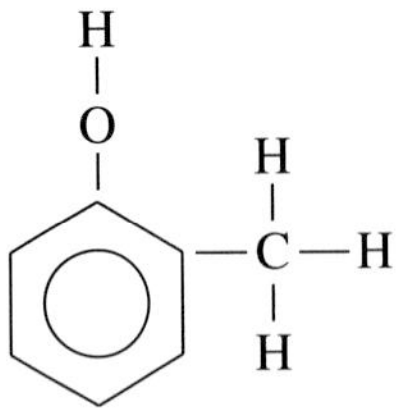

| *Ortho* cresol | *Meta* cresol | *Para* cresol |
| o-CH$_3$C$_6$H$_5$OH | m-CH$_3$C$_6$H$_5$OH | p-CH$_3$C$_6$H$_5$OH |

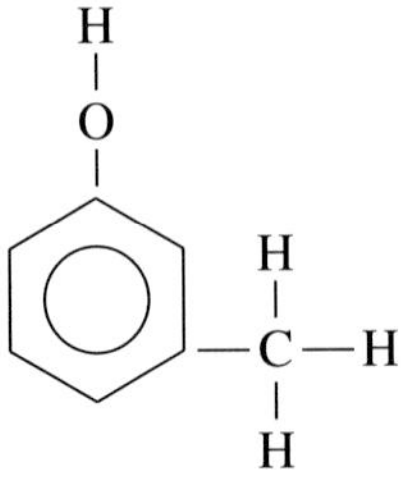

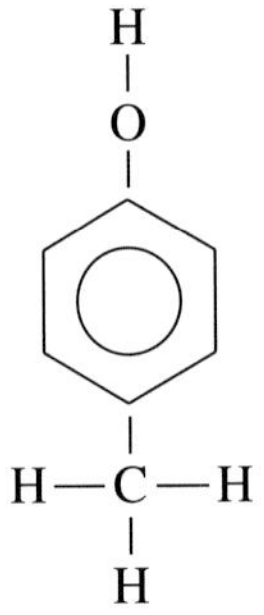

## Aldehydes

Aldehydes are highly toxic compounds that are known to be human carcinogens. Formaldehyde is toxic through inhalation and is a strong irritant. The TLV is 1 ppm in air. Acetaldehyde is toxic, with narcotic effects. The TLV is 100 ppm in air and the IDLH is 10,000 ppm. Propionaldehyde is an irritant. The toxicity hazards of the aldehydes also vary from one compound to another. Care should be taken to determine the exact hazards of any given compound.

**Acrolein, CH$_2$CHCHO,** also known as acrylaldehyde, is an aldehyde hydrocarbon derivative and it is a colorless or yellowish liquid with a disagreeable, suffocating odor. Acrolein is soluble in water with a specific gravity of 0.84, which is lighter than water. It polymerizes readily, is very reactive, and is not shipped or stored without an inhibitor. Acrolein is toxic by inhalation and ingestion, and is a strong irritant to the skin and eyes. The TLV is 0.1 ppm in air and the IDLH is 5 ppm. The target organs are the heart, eyes, skin, and respiratory system. Acrolein is a dangerous fire and explosion risk with a wide flammable range of 2.8 to 31% in air. The four-digit UN identification number for acrolein is 1092. The NFPA 704 designation is health 4, flammability 3, and reactivity 3. The primary uses of acrolein are in the manufacture of polyester resins, polyurethane resins, pharmaceuticals, and as an herbicide. The structure and molecular formula for acrolein are shown in the following example:

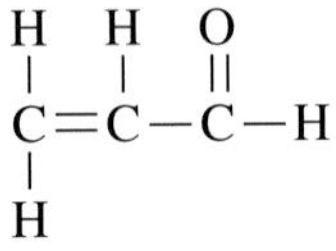

Acrolein
Acrylaldehyde
CH$_2$CHCHO

Agricultural feed store, a primary source of restricted-use agricultural pesticides.

## Organic Acids

Organic acids are toxic and corrosive. Corrosivity is a form of toxicity to the tissues that the acid contacts. However, the organic acids have other toxic effects. Formic acid is corrosive to skin and tissue. It has a TLV of 5 ppm in air and an IDLH of 30 ppm. Pure acetic acid is toxic by ingestion and inhalation. It is a strong irritant to skin and tissue. The TLV is 10 ppm in air and the IDLH is 1000 ppm. Propionic acid is a strong irritant with a TLV of 10 ppm in air. Butyric acid is a strong irritant to skin and tissue. The degree of toxicity varies with the different organic acid compounds. Reference sources and MSDS sheets will have to be reviewed to determine the exact hazards of specific acids.

## MISCELLANEOUS TOXIC MATERIALS

A wide variety of toxic liquids and solids do not fit neatly into any of the families discussed so far. Some of the most dangerous and more common ones will be listed here, but this will by no means be a comprehensive listing. The intent is to foster familiarity with the many different types of toxic chemicals that may be encountered in the real world, both in transportation and fixed facilities. The chemicals presented will be drawn from the DOT hazardous materials tables and the NFPA Hazardous Chemicals Data listing.

**Chloropicrin, $CCl_3NO_2$,** is a slightly oily, colorless, refractive liquid. It is relatively stable and slightly water soluble. The specific gravity is 1.65, which is heavier than air. It is toxic by ingestion and inhalation and is a strong irritant. The TLV is 0.1 ppm in air. The four-digit UN identification number is 1580. The

NFPA 704 designation is health 4, flammability 0, and reactivity 3. The primary uses of chloropicrin include dyestuffs, fumigants, fungicides, insecticides, rat exterminator, and tear gas. The structure and molecular formula for chloropicrin are shown in the following example.

$$Cl-\underset{\underset{Cl}{|}}{\overset{\overset{Cl}{|}}{C}}-N\overset{\diagup O}{\underset{\diagdown O}{|}}$$

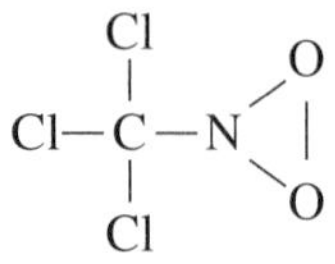

Chloropicrin
$CCl_3NO_2$

**Epichlorohydrin, $ClCH_2CHOCH_2$,** an epoxide, is a highly volatile, unstable liquid, with a chloroform-like odor. It is slightly water soluble and has a specific gravity of 1.18, which is heavier than water. Epichlorohydrin is toxic by ingestion, inhalation, and skin absorption. It is a strong irritant and a known carcinogen. The TLV is 2 ppm in air and the IDLH is 250 ppm. The target organs affected are the respiratory system, skin, and kidneys. The four-digit UN identification number is 2023. The NFPA 704 designation is health 3, flammability 3, and reactivity 2. The primary uses are as a raw material for epoxy and phenoxy resins, the manufacture of glycerol, and as a solvent. The structure and molecular formula for epichlorohydrin are illustrated in the following example:

$$H-\underset{}{\overset{\overset{H}{|}}{C}}-\underset{\underset{O}{\diagdown \diagup}}{\overset{\overset{H}{|}}{C}}-\underset{\underset{H}{|}}{\overset{\overset{H}{|}}{C}}-Cl$$

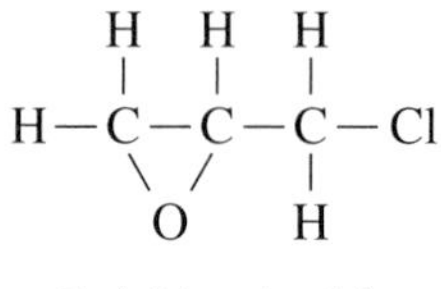

Epichlorohydrin
$CH_2CHOCH_2Cl$

**Hydrogen cyanide, HCN,** also known as hydrocyanic acid, is a water-white liquid with a faint odor of bitter almonds. The odor threshold is 0.2 to 5.1 ppm in air; however, if the odor of hydrogen cyanide is detected, it has already exceeded the allowable amount. The immediately fatal concentration is usually 250 to 300 ppm in air. Hydrogen cyanide blocks the uptake of oxygen by the cells. It is soluble in water with a specific gravity of 0.69, which is lighter than water. Hydrogen cyanide is highly toxic by inhalation, ingestion, and skin absorption. The TLV is 4.7 ppm in air and the IDLH is 50 ppm. The target organs are the central nervous system, cardiovascular system, kidneys, and liver. The four-digit UN identification number is 1051 for anhydrous (without water) and 1614 when it is absorbed in a porous material. The NFPA 704 designation is health 4, flammability 4, and reactivity 2. The primary uses are in the manufacture of acrylonitrile, acrylates, cyanide salts, rodenticides, and pesticides. The structure and molecular formula for hydrogen cyanide are shown in the following illustration:

$$H-C\equiv N$$

Hydrogen cyanide
HCN

**Methyl hydrazine, CH₃NHNH₂,** is a colorless, hygroscopic liquid with an ammonia-like odor. It is soluble in water with a specific gravity of 0.87, which is lighter than water. Methyl hydrazine is toxic by inhalation and ingestion and is a suspected human carcinogen. The TLV ceiling is 0.2 ppm in air and the IDLH is 50 ppm. The target organs are the central nervous system, respiratory system, liver, blood, eyes, and cardiovascular system. The four-digit UN identification number is 1244. The NFPA 704 designation is health 4, flammability 3, and reactivity 2. The primary uses are as a missile propellant and a solvent. The structure and molecular formula for methyl hydrazine are shown in the following illustration:

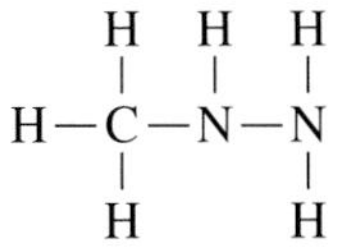

Methyl hydrazine
CH₃NHNH₂

**Methyl isocyanate, MIC, CH₃NCO,** is a colorless liquid. It is water reactive with a specific gravity of 0.96, which is lighter than water. This is the chemical that was released in Bhopal, India, that killed over 3000 people in 1984. Methyl isocyanate is toxic by skin absorption and a strong irritant. The TLV is 0.02 ppm in air and the IDLH is 20 ppm. The target organs are the respiratory system, eyes, and skin. The four-digit UN identification number is 2480. The NFPA 704 designation is health 4, flammability 3, and reactivity 2. The white section at the bottom of the diamond has a W with a slash through it, indicating water reactivity. The primary use of methyl isocyanate is as a chemical intermediate. The structure and molecular formula are shown in the following example.

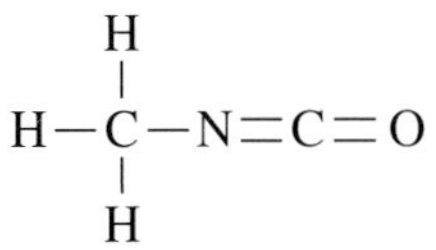

$$H-\underset{\underset{H}{|}}{\overset{\overset{H}{|}}{C}}-N=C=O$$

Methyl isocyanate
CH₃CNO

**Toluene diisocyanate, TDI, $(OCN)_2C_6H_3CH_3$,** is a water-white to pale yellow liquid with a sharp, pungent odor. It reacts with water to release carbon dioxide. The specific gravity is 1.22, which is heavier than water. TDI is toxic by inhalation and ingestion and is a strong irritant to skin and other tissue, particularly the eyes. The TLV is 0.005 ppm in air and the IDLH is 10 ppm. The target organs are the respiratory system and the skin. The four-digit UN identification number is 2078. The NFPA 704 designation is health 3, flammability 1, and reactivity 3. The white section at the bottom of the diamond has a W with a slash through it, indicating water reactivity. The primary uses of TDI are in the manufacture of polyurethane foams, elastomers, and coatings. The structure and molecular formula for TDI are illustrated in the following example:

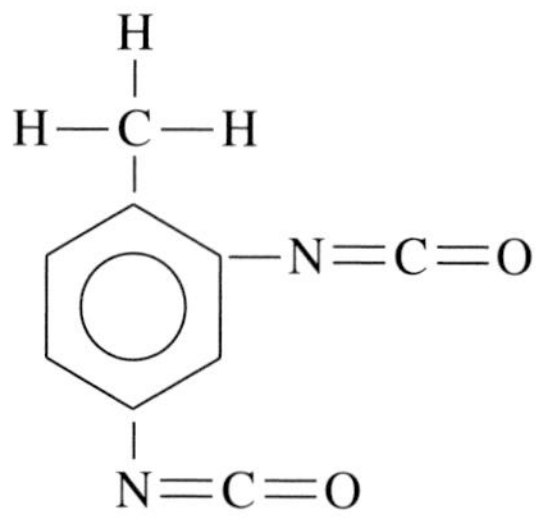

Toluene diisocyanate
TDI
$CH_3C_6H_3(NCO)_2$

## PESTICIDES

The EPA estimates that there are 45,000 accidental pesticide poisonings in the United States each year, where more than 1 billion pounds are manufactured annually. Pesticides can be found in manufacturing facilities, commercial warehouses, agricultural chemical warehouses, farm supply stores, nurseries, farms, supermarkets, discount stores, hardware stores, and other retail outlets. "Pesticide" is a generic name for toxic materials used to destroy different types of pests, including insects (insecticides), fungi (fungicides), rodents (rodenticides), or plants (herbicides). The definition of a pesticide from the Federal Insecticide, Fungicide, and Rodenticide Act (FIFRA) is "a chemical or mixture of chemicals or substances used to repel or combat an animal or plant pest. This includes insects and other invertebrate organisms; all vertebrate pests, e.g., rodents, fish, pest birds, snakes, gophers; all plant pests growing where not wanted, e.g., weeds; and all microorganisms which may or may not produce disease in humans. Household germicides, plant growth regulators, and plant root destroyers are also included."

Common pesticide families are organophosphates, carbamates, chlorophenols, and organocholorines. Organophosphates are derivatives of phosphoric acid and are acutely toxic, but are not enduring. They break down rapidly in the

Agricultural spray plane used to apply pesticides to food crops.

environment and do not accumulate in the tissues. Organophosphates are associated with more human poisonings than any other pesticide and are closely related to some of the most potent nerve gases. They function by overstimulating then inhibiting neural transmission, primarily in the nervous, respiratory, and circulatory systems. Examples include malathion, methyl parathion, thimete, counter, lorisban, and dursban. The chemical formulas of the organophosphates contain carbon, hydrogen, phosphorus, and at least one sulfur atom, and some may contain at least one nitrogen atom.

Carbamates are derivatives of carbamic acid, and are among the most widely used pesticides in the world. Most are herbicides and fungicides, such as 2,4,D, paraquat, and dicamba and function by inhibiting nerve impulses. The formula will contain carbon, hydrogen, nitrogen, and sulfur. Other examples include furadan, temik, and sevin.

Organochlorines are chlorinated hydrocarbons. The formula contains carbon, hydrogen, and chlorine. They are neurotoxins, which function by overstimulating the central nervous system, particularly the brain. Examples are aldrin, endrin, hesadrin, thiodane, and chlordane. The best known organochlorine is DDT, which has been banned for use in the United States because of its environmental persistence. Organophosphates do not break down in the environment and accumulate in the tissues, affecting the food chain.

Chlorophenols contain carbon, hydrogen, oxygen, and chlorine. They affect the central nervous system, kidneys, and liver.

Pesticide labels contain valuable information for the emergency responder and medical personnel treating a patient exposed to pesticides. This information includes product name, "signal word" ("danger," "caution," or "warning"), a statement of practical treatment, EPA registration number, a note to physician,

**PESTICIDE SIGNAL WORDS**

| SIGNAL WORD | TOXICITY | LETHAL DOSE |
|---|---|---|
| Danger/Poison* | Highly Toxic | Few Drops to 1** Teaspoon |
| Warning | Moderately Toxic | 1 Teaspoon to 1 Tablespoon |
| Caution | Low Toxicity | 1 Ounce to More Than a Pint |

\* Skull and crossbones symbol included
\*\* Less for a child or person weighing less than 160 pounds

**Figure 7.12**

and a statement of chemical hazards. Other information includes active and inert ingredients. "Inert" does not necessarily mean that the ingredients do not pose a danger; it means only that the inert ingredients do not have any action on the pest for which the pesticide was designed. Many times the inert ingredient is a flammable or combustible liquid.

The label also contains information about treatment for exposure. This information should be taken to the hospital when someone has been contaminated with a pesticide. Do not however, take the pesticide container to the hospital. Take the label or write the information down, take a Polaroid picture of the label or use a pesticide label book. (Label books are available from agricultural supply dealers.)

The three signal words indicating the level of toxicity of a pesticide are "danger," "caution," and "warning" (Figure 7.12). Highly toxic materials bear the word "danger" with a skull-and-crossbones symbol and the word "poison" printed on the label. The lethal dose may be a few drops to 1 tsp. Moderately toxic pesticides have the word "warning," and the lethal dose is 1 tsp. to 1 tbsp. Low toxicity pesticides carry the word "caution," and the lethal dose is 1 oz. to 1 pt. (The National Pesticide Network, located in Texas, provides emergency information through a toll-free telephone number, 800-858-7378, from 8 a.m. to 6 p.m. Central Standard Time.)

## TOP 50 INDUSTRIAL CHEMICALS

**Ethylene glycol, $CH_2OHCH_2OH$,** an alcohol hydrocarbon derivative, is the thirty-second most produced industrial chemical, with 5.23 billion lbs. in 1995. It is a clear, colorless, syrupy liquid with a sweet taste. Ethylene glycol is soluble in water and has a specific gravity of 1.1, which is slightly heavier than water. It is toxic by inhalation and ingestion. The lethal dose is reported to be 100 cc and the TLV is 50 ppm. Ethylene glycol has not been assigned a four-digit UN identification number. The NFPA 704 designation is health 1, flammability 1, and reactivity 0. The primary uses are as coolants and antifreeze, brake fluids, low-freezing dynamite, a solvent, and a deicing fluid for airport runways. The structure and molecular formula for ethylene glycol are shown in the following illustration:

$$
\begin{array}{ccc}
& H & H \\
& | & | \\
H - & C - & C - H \\
& | & | \\
& O & O \\
& | & | \\
& H & H
\end{array}
$$

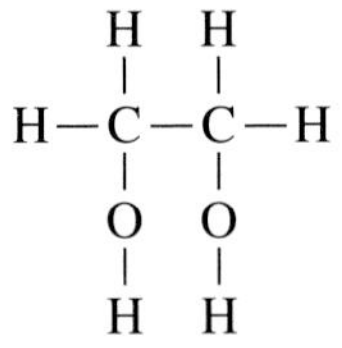

Ethylene glycol
$CH_2OHCH_2OH$

**Phenol, $C_6H_5OH$,** also known as carbolic acid, is the thirty-fourth most produced industrial chemical, with 4.16 billion lbs. in 1995. It is a white crystalline mass that turns pink or red if not perfectly pure or if exposed to light. Phenol absorbs water from the air and liquefies; it may also be found in transport as a molten material. It has a distinctive odor and a sharp, burning taste, but in a weak solution it has a slightly sweet taste. Phenol is soluble in water with a specific gravity of 1.07, which is heavier than water. It is toxic by ingestion, inhalation, and skin absorption and is a strong irritant to tissue. The TLV is 5 ppm in air and the IDLH is 250 ppm. The target organs are the liver, kidneys, and skin. The four-digit UN identification number is 1671 for solids, 2312 for molten materials, and 2821 for solutions. The NFPA 704 designation is health 4, flammability 2, and reactivity 0. The primary uses are phenolic resins, epoxy resins, 2-4-D herbicides, solvents, pharmaceuticals, and as a general disinfectant. The structure and molecular formula for phenol are shown in the following example:

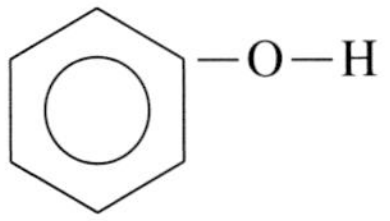

Phenol
$C_6H_5OH$

**Caprolactam, $CH_2CH_2CH_2CH_2CH_2NHCO$,** was the forty-seventh most produced industrial chemical, with 1.68 billion lbs. in 1994; however, it dropped from the Top 50 in 1995. It is a solid material composed of white flakes. Caprolactam is soluble in water and has a specific gravity (in a 70% solution) of 1.05, which is heavier than water. It may also be encountered as a molten material. Caprolactam is toxic by inhalation with a TLV of (vapor) 5 ppm in air and (dust) 1 mg/m³ of air. The primary uses are in the manufacture of synthetic fibers, plastics, film, coatings, and polyurethanes. The structure and molecular formula for caprolactam are shown in the following illustration:

$$
\begin{array}{c}
\text{H} \quad \text{H H} \quad \text{H} \\
\backslash\, / \quad \backslash\, / \\
\text{H} \quad \diagup \text{C} - \text{C} - \text{C} = \text{O} \\
\diagdown\, \text{C} \quad \quad | \\
\text{H} \diagup \quad \diagdown \text{C} - \text{C} - \text{N} - \text{H} \\
\diagup \backslash \quad \diagup \backslash \\
\text{H} \quad \text{H H} \quad \text{H}
\end{array}
$$

Caprolactam
$CH_2CH_2CH_2CH_2CH_2NHCO$

**Nitrobenzene** entered the Top 50 in 1995 as the forty-ninth most produced chemical, with 1.65 billion lbs. annually. It is a greenish-yellow crystal or yellow oily liquid and is slightly soluble in water. The primary hazard of nitrobenzene is toxicity; however, it is also combustible. The boiling point is about 410° F, the flash point is 190° F, and the ignition temperature is 900° F. The specific gravity is 1.2, which is heavier than water and the material will sink to the bottom. The vapor density is 4.3, which is heavier than air. Nitrobenzene is toxic by ingestion, inhalation, and skin absorption with a TLV of 1 ppm in air. The four-digit UN identification number is 1652. The NFPA 704 designation is health 3, flammability 2, and reactivity 1. Nitrobenzene is a nitro hydrocarbon derivative but it is not very explosive. The primary uses are as a solvent, an ingredient of metal polishes and shoe polishes, and in the manufacture of aniline. The structure and molecular formula are shown in the following example:

Nitrobenzene
$C_6H_5NO_2$

## MILITARY CHEMICAL WARFARE AGENTS

The Federal government recently released previously classified information concerning the location in the United States of chemical warfare agents that are scheduled for destruction by the year 2004. The destruction is required by a treaty banning the use of chemical weapons as warfare agents. The agents include nerve and mustard gases.

The Centers for Disease Control (CDC) have determined that the chemicals pose more of a threat in storage than during the destruction process. These agents are located in stockpiles in Oregon, Utah, Colorado, Indiana, Kentucky, Arkansas, Alabama, and Maryland (Figure 7.13). The containers and munitions holding

## U.S. Chemical Weapons Stockpile Storage Sites

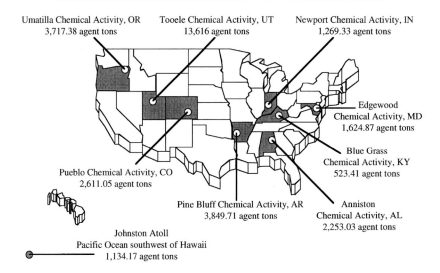

Umatilla Chemical Activity, OR
3,717.38 agent tons

Tooele Chemical Activity, UT
13,616 agent tons

Newport Chemical Activity, IN
1,269.33 agent tons

Edgewood
Chemical Activity, MD
1,624.87 agent tons

Blue Grass
Chemical Activity, KY
523.41 agent tons

Pueblo Chemical Activity, CO
2,611.05 agent tons

Pine Bluff Chemical Activity, AR
3,849.71 agent tons

Anniston
Chemical Activity, AL
2,253.03 agent tons

Johnston Atoll
Pacific Ocean southwest of Hawaii
1,134.17 agent tons

**Figure 7.13**

these chemical agents have been around, in some cases, since World War I. Small leaks occur on a frequent basis at many of the facilities. Though these chemicals are stored on government facilities, it is possible that by an accident the materials could be released and affect civilian populations. Congress has mandated that the public have maximum protection against these aging stockpiles of chemical agents. Listed below are some of the more common agents in the storage facilities and their physical and chemical characteristics. These agents are stored in ton cylinders and various types of munitions. Ton containers hold approximately 1500 to 1800 lbs. of mustard agent. The munitions include M-55 rockets, mines, shells, and bombs with small amounts of nerve and mustard agents. These materials are a fixed-storage hazard and will not be allowed to be transported at the present time by the Army.

**Blister Agent HT, Sulfur Mustard,** is a chlorinated sulfur compound. It has a chemical name of H-sulfide, bis(2-chloroethyl), T-bis-(2-(2-chloroethylthio)-ethyl) ether. This material would fit the definition of a DOT Hazard Class 6.1 poison liquid, inhalation hazard, if it were allowed to be transported. The UN identification number would be 1955. It is estimated that the NFPA 704 designation would be health 4, flammability 1, and reactivity 1. Mustard agent is a known carcinogen.

**Blister Agent L, Lewisite,** is an arsenical compound. The chemical name is chlorovinylarsine dichloride. The DOT Hazard Class would be 6.1 poison inhalation hazard if it were allowed to be transported. The four-digit UN identification number would be 1955. It is estimated that the NFPA 704 designation would be health 4, flammability 1, and reactivity 1.

**Nerve Agent VX** is a sulfinated organophosphorous compound. The chemical name is phosphonothioic acid, methyl-, S-(2-bis(1-methylethylamino)ethyl) ethyl) O-ethyl ester. The DOT Hazard Class would be 6.1 poison liquid inhalation hazard. The four-digit identification number would be 2810 if the material were allowed to be transported. The estimated NFPA 704 designation would be health 4, flammability 1, and reactivity 1.

**Blister Agent H, Mustard,** is a chlorinated sulfur compound. The chemical name is sulfide, bis(2-chloroethyl). The DOT Hazard Class would be 6.1 poison liquid inhalation hazard. The four-digit UN identification number would be 2810 if the material were allowed to be transported. The estimated NFPA 704 designation would be health 4, flammability 1, and reactivity 1. Mustard agent is a known carcinogen.

**Nerve Agent GA, Tabin,** is an organophosphate compound. The chemical name is ethyl N,N-dimethylphosphoramidocyanidate. Fires may result in the release of hydrogen cyanide. The DOT Hazard Class would be 6.1 poison liquid inhalation hazard. The four-digit UN identification number would be 2810 if it were allowed to be transported. The estimated NFPA 704 designation would be health 4, flammability 2, and reactivity 1.

**Nerve Agent GB, Sarin,** is a fluorinated organophosphorous compound. The chemical name is phosphonofluoridic acid, methyl-, isopropyl ester. The DOT Hazard Class is 6.1 poison liquid inhalation hazard. If the material were allowed to be transported the four-digit UN identification number would be 2810. The estimated NFPA 704 designation is health 4, flammability 1, and reactivity 1. This is the same agent that was released in the recent subway incident in Japan that injured and killed several people.

## CHEMICAL WARFARE AGENTS

Chemical warfare agents were introduced during World War I and were used as recently as the Gulf War with Iraq. The United States stockpile of unitary chemical warfare agents has been slated for destruction as a result of international treaty by the year 2004. Unitary agents are those that act on their own without having to be mixed with another chemical to be activated. Chemical agents are stored in bulk and in munitions at seven sites on the United States mainland and on Johnston Island in the South Pacific. Destruction through incineration is already in progress at the Johnston Island site. Destruction of the mainland stockpiles is in various stages of planning and implementation of destruction procedures. Methods of destruction under consideration range from incineration to various neutralization procedures. Many of the aging munitions have experienced leaks of chemical agents at some of the facilities, although to this date no off-post releases of any significance have occurred.

The Army and the Federal Emergency Management Agency (FEMA) are coordinating efforts to ensure public safety and protection in the event an accident occurs involving any of the chemical agents at the stockpile sites. This cooperative

A rail tank car of organophosphate pesticide, the same type of material used as a military nerve agent.

Toxic materials are the primary reason decontamination is conducted at HazMat incidents.

effort is known as the Chemical Stockpile Emergency Preparedness Program (CSEPP). Emergency responders in areas likely to be affected by a chemical agent release are trained in chemical agent awareness, emergency medical treatment, and response procedures, including personal protective equipment (PPE) and decontamination, through training programs developed jointly by FEMA and the Army.

In addition to the stockpiles, chemical agents have been found at various locations around the country in unexploded munitions buried on and off present and past military installations. Such an incident occurred in Washington D.C. in a new housing development near American University several years ago. No serious accidents involving these munitions have occurred.

The terrorist incidents that occurred in Tokyo, Japan during March of 1995 and prior to that in Matsumoto, Japan have added a new dimension to the threat of chemical agents. In the hands of terrorists these agents pose a real threat to the general population, and almost any public arena is vulnerable. Responders may be faced with hundreds and even thousands of casualties. Emergency responders, with the possible exception of those in the seven CSEPP locations around the country, may be totally unprepared to deal with chemical agents released through a terrorist attack.

## Mustard Agents (Vesicants)

There are two primary types of chemical agents, mustard and nerve agents. Mustard agents are also referred to as vesicants or blister agents, because they cause blistering when they contact the skin. There are two types of mustard agent, sulfur mustard and nitrogen mustard. Nitrogen mustard is similar to sulfur mustard in its health effects, but has slightly more systemic effects (affecting the entire body).

Mustard agent is often incorrectly referred to as "mustard gas." Mustard agent is actually a viscous oily liquid, light yellow to brown in color, with an onion, garlic, or mustard smell; it freezes at 57°F. Mustard agent is usually nonvolatile but may evaporate and produce a vapor hazard in warm weather or when involved in a fire. The vapor is heavier than air, with a density of 5.4. Mustard agent is heavier than water and nonsoluble in water. There are three basic types of blister agents: sulfur mustard, lewisite, and phosgene oxime. Mustard agent is designed to function through skin and tissue contact and is generally not a major inhalation hazard under normal conditions.

Vesicants are persistent agents, which means they do not readily vaporize and will remain as a contaminant for a long time. Mustard agents do not cause pain on contact and do not act immediately, but rather have a dormant period of from 1 to 24 hours before symptoms present themselves. Symptoms of mustard exposure include erythema (redness of the skin), blisters, conjunctivitis (eye inflammation), and upper respiratory distress; the symptoms may worsen over several hours. Once mustard has contacted tissues the damage begins immediately, although there may not be any indication that contact has occurred. Mustard is highly soluble in fat, which results in rapid skin penetration.

The lethal dose of liquid mustard applied to the skin is approximately 7 g spread over 25 percent of the body surface area. The threshold for skin erythema and blistering is 10 µg (a microgram is one millionth of a gram) per square centimeter deposited on the skin. The median lethal dose ($LD_{50}$; lethal dose for 50% of the population) for ingestion of mustard is estimated as 0.7 mg/kg of body weight. Damage occurs primarily to the skin, eyes, and respiratory tract.

Vapor threshold doses of mustard that cause effects to the eyes are 0.1 mg/m$^3$ over 10 to 30 minutes of continuous exposure. Doses of 200 mg-min/m$^3$ may cause corneal edema (swelling and fluid build-up within the cornea), keratitis (inflammation of the cornea), and blepharosphasm (uncontrolled winking), leading to temporary blindness. Irritation of the nasal mucous membranes and hoarseness first occur at doses ranging from 12 to 70 mg-min/m$^3$. Lower respiratory effects, such as tracheobronchitis (bronchitis or inflammation of the trachea), tachypnea (excessively rapid respirations), cough, and bronchopneumonia (inflammation of the lungs), begin to occur with doses exceeding 200 mg-min/m$^3$. Mild skin erythema may be seen with doses of 50 mg-min/m$^3$. Severe erythema, followed by blistering, may begin at concentration-time profiles exceeding 300 mg-min/m$^3$. Warm, humid environments may cause the earlier development of erythema and blistering at lower doses. The maximum safe doses for mustard have been established as 5 mg-min/m$^3$ for skin exposure and 2 mg-min/m$^3$ for eye exposure. The threshold limit value–time weighted average (TLV-TWA) for mustard as established by the U.S. Surgeon General is 0.003 mg/m$^3$. The IDLH for mustard has been recommended at 1.67 mg/m$^3$; however, as of this writing, the Office of the U.S. Army Surgeon General has not formally established IDLH (Immediately Dangerous to Life and Health) levels for HD or L. There is no known antidote for mustard agent. Mustard has been established as a human carcinogen.

Unless decontamination occurs within seconds after exposure to mustard, the results will be only minimal, if any. Hypochlorite solutions or large volumes of water are used to attempt to flush the agent from the affected tissues. If decontamination is begun after symptoms start, it will have no effect at all, as the damage has already occurred. Mustard enters the body through the cells of the skin or mucous membrane and produces biochemical damage within seconds to minutes and no known procedure can reverse the process. Treatment of mustard exposure is much the same as for thermal burns and involves managing the symptoms and lesions (blisters). Mustard is rapidly transformed when it contacts the tissues in the body but is not found to be present in the blood, blister fluid, tissue, or urine. Mustard exposures are rarely fatal and patients recover over a period of months. In World War I, one third of all U.S. casualties were the result of chemical exposure (sulfur mustard), but the fatality rate was only 5%. The military has four designations for sulfur mustard: H, HD (distilled), HS, and HT, which is mixed with a thickener.

Lewisite is a vesicant that is believed to have been used by Japan against China in the 1930s. It has no other known battlefield uses and the military designation is L. Phosgene oxime is also a vesicant; it has not been used on the battlefield and has a military designation of CX.

Mustard is found in artillery shells, mortar shells, land mines, and ton containers, which are similar to the ton containers used to ship and store chlorine. It is not likely that mustard would be used in terrorist activity because of the time it takes for mustard to act and its low volatility; however, if it were to be used, many hours would pass before the symptoms began to surface. This time delay would make diagnosis and determination of the incident source very difficult. For mustard to be an effective vapor hazard there would have to be heat, fire, or explosion to disperse the vapor, or the mustard would have to be sprayed over an area with a pesticide spray plane or other source.

## Nerve Agents

Nerve agents are the most toxic of all chemical agents used in weapons. Nerve agents are nothing more than close relatives of organophosphate pesticides; in fact, many of the nerve agents were discovered by chemists trying to make new pesticides. However, military nerve agents are much more toxic. Pesticides generally work by interfering with the functions of the central nervous system and nerve agents act by the same means. Some common organosphosphate pesticides include Malathion® and Parathion®. Another group of pesticides, called carbamates, cause the same type of central nervous system damage as the nerve agents although they are not related chemically. A common carbamate is the pesticide Sevin®. Nerve agents were first discovered in the 1800s; however, the toxicity of the agents was not realized until the early 1900s.

Nerve agents under moderate temperature conditions are liquids. The agents are clear, colorless, and tasteless, and most are odorless. GB agent is the most volatile even though the evaporation rate is less than that of water. GD has a greater evaporation rate then GA, and GA has a greater rate than GF. It is unlikely that GD would be used as a terrorist weapon because of the complexity of its manufacture. VX has a viscosity similar to light motor oil, and although it produces a slight vapor it generally is not considered to be a vapor hazard unless the ambient temperature is very warm. All nerve agents, in addition to being inhalation hazards, are also absorbed through the skin and will travel through ordinary clothing. The $LD_{50}$ (lethal dose for 50% of those exposed through skin contact) of VX is much smaller than for GB, BD, BA, and GF; however, this is because the "G-agents" will evaporate from the skin before they can penetrate. The primary military nerve agents of importance are GA (Tabun), GB (Sarin, which was used in the terrorist attacks in Japan), GD (Soman), GF, and VX.

Tabun was first made in the 1930s by a German chemist. Sarin was discovered in Germany about two years after Tabun. The German government manufactured large amounts of these nerve agents and stockpiled them during World War II. However, for unknown reasons, they never used them. The Allies were not aware of the existence of the chemical nerve agents and had no protection or antidotes against them. After World War II, Soman, GF, and VX were discovered, manufactured, and stockpiled by the United States and the Soviet Union. These stockpiles are now in the process of being destroyed. The only known battlefield use of nerve agents was during the Iraq/Iran war when Iraq used nerve agents against

Iran. The major concern presently with nerve agents is their manufacture and use by terrorist groups.

Unlike the slow-acting vesicants, vapor exposure to nerve agents will cause symptoms within seconds to several minutes after contact. Large amounts of nerve agent in vapor form will cause loss of consciousness and convulsions within seconds after one or two breaths. Nerve agent exposures on the skin will not present symptoms for varying periods of time depending on the amount of the exposure. Effects may occur from several minutes to as much as 18 hours after exposure. It takes a period of time for the agent to penetrate the skin and reach the target organ(s).

Nerve agents act by inhibiting the enzyme cholinesterase. The function of cholinesterase is to destroy the neurotransmitter acetylcholine. Neurotransmitters are chemical substances released by a nerve impulse at the nerve ending. When released they travel to the organ that the nerve stimulates. Once it arrives at the organ, the neurotransmitter combines with the receptor site on the organ to cause an effect on the organ. For example, to move a muscle anywhere in your body, an electrical impulse originates in the brain and travels down appropriate nerves to the nerve ending near that muscle. The electrical impulse does not go to the muscle. The electrical impulse causes the release of a neurotransmitter that then travels across the very tiny gap between the nerve ending and the muscle to stimulate the muscle. The muscle reacts to this stimulation by moving. The neurotransmitter is then destroyed to prevent the stimulation of the muscle again. If additional muscle movement is required, another nerve impulse causes the release of more neurotransmitter. The neurotransmitter acetylcholine is released by nerve endings and stimulates the intended organ. It is then destroyed by the enzyme acetylcholinesterase. As long as the acetylcholinesterase is intact, the body functions normally. If the cholinesterase is inhibited, the acetylcholine builds up and overstimulates the muscles, glands, and other nerves, which produces the symptoms exhibited by nerve agent exposures. The cholinergic nervous system stimulates skeletal muscles (those that are voluntarily moved, such as the arms, legs, trunk, and face). Additionally the exocrine glands (lacrimal glands, nasal glands, salivary glands, sweat glands, and the glands that line the airways and gastrointestinal tract) are stimulated by acetylcholine. Smooth muscles (of primary importance are the muscles that surround the airways and the gastrointestinal tract) are also stimulated by the neurotransmitters. When the acetylcholinesterase is inhibited, the excess acetylcholine overstimulates all of these structures to cause involuntary movement in the skeletal muscles. Excess secretions develop from the lacrimal, nasal, salivary, and sweat glands. Excess secretions continue into the airways and gastrointestinal tract with resulting constriction of muscles in the airways that cause bronchoconstriction, similar to asthma. Constriction in gastrointestinal tract causes cramps, vomiting, and diarrhea.

Unlike mustard, the effects of nerve agents can be reversed by the use of atropine as an antidote. Atropine blocks the effects of the excess acetylcholine. Atropine is most effective for smooth muscles and glands and does not help the skeletal muscles. Another drug call pralidoxime chloride (2-PAMCl) is used in conjunction with atropine to treat the skeletal muscles. Convulsions may also occur with exposure to nerve agents, and diazepam (Valium®) is administered in some instances to help control the convulsions. Local protocols and procedures should be established for administration of antidotes based upon advice from the local EMS Medical Director.

Mustard and nerve agents are chemicals and therefore require an appropriate level of chemical protective clothing just like any other chemical. The military has protective equipment for battlefield protection against chemical agents. This PPE is composed of a charcoal suit, protective hood made of butyl rubber protected cloth, butyl rubber gloves with thin cotton inserts, vinyl boots, and a powered air-purifying respirator (PAPR). The PAPR is used instead of SCBA because of the length of time that the respiratory protection may be needed and the difficulty in changing SCBA bottles in a hazardous atmosphere or under battlefield conditions. The PAPR will remove up to 0.5 mg/m³ of nerve agent GB for up to 16 hours based upon the longest time an emergency responder has used the PAPR (tests indicate that the filters and cartridges have the shortest service life against nerve agent GB compared to other lethal chemical agents). The actual expected time of cartridge usage during an incident is up to 12 hours.

One of the major problems that occurs with any type of chemical protective clothing is heat stress. Because of the extended time the PAPR can be worn the danger of heat stress is great. In reality responders will be limited in the amount of time spent in the PPE by heat stress potential rather than the limitations of the air supply. Most of the chemical agents require warm temperatures for them to be a vapor problem. So if there is vapor, heat stress will be a problem. When temperatures are from 50 to 70°F, the recommended work time in the PPE is 30 to 45 minutes with a 10 to 15 minute rest time. Temperatures from 70 to 80°F require a recommended work time in the PPE of 20 to 30 minutes with a rest time of 40 to 60 minutes. Temperatures from 85 to 100°F require a recommended work time in the PPE of 15 to 20 minutes with an indefinite rest time. The PAPR should not be used in Immediately Dangerous to Life and Health (IDLH) atmospheres or in atmospheres where the oxygen concentration is less than 16.5%. Airborne agent concentration IDLH values have been established for the following nerve agents: GA/GB 0.2 mg/m³, GD 0.06 mg/m³, VX 0.02 mg/m³. The PAPR uses a battery-operated blower that is designed to deliver essentially decontaminated air at a slight positive pressure into a full facepiece. The blower draws ambient air through two or three air-purifying elements (filters or chemical cartridges) that remove specific contaminants and deliver the subsequent air through a corrugated breathing tube into a facepiece assembly on the face of the respirator wearer. There is also commercially available chemical protective clothing that has been tested against live agents and passed for use by emergency responders. The Army PPE is available for civilian use by emergency responders in the CSEPP-affected communities. However, many of the emergency responders in those areas have chosen to use a commercial chemical suit with the PAPR rather than the military battle suit. The butyl rubber hood is still required even with the commercial chemical suit.

Unlike most hazardous materials incidents, decontamination for chemical agent releases will focus primarily on contaminated victims. There are likely to be hundreds and even thousands of exposed citizens who may need various levels of decontamination. The process will be very labor intensive and will be unlike any other decontamination process responders have previously faced. In some cases citizens will need to be decontaminated simply for their own peace of mind. Vapor exposures do not require as extensive decontamination as do liquid exposures. In reality, little contamination occurs from a vapor exposure. However, the public may perceive a need for decontamination, which may have to be performed, even if only for psychological reasons. Decontaminating hundreds and thousands of potentially contaminated victims could create a logistical nightmare. Responders in the Chemical Stockpile Program have acquired specially designed decontamination trailers for treating large numbers of citizens. These trailers would work well for mass exposures from terrorist incidents; unfortunately, only the CSEPP states may have them available at the present time. During the incident in Tokyo, Japan no decontamination was performed. Medical personnel and responders did not know what the source of the medical problems was for a period of time.

When large numbers of casualties start showing up from apparently unknown sources, chemical or biological terrorism should be one of the considerations. These types of incidents will tax the emergency response system as never before. Responders need to plan and train for chemical as well as nuclear and biological terrorism attacks. Even with planning these will be difficult incidents to respond to. The United States Department of Veterans Affairs has a program available to provide chemical and biological weapons training and assistance. The agency can provide Disaster Medical Assistance Teams (DMATs) to assist local medical personnel at the scene of terrorist attacks or any other disaster. Most local hospitals have only a limited supply at best of antidotes to deal with chemical agent exposures. The VA's Office of Medical Preparedness (OMP) has a stockpile of antidotes for chemical agent exposures. The DMAT can be airlifted with antidotes directly to a disaster scene. For additional information on chemical warfare agent training, technical information, or contacts, write the author in care of *Firehouse Magazine*.

## CHEMICAL RELEASE STATISTICS

There is one Class 6.1 poison in the EPA Top 15 chemical releases. Mercury is a liquid metal at normal temperatures and pressures. It was involved in 535 accidents resulting in over 541,000 lbs. being released into the environment. Molten phenol is in the Top 25 railroad commodities with 8982 carloads shipped in 1994. In 1994, there were 20 leaks reported from tank cars carrying Class 6.1 poisons. The DOT reports four poisons in the Top 50 incidents involving hazardous materials during 1995: dichloromethane, 1,1,1,-trichloroethane, tetrachloroethylene, all alkyl halides; and poisonous materials not otherwise specified. Together these materials accounted for 484 incidents with over 40 injuries.

## SUMMARY

Class 6.1 poisons when spilled usually do not affect large segments of the population as do the Class 2.3 poison gases. However, they can still present serious dangers to emergency responders. Several protective measures may minimize the effects of toxic materials. Antidotes are available for a small number of toxic materials, but they must be administered immediately after exposure. Your body has the ability to filter out some toxic materials through the normal process of eliminating wastes.

Protect yourself from toxic materials by wearing protective clothing and avoiding contact with toxic materials. Practice contamination prevention. Establish zones, deny entry, and provide protection to responders and to the public.

## REVIEW QUESTIONS

1. A one-time short-duration exposure to a toxic material is referred to as an _____ exposure.

2. Multiple exposures including 8 hours per day, 40 hours per week are referred to as _____ exposures.

3. The four routes of exposure in which toxic materials may enter the body are _____, _____, _____, and _____.

4. Long-term effects of exposure to toxic materials may cause any of the following: _____, _____, or _____.

5. Etiologic agents are _____.

6. Which of the following are rates of exposure for toxic materials?
   A. Curies
   B. TLV-TWA
   C. RADs
   D. PEL
   E. STEL
   F. Isomers

7. Protective measures against toxic materials may be _____ or _____.

8. Provide the names, structures, and hazards for the following toxic hydrocarbon and hydrocarbon derivative compounds.

   $C_7H_8$     $C_2H_5F$     $CH_3OH$     $HCOOH$     $t\text{-}C_4H_9OH$

9. What are the three "signal words" that may appear on the label of a pesticide?

10. Concentrations for toxic materials are usually expressed in any of the following terms: _____, _____, _____, and _____.

11. The dose of a toxic material is the _____ that enters the body and the response is the _____ effect.

# 8             RADIOACTIVE MATERIALS

Radioactivity is caused by changes in the nucleus of the atom. Radioactivity is not a chemical activity, but rather it is a nuclear event. Chemical activity involves electrons orbiting around the nucleus of an atom, particularly the outer-shell electrons. It is within these outer-shell electrons that chemical reactions take place. Radioactivity, on the other hand, involves the nucleus of the atom. There is normally a "strong force" that holds the nucleus of an atom together. There are some nuclei of elements that the force cannot hold together, and the nuclei begin to disintegrate. A basic law of nature says that unstable materials may not exist naturally for long. The unstable materials must do whatever they can to achieve stability. Radioactive elements throw off particles from the nucleus to reach stability. This throwing-off of particles is called radioactivity; the process is known as nuclear decay. This decay process is a random, spontaneous occurrence. There is no way that it can be shut off, nor is there any way to predict when a particular atom will begin to decay. The result of the decay is the ejection of a particle from the nucleus of the atom.

According to the Department of Transportation (DOT), a radioactive material is "any material having a specific activity greater than 0.002 microcuries per gram." Specific activity of a radionuclide means "the activity of the radionuclide per unit mass of that nuclide." Simply stated, a microcurie is a measurement of radioactivity. When a radioactive material emits more than 0.002 microcuries per gram of material, which is a term of weight, the material is then regulated in transportation by the DOT.

Radiation is ionizing energy spontaneously emitted by a material or combination of materials. A radioactive material, then, is a material that spontaneously emits ionizing radiation. There are three types of DOT labels used to mark radioactive packages. They are radioactive I, II, and III. These radioactive materials are determined by the radiation level at the package surface (Figure 8.1). Radioactive III materials are the only radioactives that require placarding on a transportation vehicle.

Elements above lead (atomic numbers 83 and above) on the Periodic Table are radioactive (Figure 8.2). Other elements may have one or more radioactive

**DETERMINATION OF RADIOACTIVE PLACARDING AND LABELING**

| | |
|---|---|
| Radioactive I: | $\leq$ 0.5 Millirem per hour |
| Radioactive II: | $\geq$ 0.5 - $\leq$ 50 Millirem per hour |
| Radioactive III: | $\geq$ 50 - $\leq$ 200 Millirem per hour |

All measurements at the surface of the package

**Figure 8.1**

isotopes. Some elements occur naturally, while others are man-made. Each symbol on the Periodic Table represents one atom of that element. An atom is made up of a nucleus with varying numbers of electrons in orbits circling around the nucleus (Figure 8.3). Located inside the nucleus are protons and neutrons. The number of protons also reflects the atomic number of that element. The number of neutrons may vary within the same type of element or from one element to another, but the number of protons must stay the same. The atom is the smallest part of an element that normally exists; so any particle of an element that is smaller than an atom is commonly referred to as a subatomic particle.

## TYPES OF RADIATION

There are two types of radiation: ionizing and nonionizing. Ionizing radiation involves particles and "waves of energy" traveling in a wave-like motion. Exam-

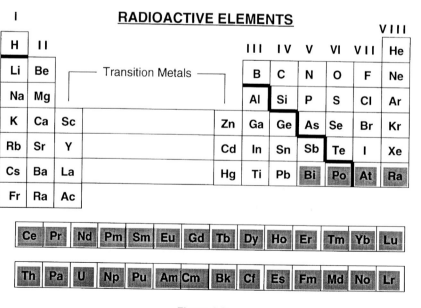

**Figure 8.2**

## THE ATOM

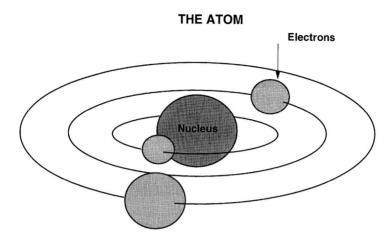

**Figure 8.3**

ples are alpha, beta, gamma, X-ray, neutron, and microwave. Nonionizing radiation is made up of "waves of energy." Examples include ultraviolet, radar, radio, visible light, and infrared light. All radioactivity travels in a straight line. There are three primary types of radioactive emissions from the nucleus of radioactive atoms. The first is the alpha particle, which looks much like an atom of helium stripped of its electrons with two protons and two neutrons remaining (Figure 8.4). The alpha particle is a positively charged particle that is emitted from the nucleus of an atom. It is large in size and, therefore, will not penetrate as much or travel as far as beta or gamma radiation. Alpha particles travel 3 to 4 in. and will not penetrate the skin. Complete turnouts, including self-contained breathing apparatus (SCBA), hood, and gloves, will protect responders from external exposure. However, if alpha emitters are ingested or enter the body through a broken skin surface, they can cause a great deal of damage to internal organs.

The beta particle is negatively charged and is smaller, travels faster, and penetrates farther than the alpha particle. It is 1/1800 the size of a proton or roughly equal to an electron in mass (Figure 8.5). The beta particle will penetrate

## ALPHA PARTICLE

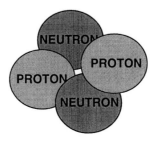

**Figure 8.4**

**BETA PARTICLE**

**ELECTRON
(NEGATRON)**                    **POSITRON**

**WEIGHT = 1/1800 OF A PROTON**

**Figure 8.5**

the skin and travel from 3 to 100 ft. Full turnouts and SCBAs *will not* provide full protection from beta particles. Particulate radiation results in contamination of personnel and equipment where the particles come to rest. Energy waves of radiation do not cause contamination.

There is a third type of radioactive particle, but it does not occur naturally. The neutron particle is the result of an atom being split in a nuclear reactor or accelerator, or it may occur in a thermal nuclear reaction. When an atom is split, neutron particles are thrown out. You would have to get inside a reactor or experience a thermonuclear explosion to be exposed to neutron particles.

Gamma radiation is a naturally occurring, high-energy electromagnetic wave that is emitted from the nucleus of an atom. It is not particulate in nature and has high penetrating power. Gamma rays have the highest energy level known and are the most dangerous of common forms of radiation. Gamma rays travel at the speed of light, or more than 186,000 miles per second, and will penetrate the skin and injure internal organs. No protective clothing can protect against gamma radiation. Shielding from gamma radiation requires several inches of lead, other dense metal, or several feet of concrete or earth. Gamma radiation does not result in any contamination because there are no radioactive particles, only energy waves. Examples of other electromagnetic energy waves include ultraviolet, infrared, microwave, visible light, radio, and X-ray (Figure 8.6). There is little differ-

## ELECTROMAGNETIC–ENERGY WAVES

- Speed = 186,000 miles per second

- The energy of gamma is greater than visible light

**Figure 8.6**

**ISOTOPES OF SELECTED ELEMENTS**

| | | |
|---|---|---|
| **CARBON 12 (Normal)** | **HYDROGEN 1** | **Protium** |
| **CARBON 13** | **HYDROGEN 2** (heavy water) | **Deuterium** |
| **CARBON 14** | | |
| | **HYDROGEN 3** | **Tritium** |

**Figure 8.7**

ence between gamma rays and X-rays; X-rays are produced by a cathode ray tube. To be exposed to X-rays, however, there has to be electrical power to the X-ray machine, and the machine has to be turned on. If there is no electrical power, there is no radiation.

## Isotopes

When the nucleus of an element contains more neutrons than the "normal" atom of that element, it is said to be an isotope of that element. In addition to neutrons, there are also protons in the nucleus. The total number of protons is also the atomic number of the element and changing the number of protons in the element produces another element.

All atoms have from 3 to 25 isotopes; the average is 10 per element (Figure 8.7). **All isotopes are not radioactive.** Hydrogen has three important isotopes: Hydrogen 1, sometimes called protonium, has 1 proton in the nucleus and no neutrons. Hydrogen 2 has 1 proton and 1 neutron in the nucleus and is called deuterium or heavy water. Hydrogen 1 and Hydrogen 2 are not radioactive. Hydrogen 3 has 1 proton and 2 neutrons in the nucleus and is called tritium. Hydrogen 3 is radioactive. Tritium is used in some "exit" signs, which gives them the glow-in-the-dark ability without batteries or any other electrical source.

Carbon has several isotopes. "Normal" carbon is known as carbon-12. Carbon 12 has six protons and six neutrons in the nucleus and is not radioactive. Carbon-13 has six protons and seven neutrons in the nucleus. It is not radioactive. There are 999 out of 1000 carbon atoms that are carbon-12, the other is carbon-13. Carbon-14 has six protons and eight neutrons and is naturally radioactive. Carbon-14 is a beta emitter that is produced in the atmosphere by the action of cosmic radiation on atmospheric nitrogen. In the process, a proton is forced from the nucleus of nitrogen, which then becomes carbon-14. The human body is made up of 1/10,000 of 1% of carbon-14. You inhale some carbon-14 every time you take a breath. It decays and is replaced in the body so there is a constant supply. Carbon-14 is sometimes used to determine the age of organic materials; it takes years to disappear. The amount of carbon-14 can be used to determine how long a person has been dead, or the age of a body or other object. There are three other carbon isotopes that are all man-made: carbon-11 has six protons and five neutrons; carbon-15 has 6 protons and 9 neutrons, and carbon-16 has 6 protons and 10 neutrons; all three are radioactive.

## RADIATION MEASUREMENTS

**Curie** = A physical amount of radioactive material.

1 Megacurie (MCi) = 1,000,000 Curies
1 Kilocurie (kCi) = 1,000 Curies
1 Millicurie (mCi) = 0.0001 Curies
1 Microcurie (uCi) = 0.000001 Curies

**Roentgen** = Ionization per $cm^3$ of dry air.

**RAD** = Radiation absorbed dose - Dosage.

**REM** = Biological effects.

**Figure 8.8**

Radioactive materials are heavily regulated in transportation by the DOT and by the Atomic Energy Commission in other situations. The design and construction of packaging for radioactive materials in transportation makes the likelihood of a release very small. The packaging undergoes rigorous testing before it is approved for use with radioactive materials. The metal casks used for high-level radioactive materials have never been involved in an accident where a serious release occurred. Most releases of radioactive materials are low level in nature and often involve radioactive isotopes used for medical purposes.

## Intensity of Radiation

Several terms are used to express the intensity of radiation (Figure 8.8). Radiation level is a term often substituted for dose rate or exposure rate. It is generally referred to as the effect of radiation on matter; i.e., the amount of radiation that is imparted from the source and absorbed by matter due to emitted radiation per unit of time. A curie is 37 billion disintegrations per second. The curie is a physical amount of material that is required to produce a specific amount of ionizing radiation; 1 millicurie = 0.001 curie and 1 microcurie = 0.000001 curie. It may take several hundred pounds of one radioactive material to produce the same amount of curies as one pound of another radioactive material. A roentgen is a measure of the amount of ionization produced by a specific material. It is the amount of X-ray or gamma radiation that produces 2 billion ionizations in 1 $cm^3$ of dry air. A RAD is the radiation absorbed dose (roughly equal to a roentgen). The radiation equivalent man (REM), also roughly equal to a roentgen, is a term for how much radiation has been absorbed or the biological effect of the dose.

If radioactive materials are released, the human senses *cannot* detect radioactivity. The only way responders will know if radioactive materials are present is with the use of instruments specially designed to detect radioactivity. There are two types of civil defense meters widely available to emergency responders: the CD V-700 and the CD V-715. Neither of these instruments can detect alpha radiation. The CD V-700 survey meter has a range of 0 to 50 mR/hour. An experienced operator can detect beta radiation with the CD V-700 through a process of elimination. If you check for radiation with the Geiger-Mueller (GM) tube on the CD V-700 with the window closed and no radiation exists beyond normal background, no gamma radiation is present. If the window is opened and another reading detects radiation, it is beta radiation. The CD V-715 survey meter has a range of 0.05 to 500 R/hour or 50 to 500,000 mR/hour. Radiation is detected through an ionization chamber much like the ionization chamber of a smoke detector. This unit is designed to detect gamma radiation from fallout in a nuclear attack. Dosimeters are used in conjunction with survey meters to monitor the levels of exposure to personnel. There are two types of civil defense dosimeters with different monitoring scales. The CD V-138 is used for monitoring relatively low levels of exposure and has a minimum scale reading of 200 mR. The CD V-742 has a range up to 200 R (200,000 mR) and is used for high levels of personnel exposure. Both meters should be worn by responders to ensure proper protection. There are also similar commercial instruments available, including those sensitive enough to detect alpha radiation. The health effects of exposure to radiation can vary (Figure 8.9). Nonionizing radiation comes from ultraviolet and infrared energy waves. This type of radiation causes sunburn types of injury. This is not a major concern for hazardous materials responders. Ionizing radiation comes from alpha, beta, and gamma sources. Ionization damage occurs at the cellular level. Four types of short-term effects on the cells can occur: no damage at all, the ionization passes through the cell; damage occurs, but the damaged cells can be repaired; there can be irreparable damage to the cells that does not cause death, but the damage is permanent; there is the destruction of the cells. There are also long-term effects from ionizing radiation. Exposures can cause cancer and birth defects of a teratogenic or mutagenic nature. Teratogenic birth defects result from the fetus being exposed to and damaged by radiation. The child is then born with some kind of birth defect as a result of the exposure. Providing no further exposures occur, future children can be born normal. Mutagenic damage occurs when the DNA or other part of the reproductive system is damaged by exposure

### HEALTH EFFECTS OF EXPOSURE

| | |
|---|---|
| 25 REM | Maximum single lifetime exposure |
| 20-100 REM | Chromosomal damage, alteration of white blood count |
| 100-200 REM | Nausea and vomiting, WBC reduction |
| 200-400 REM | Severe WBC reduction, hair loss, some death from infection |
| 600-1,000 REM | 50% Death in 30 days |
| 1,000-2,000 REM | Death within 4-14 days |
| 2,000 or more | Death (immediate) |

**Figure 8.9**

to radiation. The ability to produce normal children is lost; the damage is permanent.

## Radiation Exposure

The routes of entry for radioactive materials are much the same as for poisons. However, the radioactive source or material does not have to be directly contacted for radiation exposure to occur. Once a radioactive material enters the body, it is very dangerous because the source now becomes an internal source rather than an external one. You cannot protect yourself by distance or shielding from a source that is inside your body. Contact with a radioactive material or ingestion of a radioactive material does not make you radioactive. You may become contaminated with radioactive particles, but with proper decontamination, these can be successfully removed. After they are removed, they cannot cause any further damage to your body. There are no truly safe levels of exposure to radioactive materials.

Radiation does not cause any specific diseases. Symptoms of radiation exposure may be the same as those from exposure to cancer-causing materials. The tolerable limits for exposure to radiation that have been proposed by some scientists are very arbitrary. Scientists concur that some radiation damage can be repaired by the human body. Therefore, tolerable limits are considered acceptable risks when the activity benefits outweigh the potential risks. The maximum annual radiation exposure for an individual person in the United States is 0.1 REM. The maximum for workers in the nuclear industry is 5 REMs per year. An emergency exposure of 25 REMs has been established by The National Institute of Standards and Technology for response personnel. This type of exposure should be attempted under only the most dire circumstances and should occur only once in a lifetime.

The effects of exposure to radiation on the human body depend on the amount of material the body was exposed to, the length of exposure, the type of radiation, the depth of penetration, and the frequency of exposures. The most susceptible cells are rapidly dividing cells, such as in the bone marrow. Children are more susceptible than adults and the fetus is the most susceptible.

Radiation injuries frequently do not present themselves for quite a long time after exposure. It can be years or even decades before symptoms appear. Cancer is one of the main long-term effects of exposure to radiation. Leukemia may take from 5 to 15 years to develop. Lung, skin, and breast cancer may take up to 40 years to develop. Figure 8.10 shows exposure rates and resulting radiation sickness effects. Radiation burns are much like thermal burns, although they can be much more severe. First-degree radiation burns result from an exposure of 50 to 200 RADs, second-degree burns result from 500 RADs, and third degree burns result from 1000 RADs.

Because of the physical characteristics of radioactives, protection can be provided by taking a few simple steps, commonly referred to as time, distance, and shielding. Time refers to the length of exposure to a radioactive source and the half-life of a radioactive material. A half-life is the length of time necessary

### RADIATION SICKNESS

| | |
|---|---|
| 25 REM | No Detectable Symptoms |
| 50 REM | Temperature, blood count change |
| 100 REM | Nausea, fatigue |
| 200-250 REM | Fatal to some in 30 days, all sick |
| 500 REM | 1/2 dead in 30 days |
| 600+ REM | All will die |

**Figure 8.10**

for an unstable element or nuclide to lose one-half of its radioactive intensity in the form of alpha, beta, and gamma radiation. Half-lives range from fractions of seconds to millions of years. In 10 half-lives, almost any radioactive source will no longer put out any more radiation than normal background radiation.

Distance is the second protective measure against radiation. As was previously mentioned radiation travels in a straight line, but only for short distances. Therefore, the greater the distance from the radioactive material, the less the intensity of the exposure. There is a law in dealing with radioactivity, known as the "inverse-square law." This means that as the distance from the radioactive source is doubled, the radiation intensity drops off by one quarter. If the distance is increased 10 times, the intensity drops off to 1/100 of the original intensity (Figure 8.11).

Shielding is the third protective measure against radiation. Shielding simply means that if there is enough mass between personnel and the radiation, they will be protected from the radiation (Figure 8.12). In the case of alpha particles, your skin or a sheet of paper will produce enough shielding. Turnouts, if worn, will provide extra protection. Ingestion is the major hazard of radioactive particles

### INVERSE SQUARE LAW

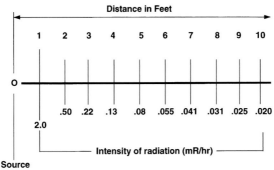

**Figure 8.11**

**SHIELDING FROM RADIATION**

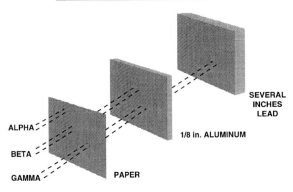

**Figure 8.12**

and wearing SCBA will prevent ingestion. Beta particles require more substantial protection from entering the body. A 1/24-in.-thick piece of aluminum will stop beta radiation. Turnouts will not provide adequate protection. Gamma radiation requires 3 to 9 in. of lead or several feet of concrete or earth. No protective clothing can protect against gamma radiation.

## RADIOACTIVE ELEMENTS AND COMPOUNDS

Radioactive materials are a part of everyday life. **Uranium, U,** is a radioactive metallic element. Uranium has three naturally occurring isotopes: uranium-234 (0.006%), uranium-235 (0.7%), and uranium-238 (99%). Uranium-234 has a half-life of $2.48 \times 10^5$ years, uranium-235 has a half-life of $7.13 \times 10^8$ years, uranium-238 has a half-life of $4.51 \times 10^9$ years. Uranium is a dense, silvery, solid material that is ductile and malleable; however, it is a poor conductor of electricity. As a powder, uranium is a dangerous fire risk and ignites spontaneously in air. It is highly toxic and a source of ionizing radiation. The TLV including metal and all compounds is 0.2 mg/m$^3$ of air. The four-digit UN identification number for uranium is 2979. Uranium is used in nuclear reactors to produce electricity and in the production of nuclear weapons systems.

## URANIUM COMPOUNDS

The primary use of uranium compounds is in the nuclear industry. Uranium has been used over the years for a number of commercial ventures, some successful and others not. Uranium dioxide was used as a filament in series with tungsten filaments for large incandescent lamps. The lamps are used in photography and motion pictures. Uranium dioxide has a tendency to eliminate the sudden surge of current through the bulbs when the light is turned on. This extends the life of the bulbs. Some alloys of uranium were used in the production of steel;

Refrigerator with radioactive isotopes inside.

however, they never proved commercially valuable. Sodium and ammonium diuranates have been used to produce colored glazes in the production of ceramics. Uranium carbide has been suggested as a good catalyst for the production of synthetic ammonia. Uranium salts in small quantities are claimed to stimulate plant growth; however, large quantities are clearly poisonous to plants.

**Uranium carbide, $UC_2$,** is a binary salt. It is a gray crystal that decomposes in water. It is highly toxic and a radiation risk. Uranium carbide is used as nuclear reactor fuel.

**Uranium dioxide,** also known as yellow cake, has a molecular formula of **$UO_2$.** It is a black crystal that is insoluble in water. It is a high radiation risk and ignites spontaneously in finely divided form. It is used to pack nuclear fuel rods.

**Uranium hexafluoride** has a molecular formula of **$UF_6$.** It is a colorless volatile crystal that sublimes and reacts vigorously with water. It is highly corrosive and is a radiation risk. The four-digit UN identification number for fissile material containing more than 1% of U-235 is 2977; for lower specific activity the number is 2978. Uranium hexafluoride is used in a gaseous diffusion process for separating isotopes of uranium.

**Uranium hydride** has a molecular formula of **$UH_3$** and is a brown-gray to black powder that conducts electricity. It is highly toxic and ignites spontaneously in air.

**Uranium tetrafluoride,** with the molecular formula of **$UF_4$,** is a green, nonvolatile, crystalline powder, that is insoluble in water. It is highly corrosive and is also a radioactive poison.

Medical uses of radioactive sources include sterilization, implants using radium, scans using iodine, and therapy using cobalt. X-rays are used in diagnostic

Nuclear reactor at power plant can be a source of highly radioactive materials.

medical procedures. In addition to medical facilities, radioactive materials may be found in research laboratories, educational institutions, industrial applications, and hazardous waste sites.

## RADIUM COMPOUNDS

**Radium, Ra,** is a radioactive metallic element. There are 14 radioactive isotopes; however, only radium-226 with a half-life of 1620 years is usable. It is a brilliant, white solid that is luminescent and turns black on exposure to air. Radium is water soluble and contact with water evolves hydrogen gas. It is in the alkaline earth metal family and, like calcium, it seeks the bones when it enters the body. It is highly toxic and emits ionizing radiation. Radium is destructive to living tissue. It is used in the medical treatment of malignant growths and industrial radiography.

Compounds formed with radium all have the same hazards as radium itself. Most are used in the treatment of cancer and for radiography in the medical and industrial fields. The compounds are all solids and the degree of water solubility varies. **Radium bromide** has a molecular formula of $RaBr_2$; it is composed of white crystals that turn yellow to pink. It sublimes at about 1650°F and is water soluble. The hazards are the same as for radium. It is used in the medical treatment of cancers.

**Radium carbonate,** with the molecular formula of $RaCO_3$, is an amorphous radioactive powder that is white when pure. Because of impurities, radium carbonate is sometimes yellow or pink. It is insoluble in water.

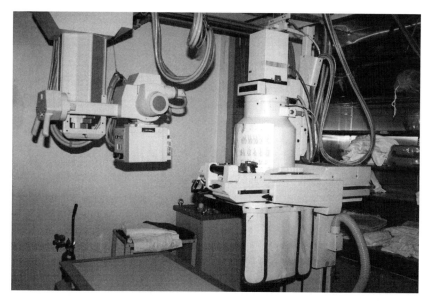

X-ray machine at hospital is a source of radiation only when the machine is turned on. There is no radioactive source when the power is off.

**Radium chloride** with the molecular formula of **$RaCl_2$,**, is a yellowish-white crystal that becomes yellow or pink upon standing. It is radioactive and soluble in water. It is used in cancer treatment and physical research.

## COBALT

**Cobalt, Co,** is a metallic element. Cobalt-59 is the only stable isotope. The common isotopes are cobalt-57, cobalt-58, and the most common, cobalt-60. Cobalt is a steel gray, shining, hard, ductile, and somewhat malleable metal. It has magnetic properties and corrodes readily in air. Cobalt dust is flammable and toxic by inhalation, with a TLV of 0.05 mg/m$^3$ of air. It is an important trace element in soils and animal nutrition. Cobalt-57 is radioactive. It has a half-life of 267 days. It is a radioactive poison and is used in biological research. Cobalt-58 is also radioactive and has a half-life of 72 days. It is a radioactive poison and it is used in biological and medical research. Cobalt-60 is one of the most common radioisotopes. It has a half-life of 5.3 years, is available in larger quantities, and is cheaper than radium. It is a radioactive poison and is used in radiation therapy for cancer and radiographic testing of welds and castings in industry. Compounds of cobalt are not radioactive.

## IODINE

**Iodine, I,** is a nonmetallic element of the halogen family. There is only one natural stable isotope, iodine-127. There are many artificial radioactive isotopes.

Radioactive isotopes are kept in special lead containers to prevent contact with the material.

Iodine is a heavy, grayish-black solid or granules having a metallic luster and characteristic odor. It is readily sublimed to a violet vapor and is insoluble in water. Iodine is toxic by ingestion and inhalation and is a strong irritant to the eyes and skin with a TLV ceiling of 0.1 ppm in air. Iodine-131 is a radioactive isotope of iodine and has a half-life of 8 days. It is used in the treatment of goiter, hyperthyroidism, and other disorders. It is also used as an internal radiation therapy source. Most iodine compounds are not radioactive.

**Krypton, Kr,** is an elemental colorless, odorless, inert gas. It is noncombustible, nontoxic, and nonreactive; however, it is an asphyxiant gas and will displace oxygen in the air. Krypton-85 is radioactive and has a half-life of 10.3 years. The four-digit UN identification number for krypton is 1056 as a compressed gas and 1970 as a cryogenic liquid. These forms of krypton are not radioactive. Radioactive isotopes of krypton would be shipped under radioactive labels and placards as required. Its primary uses are in the activation of phosphors for self-luminous markers, detecting leaks, and in medicine to trace blood flow.

**Radon, Rn,** is a gaseous radioactive element from the noble gases in family eight on the Periodic Table. There are 18 radioactive isotopes of radon all of which have short half-lives. Radon-222 has a half-life of 3.8 days. Radon is a colorless gas that is soluble in water. It can be condensed to a colorless transparent liquid and to an opaque, glowing solid. Radon is the heaviest gas known, with a density of 9.72 g/L at 32°F. Radon is derived from the radioactive decay of radium. It is highly toxic and emits ionizing radiation. Lead shielding must be used in handling and storage. Radon has appeared naturally in the basements of homes, causing some concern for the residents. The primary uses are as a cancer treatment, a tracer in leak detection, in radiography, and chemical research.

## CHEMICAL RELEASE STATISTICS

According to EPA statistics, radioactive materials were the eighth most released hazardous material by number of accidents from 1988 to 1992. There were over 1000 accidents resulting in the release of over 129 million lbs. of radioactive material. These primarily occurred at fixed facilities. The DOT reports nine transportation incidents during 1995 involving radioactive materials with no deaths or injuries.

Radioactive materials are often found in transportation. They are heavily regulated and the containers are well constructed. Most radioactive incidents are not handled by local emergency responders. Agencies other than fire, police, and EMS are responsible for response and handling of radioactive emergencies. Emergency responders must, however, be aware of radioactive materials and know how to protect themselves. Each state has radiological response teams for radioactive emergencies. They may be a part of the emergency management agency, the health department, the department of environment, or some other agency. Federal interests are represented by the Nuclear Regulatory Commission (301) 492-7000. Incidents involving weapons are handled by the Department of Defense Joint Nuclear Accident Center (703) 325-2102.

Radioactive materials may be found in many types of transportation vehicles, including taxi cabs, delivery trucks, and private cars.

Storage areas for radioactive materials are marked with this type of sign.

## REVIEW QUESTIONS

1. Chemical activity takes place with the _____ of the atom and radio-activity involves the _____.

2. An isotope is an element that has _____ in the nucleus than a normal atom of that element.

3. The three major types of radiation include _____, _____, and _____.

4. _____ and _____ are forms of particulate radiation.

5. _____ is an electromagnetic energy wave.

6. Radiation always travels in the following manner _____.
   A. In a curving motion
   B. In a straight line
   C. Around objects
   D. Straight up
   E. None of the above

7. The two types of radiation are _____ and _____.

8. Match the following types of radiation with the correct statement.

    _____ Beta        A.  High energy wave

    _____ Alpha       B.  Small, travels faster, and penetrates farther

    _____ Gamma    C.  Large, travels short distance

9. Protective measures against radiation include _____, _____, and _____.

10. Which of the following radioactive-labeled materials require placarding on the transport vehicle regardless of the quantity?

    A.  Radioactive I

    B.  Radioactive II

    C.  Radioactive III

# 9

CORROSIVES

Corrosives are the largest volume class of chemicals used by industry. It therefore stands to reason that they would frequently be encountered in transportation and fixed facilities. The Department of Transportation (DOT) Hazard Class 8 materials are corrosive liquids and solids. There are no DOT subclasses of corrosives. There are, however, two types of corrosive materials found in hazard Class 8: acids and bases. Acids and bases are actually two different types of chemicals that are sometimes used to neutralize each other in a spill. They are grouped together in Class 8, because the corrosive effects are much the same on tissue and metals, if contacted. It should be noted, however, that the correct terminology for an acid is *corrosive* and for a base is *caustic*. The DOT, however, does not differentiate between the two when placarding and labeling. The DOT definition for a corrosive material is "a liquid or solid that causes visible destruction or irreversible alterations in human skin tissue at the site of contact, or a liquid that has a severe corrosion rate on steel or aluminum. This corrosive rate on steel and aluminum is 0.246 inches per year at a test temperature of 131°F."

A definition for an acid from the *Condensed Chemical Dictionary* is "a large class of chemical substances whose water solutions have one or more of the following properties: sour taste, ability to make litmus dye turn red and to cause other indicator dyes to change to characteristic colors, the ability to react with and dissolve certain metals to form salts, and the ability to react with bases or alkalis to form salts." It is important to note here that tasting any chemical is not an acceptable means of identification for obvious reasons.

In addition to being corrosive, acids and bases can explode or polymerize; they can also be water reactive, toxic, flammable (applies to organic acids because inorganic acids do not burn), very reactive and unstable oxidizers.

There are two basic types of acid, organic and inorganic (Figure 9.1). Inorganic acids are sometimes referred to as mineral acids. As a group, organic acids are generally not as strong as inorganic acids. The main difference between the two is the presence of carbon in the compound: inorganic acids do not contain carbon. Inorganic acids are corrosive, but they do not burn. They may, however, be oxidizers and support combustion, or may spontaneously combust with organic

## TYPES OF ACIDS

| INORGANIC | ORGANIC |
|---|---|
| Sulfuric $H_2SO_4$ | Formic HCOOH |
| Hydrochloric HCl | Acetic $C_2H_5COOH$ |
| Nitric $HNO_3$ | Propionic $C_2H_5COOH$ |
| Phosphoric $HPO_4$ | Acrylic $C_2H_3COOH$ |
| Perchloric $HClO_4$ | Butyric $C_3H_7COOH$ |

**Figure 9.1**

material. Inorganic acid molecular formulas begin with hydrogen (H). For example, $H_2SO_4$ is the molecular formula for sulfuric acid, HCl is hydrochloric acid, and $HNO_3$ is nitric acid. Organic acids are hydrocarbon derivatives, therefore, they have carbon in the compound, and the name will begin with the prefix indicating the number of carbons. For example, the prefix for a one-carbon compound with the organic acids is "form", so a one-carbon acid is called formic acid; a two-carbon acid is acetic acid; a three-carbon acid is propionic acid, etc. Organic acids are corrosive, may polymerize, and some may burn.

## INORGANIC ACIDS

Acids are materials that release hydrogen ions, $H^+$, when placed in water. Inorganic acids can generally be identified by a hydrogen at the beginning of the formula (Figure 9.2), because very few other compounds begin with hydrogen. The hydrogen ion, $H^+$, consists of just a hydrogen nucleus, without electrons, and is composed of just one proton. Acids that supply just one $H^+$ are often referred to as monoprotic acids, e.g., HCl and $HNO_3$. Acids that can supply more than one $H^+$ are referred to as polyprotic acids; more specifically $H_2CO_3$ and $H_2SO_4$ are referred to as diprotic acids and $H_3PO_4$ as triprotic acids. There are two general types of inorganic acids: binary and oxyacids. Binary acids are composed of just two elements, hydrogen and some other nonmetal, e.g., HCl and $H_2S$. These acids are named by placing the prefix "hydro" before and the suffix "ic" after the nonmetal element; and the compound ends with the word acid. For example, when hydrogen is combined with chlorine, the "ine" is dropped from chlorine, and the

## INORGANIC ACIDS BEGIN WITH HYDROGEN IN FORMULA

| Binary Acids | | Ternary Acids | |
|---|---|---|---|
| Hydrofluoric | HF | Nitric | $HNO_3$ |
| Hydrochloric | HCl | Perchloric | $HClO_4$ |
| Hydrobromic | HBr | Sulfuric | $H_2SO_4$ |
| Hydrodic | HI | Phosphoric | $H_3PO_4$ |
| Hydrosulfuric | $H_2S$ | Carbonic | $H_2CO_3$ |

**Figure 9.2**

prefix "hydro" and suffix "ic" are added: hydrochloric acid; hydrogen combined with sulfur is called hydrosulfuric acid.

Acids that contain hydrogen, oxygen, and some other nonmetal element are called oxyacids, e.g., $H_2SO_4$, $HNO_3$, and $HClO_4$ (note the similarities to the oxyradicals). Like the oxysalts, these acids are named according to the numbers of oxygens in the compound. The acid with the largest number of oxygens in a series ends with the suffix "ic" and the one with the fewest number of oxygens takes the suffix "ous" (similiar to the alternate naming of the transitional metal salts discussed in Chapter 1). For example, when hydrogen is combined with sulfur, the base state of the compound is $SO_4$ and the acid, $H_2SO_4$, is called sulfur*ic* acid. If there is one less oxygen present in the compound, such as $SO_3$, the ending changes to "ous" and the acid, $H_2SO_3$, is called sulfur*ous* acid. $HNO_3$ is nitric acid and $HNO_2$ is nitrous acid, etc. When halogens are present in the acid the compound with the most oxygens in the base state ends in "ic," such as chloric acid, $HClO_3$. If the oxygen is increased by one to $HClO_4$, the prefix "per" is added, yielding the name *per*chlor*ic* acid. The acid compound with the least number of oxygens ends with "ous", such as chlorous acid, $HClO_2$. If the oxygen is reduced by one to $HClO$, the prefix "hypo" is added yielding the name *hypo*chlor*ous* acid.

## STRENGTH AND CONCENTRATION

Most inorganic acids are produced by the dissolving of a gas or a liquid in water, e.g., hydrochloric acid is derived from dissolving hydrogen chloride gas in water. All acids contain hydrogen. This hydrogen is the form of an ion ($H^+$) and can be measured by using the pH scale (Figure 9.3). In simple terms, the pH

## pH SCALE OF COMMON MATERIALS

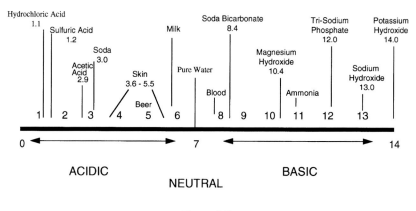

**Figure 9.3**

## IONIZATION OF COMMON ACIDS AND BASES

| Completely Ionized | Moderately Ionized | Slightly Ionized |
|---|---|---|
| Nitric | Oxalic | Hydrofluoric |
| Hydrochloric | Phosphoric | Acetic |
| Sulfuric | Sulfurous | Carbonic |
| Hydriodic | | Hydrosulfuric |
| Hydrobromic | | (Most others) |
| | | |
| Potassium Hydroxide | | Ammonium |
| Sodium Hydroxide | | Hydroxide |
| Barium Hydroxide | | (All others) |
| Strontium Hydroxide | | |
| Calcium Hydroxide | | |

**Figure 9.4**

scale measures the hydrogen ion concentration of a solution. Concentrated acids and bases measure off the pH scale. The pH scale only measures acids and bases in solutions. To determine if a concentrated material is an acid or base, litmus paper is used; however, this does not yield a numerical value. As a group, acids have high hydrogen ion concentrations. Bases have very low hydrogen ion concentrations and high hydroxyl ($OH^-$) concentrations. The strength or weakness of an acid or base is the amount of hydrogen ions or hydroxyl ions that are produced as the acid or base is created. If the hydrogen ion concentration of an acid is high, the acid is concentrated. If the hydroxyl concentration is high, it is a concentrated base. In both cases, there is almost total ionization of the material dissolved in water to make the strong acid and base (Figure 9.4). For example, hydrochloric acid is a strong acid because practically all of the hydrogen chloride gas is ionized in the water. Acetic acid is a weak acid because only a few molecules ionize in producing the acid.

Another term associated with corrosives is *concentration*. Concentration has to do with the amount of acid that is mixed with water and is often expressed in terms of percentages. A 98% concentration of sulfuric acid is 98% sulfuric acid and 2% water, a solution of 50% nitric acid is 50% nitric acid and 50% water. In the 50% concentration, the solution has only half the $H^+$ ions that the 100% concentration would have. A 50% concentration of nitric acid is a solution diluted to 50% of the original acid.

## pH

The pH scale measures the acidity or alkalinity of a *solution*. The pH scale cannot measure some strong acids and bases that are not in solution because they have values less than 0 or greater than 14. Acid solutions are considered acidic and base solutions are considered alkaline. Acid solutions have a value on the pH scale from 1 to 6.9. Materials with a pH value of 7 are considered to

## Exponential Logarithm of pH Values

| pH Value | H + and OH - Concentration |
|:---:|:---:|
| 1 | 1,000,000 |
| 2 | 100,000 |
| 3 | 10,000 |
| 4 | 1,000 |
| 5 | 100 |
| 6 | 10 |
| 7 | 1 |
| 8 | 10 |
| 9 | 100 |
| 10 | 1,000 |
| 11 | 10,000 |
| 12 | 100,000 |
| 13 | 1,000,000 |

**Figure 9.5**

be neutral, i.e., they are neither acidic or basic. Base solutions have values on the scale from 7.1 to 14. It is not important for emergency responders to understand or know how the pH scale measures corrosivity or the specific values of any given acid or base. It is important, however, for responders to know that numerical values lower than 7 are acids and values higher than 7 are bases. Usually, when dealing with numerical values, the higher the number, the greater the value that is being measured and the number 2 is twice the value of 1. With the pH scale, however, the lower the pH number, the more acidic an acid solution is, and an acid solution with a pH of 1 is 10 times more acidic than an acid solution with a pH of 2 and so on (Figure 9.5). The ratio and the intervals between the numbers are exponential, e.g., a pH of 5 is 10 times more acidic than a pH of 6, etc. The result of this exponential ratio is that on the full scale, a solution with a pH of 1 is 1,000,000 times more acidic than an acid with a pH of 7. So the difference between individual values on the pH scale is very great and is one of the reasons why dilution and neutralization are not as simple as they might sound. Those terms will be discussed further under the section Dilution vs. Neutralization.

If the chemical name of a hazardous material is known and it is determined to be a corrosive, looking up the chemical name in reference sources will identify whether the material is an acid or a base. It will not be necessary for responders to get a pH measurement of the material unless it is to verify the reference information. The use of pH measurements can be useful when a material has not been positively identified. The pH measurement can be used to narrow the chemical family possibilities in the identification process.

There are a number of ways for emergency responders to measure the pH of a corrosive material. First of all, the proper chemical protective clothing must be worn when working around the corrosive material. The simplest and least expensive method of determining pH is the use of pH paper, which changes color based on the type and strength of corrosive material that is present. The colored paper is then compared to a chart on the pH paper container. The chart will indicate numerical pH values much the same as do expensive measuring instruments. Although not as accurate as a pH meter, the numbers will give a "ballpark" measure of the pH of the material. There are also commercially available pH meters from hand-held to sophisticated laboratory instruments. This equipment can be expensive and pH paper is accurate enough for emergency response identification purposes. If the only information needed is whether a material is an acid or a base, litmus paper can be used. Litmus paper turns blue if the corrosive material is a base, red if the corrosive material is an acid. The litmus paper will not give actual pH numerical values.

The definition of a base from the *Condensed Chemical Dictionary* is "a large class of compounds with one or more of the following properties: bitter taste, slippery feeling in solution, ability to turn litmus blue, and to cause other indicators to take on characteristic colors, and the ability to react with (neutralize) acids to form salts." It is important to note that while the definition of acid and base mentions the taste and feeling of the materials, these are dangerous chemicals and can cause damage to tissues upon contact. Therefore, it is *NOT* recommended that responders come in contact with these materials through taste or touch! Ionization occurs with the bases, just as with the acids, as they are made. Most bases are produced by dissolving a solid, usually a salt, in water. However, with the bases the ion produced is the hydroxyl ion ($OH^-$). The base is considered strong or weak depending on the number of hydroxyl ions produced as a corrosive material is dissolved in water. A large $OH^-$ concentration produces a strong base; a small $OH^-$ concentration produces a weak base. Sodium and potassium hydroxide are strong bases, calcium hydroxide (hydrated lime) is a weak base. Bases will have a pH from 7.1 to 14 on the pH scale. The degree of alkalinity increases from 7.1 to 14, with 7.1 being the least basic and 14 the most basic. The amount of alkalinity between the numerical values on the pH scale is exponential just as with the acids. A base with a pH of 9 is 10 times more basic than one with a pH of 8 and so on.

Corrosivity is not the only hazard of Class 8 materials. In addition to being corrosive, they may have other hazards, such as toxicity, flammability, or oxidation. Many corrosives, especially acids, can be violently water reactive. Contact with water may cause splattering of the corrosive, produce toxic vapors, and evolve heat that may ignite nearby combustible materials. Some of the water may be turned to steam by the heat produced in the reaction. This can cause overpressurization of the container. Many corrosives may also be unstable, reactive, may explode, polymerize, or decompose, and produce poisons.

**Picric acid, $C_6H_2(NO_2)_3OH$,** for example, becomes a high explosive when dried out and is sensitive to shock and heat. The hazard class for picric acid is 4.1 flammable solid. It is considered a wetted explosive. The name would indicate acid; however, the corrosivity of picric acid is far outweighed by its explosive dangers. The slightest movement of dry picric acid may cause an explosion. Picric acid, when shipped, is mixed with 12 to 20% water to keep it stable. When this water evaporates in storage, over time, the material becomes explosive.

**Perchloric acid, HClO$_4$** is a colorless, volatile, fuming liquid that is unstable in its concentrated form. It is a strong oxidizing agent and will spontaneously ignite in contact with organic materials. It is corrosive, with the highest concentration of 70%. Contact with water produces heat; when shocked or heated, it may detonate. The boiling point is 66°F and it is soluble in water, with a specific gravity of 1.77, which is heavier than water. The vapor density is 3.46, which is heavier than air. Perchloric acid is toxic by ingestion and inhalation. It is used in the manufacture of explosives and esters; in electro-polishing and analytical chemistry; and as a catalyst.

**Hydrocyanic acid, HCN,** is corrosive in addition to being a very toxic material. It is also a dangerous fire and explosion risk. It has a wide flammable range of 6 to 41% in air. The boiling point is 79°F, the flash point is 0°F, and the ignition temperature is 1004°F. It is toxic by inhalation, ingestion, and through skin absorption. The TLV of hydrocyanic acid is 10 ppm in air. It is used in the manufacture of acrylonitrile, acrylates, cyanide salts, dyes, rodenticides, and other pesticides.

## ORGANIC ACIDS

Organic acids are hydrocarbon derivatives. They are flammable, corrosive, and may polymerize by exposure to heat or sudden shock. Organic acids are "super-duper" polar materials; they are the most polar of the hydrocarbon derivatives. Organic acids have hydrogen bonding and a carbonyl that gives them a double dose of polarity. The functional group is represented by a carbon, two oxygens, and a hydrogen. The general formula is **R–C–O–O–H.** One radical is attached to the carbon of the functional group. Organic acids use the alternate prefix for one- and two-carbon compounds. When naming them, all of the carbons, including the one in the functional group, are counted to determine the hydrocarbon prefix name. To represent an acid, "ic" is added to the hydrocarbon prefix and the name ends in acid, e.g., a one-carbon acid uses the alternate prefix name "form"; "ic" is added to "form", making it formic, and acid is added to the end: formic acid. A two-carbon acid uses the alternate name for two carbons, which is "acet", plus "ic", and ends with acid: acetic acid. Naming three- and four-carbon acids reverts back to the normal prefixes for three- and four-carbon radicals with some minor alterations to make the names flow more smoothly, which really has nothing to do with the naming rules or the ultimate compound name. For example, a three-carbon acid uses the prefix "prop", indicating three carbons, the letters "ion" are then added to make the name flow smoothly, the ending "ic" is added to the radical and the word acid for the compound name: propionic acid. A four-carbon organic acid begins with the radical prefix "but", the filler letters "yr" are attached, the radical ends with "ic", acid is added and the name for a four-carbon organic acid is butyric acid. The structures and molecular formulas for organic acids are illustrated with one through four carbons. Note that the carbon in the functional group is counted when determining which hydrocarbon radical is used in naming it.

$$
\begin{array}{c}
O \\
\parallel \\
H-C-O-H
\end{array}
$$

Formic acid
HCOOH

$$
\begin{array}{c}
H \quad O \\
| \quad \parallel \\
H-C-C-O-H \\
| \\
H
\end{array}
$$

Acetic acid
CH$_3$COOH

$$
\begin{array}{c}
H \quad H \quad O \\
| \quad | \quad \parallel \\
H-C-C-C-O-H \\
| \quad | \\
H \quad H
\end{array}
$$

Propionic acid
C$_2$H$_5$COOH

$$
\begin{array}{c}
H \quad H \quad H \quad O \\
| \quad | \quad | \quad \parallel \\
H-C-C-C-C-O-H \\
| \quad | \quad | \\
H \quad H \quad H
\end{array}
$$

Butyric acid
C$_3$H$_7$COOH

It is also possible to add double-bonded radicals to the organic acid functional group. For example, when the vinyl radical is attached to the carbon in the functional group, a three-carbon double-bonded radical is created. The acryl radical is used for the three carbons with a double bond, the ending "ic" is added to the radical, and the word acid to the end. The compound formed is acrylic acid. The double bond between the carbons can come apart in a polymerization reaction. Generally, materials that have double bonds are reactive in some manner. If polymerization occurs inside a container, an explosion may occur that can produce heat, light, fragments, and a shock wave.

**Formic acid, HCOOH,** is a colorless, fuming liquid with a penetrating odor. The highest concentration is 90%. It is a polar material that is soluble in water and has a specific gravity of 1.2, which is heavier than water. As with many of the organic acids, formic acid is flammable. The boiling point is 213°F, the flash point is 156°F, and the flammable range is 18 to 57%. The ignition temperature is 1004°F and the vapor density is 1.6, which is heavier than air. Formic acid is toxic, with a TLV of 5 ppm in air. The four-digit UN identification number is 1779. The NFPA 704 designation is health 3, flammability 2, and reactivity 0. It is used in dyeing and finishing of textiles, the treatment of leather, and the manufacture of esters, fumigants, insecticides, refrigerants, etc.

**Propionic acid, C$_2$H$_5$COOH,** is a colorless, oily liquid with a rancid odor. It is a polar compound and soluble in water. Propionic acid is flammable, with a flammable range of 2.9 to 12% in air, a boiling point of 286°F, a flash point of 126°F, and an ignition temperature is 955°F. Polar solvent foam will have to be used to extinguish fires. It is toxic, with a TLV of 10 ppm in air. The four-digit UN identification number for propionic acid is 1848. The NFPA 704 designation is health 3, flammability 2, and reactivity 0. It is used as a mold inhibitor in bread and as a fungicide, an herbicide, a preservative for grains, in artificial fruit flavors, pharmaceuticals, and others.

**Butyric acid,** also known as butanoic acid, is a colorless liquid with a penetrating, obnoxious odor. It is miscible in water, with a specific gravity of 0.96, which makes it slightly lighter than water. Butyric acid is flammable, with a boiling point of 326°F and a flash point of 161°F. The flammable range is 2 to 10 percent in air and the ignition temperature is 846°F. Butyric acid is a strong irritant to skin and eyes. The four-digit UN identification number is 2820. The NFPA 704 designation is health 3, flammability 2, and reactivity 0. The primary uses of butyric acid are in the manufacture of perfume, flavorings, pharmaceuticals, and disinfectants.

**Acrylic acid, $C_2H_3COOH$,** is a colorless liquid with a acrid odor. It polymerizes readily and may undergo explosive polymerization. The boiling point is 509°F, the flash point is 122°F, and the ignition temperature is 820°F. The flammable range is 2.4 to 8% in air. Acrylic acid is miscible with water and has a specific gravity of 1.1, which is slightly heavier than water. The vapor density is 2.5, which is heavier than air. It is an irritant and corrosive to the skin, with a TLV of 2 ppm in air. The four-digit UN identification number is 2218. Acrylic acid must be inhibited when transported. The NFPA 704 designation is health 3, flammability 2, and reactivity 2. The primary uses are as a monomer for polyacrylic and polymethacrylic acids and other acrylic polymers. The structure and molecular formula for acrylic acid are shown as follows:

$$
\begin{array}{ccc}
H & H & O \\
| & | & || \\
C{=}C & {-}C & {-}O{-}H \\
| & & \\
H & &
\end{array}
$$

Acrylic acid
$C_2H_3COOH$

**Phosphorus trichloride, $PCl_3$,** is a clear, colorless, fuming corrosive liquid. It decomposes rapidly in moist air and has a boiling point of about 168°F. $PCl_3$ is corrosive to skin and tissue and reacts with water to form hydrochloric acid. The TLV is 0.2 ppm and the IDLH is 50 ppm in air. The four-digit UN identification number is 1809. The NFPA 704 designation is health 4, flammability 0, and reactivity 2. The white section at the bottom of the diamond contains a W with a slash through it, indicating water reactivity. The primary uses are in the manufacture of organophosphate pesticides, gasoline additives, dye stuffs, as a chlorinating agent, a catalyst, and in textile finishing.

Corrosives in contact with poison may produce poison gases as the poison decomposes. In responding to an incident involving corrosives, the toxicity of the vapors could be much more of a concern for personnel than the corrosivity. When acids come in contact with cyanide, hydrogen cyanide gas which is very toxic, with a TLV of 10 ppm in air, is produced.

When strong corrosives contact flammable liquids, the chemical reaction that occurs may produce heat. The heat produced will cause more vapor to be produced and, if an ignition source is present, combustion may occur. Corrosives may also be strong oxidizers. If they come in contact with particulate combustible solids, spontaneous combustion may occur. Once ignition has occurred, the corrosive will act as an oxidizer and accelerate the rate of combustion. Nitric acid in contact with combustible organic materials containing cellulose will produce a chemical reaction. This reaction will produce nitrocellulose, which is a dangerous fire and explosion risk. Toxic vapors may also be produced when the cellulose burns.

## TOP 50 INDUSTRIAL CHEMICALS

After flammable liquids and gases, corrosives are the next most common hazardous material encountered by emergency responders. There are 11 corrosive materials in *Chemical and Engineering News* magazine's Top 50 industrial chemicals for 1995. Combined, these 11 account for over 250 billion lbs. of corrosives produced annually.

**Sulfuric acid, H$_2$SO$_4$,** is an inorganic acid and is the number one produced and most widely used industrial chemical at 95.36 billion lbs. in 1995. It is a strong corrosive with a solution pH of 1.2. It is a dense, oily liquid, colorless to dark brown, depending on purity. Sulfuric acid is miscible in water, but violently water reactive, producing heat and explosive splattering if water is added to the acid. The boiling point is 626°F and the specific gravity is 1.84, which is heavier than water. Sulfuric acid is very reactive and dissolves most metals. When in contact with metals, hydrogen gas is released. The vapors are toxic by inhalation and the TLV is 1 ppm in air. Sulfuric acid is incompatible with potassium chlorate, potassium perchlorate, potassium permanganate, and similar compounds of other

Bulk container of sulfuric acid.

Railcar with liquid sulfuric acid, the number 1 industrial chemical produced in 1995.

light metals. Sulfuric acid has a four-digit UN identification number of 1830. The NFPA 704 designation is health 3, flammability 0, and reactivity 2. The white section contains a W with a slash through it, indicating water reactivity. Sulfuric acid is used in batteries for cars and other vehicles. It is also used in the manufacture of fertilizers, chemicals, and dyes; as an enchant and a catalyst; in electroplating baths and explosives; in or for pigments, and many other uses.

**Fuming sulfuric acid** is also called oleum, which is a trade name. Fuming sulfuric acid is a solution of sulfur trioxide in sulfuric acid. The sulfur trioxide is forced into solution with the sulfuric acid to the point that the solution cannot hold anymore. As soon as the solution is exposed to air, the fuming begins, forming dense vapor clouds. It is violently water reactive, as are most acids. The four-digit UN identification number is 1831.

**Lime, CaO** (also known as calcium oxide, quicklime, hydrated lime, and hydraulic lime), a binary oxide salt, is the fifth most produced industrial chemical, with 41.23 billion lbs. produced in 1995. It is a white or grayish-white material in the form of hard clumps. It may have a yellowish or brownish tint due to the presence of iron. Lime is odorless and crumbles on exposure to moist air. It is a corrosive caustic that yields heat and calcium hydroxide when mixed with water and is a strong irritant with a TLV of 2 mg/m$^3$ of air. The four-digit UN identification number for calcium oxide is 1910. The primary uses are in the manufacture of other chemicals, like calcium carbide; pH control; neutralization of acid waste, insecticides, and fungicides.

**Sodium hydroxide, NaOH,** also known as caustic soda, is the eighth most produced industrial chemical, with 26.19 billion lbs. in 1995. Sodium hydroxide is a hydroxide salt. Sodium hydroxide is a strong base and is severely corrosive with a solution pH of 13. It is the most important industrial caustic material.

Rail tank car of sodium hydroxide solution, a corrosive material.

Sodium hydroxide is a white, deliquescent solid found in the form of beads or pellets. It is also found in solutions with water of 50 and 73%. Sodium hydroxide is water soluble, water reactive, and absorbs water and carbon dioxide from the air. The specific gravity is 2.8, which is heavier than water. It is corrosive to tissue in the presence of moisture and is a strong irritant to eyes, skin, mucous membranes, and is toxic by ingestion. The TLV ceiling is 2 mg/m$^3$ of air. The four-digit UN identification number is 1823 for dry materials and 1824 for solutions. The NFPA 704 designation is health 3, flammability 0, and reactivity 1. It is used in the manufacture of chemicals, as a neutralizer in petroleum refining, in metal etching, electroplating, and as a food additive.

**Phosphoric acid, H$_3$PO$_4$,** an inorganic acid, is the seventh most heavily produced industrial chemical, with 26.19 billion lbs. in 1995. It is a colorless, odorless, sparkling liquid or crystalline solid, depending on concentration and temperature. Phosphoric acid has a boiling point of 410°F; at 20°C, the 50 and 75% strengths are mobile liquids. The 85% concentration has a syrupy consistency and the 100% acid is in the form of crystals. Phosphoric acid is water soluble and absorbs oxygen readily, and the specific gravity is 1.89, which is heavier than water. It is toxic by ingestion and inhalation and an irritant to the skin and eyes, with a TLV of 1 mg/m$^3$ of air. The four-digit UN identification number is 1805. The NFPA 704 designation is health 3, flammability 0, and reactivity 0. The primary use of phosphoric acid is in chemical analysis and as a reducing agent.

**Sodium carbonate, Na$_2$CO$_3$,** also known as soda ash and sodium bicarbonate, is an oxysalt, and is the eleventh most produced industrial chemical, with

22.28 billion lbs. in 1995. It is a base with a pH of 11.6. It is not particularly hazardous and is used to neutralize acid spills.

**Nitric acid, HNO₃,** an inorganic acid, is the fourteenth most heavily produced industrial chemical, with 17.24 billion lbs. in 1995. It is a colorless, transparent, or yellowish, fuming, suffocating, corrosive liquid. Nitric acid will attack almost all metals. The yellow color results from the exposure of the nitric acid to light. Nitric acid is a strong oxidizer, is miscible in water, and has a specific gravity of 1.5, which is heavier than water. It may be found in solutions of 36, 38, 40, 42, degrees B`e and concentrations of 58, 63, and 95%. Nitric acid is a dangerous fire risk when in contact with organic materials. It is toxic by inhalation and is corrosive to tissue and mucous membranes. The TLV is 2 ppm in air. Nitric acid is incompatible with acetic acid, hydrogen sulfide, flammable liquids and gases, chromic acid, and aniline. The four-digit UN identification number for nitric acid at <40% is 1760. The NFPA 704 designation for nitric acid at <40% concentration is health 3, flammability 0, and reactivity 0. There is not any information in the white area of the diamond for <40% concentrations. Below 40% concentration, nitric acid is not considered an oxidizer. The four-digit UN identification number for nitric acid at >40% is 2031. It is placarded as a Class 8 corrosive; however, individual containers are labeled corrosive, oxidizer, and poison. The NFPA 704 designation for nitric acid at >40% concentration is health 4, flammability 0, and reactivity 0. The prefix "oxy" appears in the white section of the diamond. Nitric acid above 40% concentration is an oxidizer. Nitric acid is used in the manufacture of ammonium nitrate fertilizer and explosives, in steel etching, and reprocessing spent nuclear fuel.

There are two types of fuming nitric acid. **White fuming nitric acid** is concentrated with 97.5% nitric acid and less than 2% water. It is a colorless to pale yellow liquid that fumes strongly. It is decomposed by heat and exposure to light and becomes red in color from nitrogen dioxide. **Red fuming nitric acid** contains more than 85% nitric acid, 6 to 15% nitrogen dioxide, and 5% water. The four-digit UN identification number for red fuming nitric acid is 2032. The NFPA 704 designation is health 4, flammability 0, and reactivity 1. The prefix "oxy" appears in the white section of the diamond. Red fuming nitric acid is considered an oxidizer. Both white and red fuming acids are toxic by inhalation, strong corrosives, and dangerous fire risks that may explode on contact with reducing agents. They are used in the production of nitro compounds, rocket fuels, and as laboratory reagents.

**Terephthalic acid, TPA, C₆H₄(COOH)₂,** is an organic acid. It is the twenty-fifth most produced industrial chemical, with 7.95 billion lbs. in 1995. It is a white crystalline or powdered material that is insoluble in water. It undergoes sublimation above 572°F. In addition to being corrosive, it is also combustible. The primary uses are in the production of polyester resins, fibers, and films; it is also an additive to poultry feeds. The structure and molecular formula for terephthalic acid are as follows:

$$H-O-C-\underset{}{\bigcirc}-C-O-H$$

Terephthalic acid
$C_6H_4(COOH)_2$

The naming of terephthalic acid does not follow any of the rules of naming organic acids under the trivial naming system. However, the formula and the structure indicate an organic compound and the name indicates acid. The hazards of the acids, except for flammability, are similiar. The fact that the name indicates acid should lead you to assume flammability and toxicity in addition to corrosiveness, until other information is known.

**Hydrochloric acid, HCl,** an inorganic acid, is the twenty-seventh most produced industrial chemical, with 7.33 billion lbs. in 1995. It is a colorless or slightly yellow, fuming, pungent liquid produced by dissolving hydrogen chloride gas in water. Hydrochloric acid in solution has a pH of 1.1. The specific gravity is 1.19, which is heavier than water. It is water soluble, a strong corrosive, and toxic by ingestion and inhalation. It is an irritant to the skin and eyes. The four-digit UN identification number is 1050 for anhydrous and 1789 for solution. Hydrochloric acid is used in food processing, pickling, and metal cleaning, as an alcohol denaturant, and as a laboratory reagent.

**Acetic acid, CH₃COOH,** an organic acid, also known as ethanoic acid and vinegar acid, is the thirty-third most produced industrial chemical, with 4.68

Hydrofluoric acid in 55-gal drums.

billion lbs. in 1995. Acetic acid is a clear, colorless, corrosive liquid with a pungent odor. In solution, acetic acid has a pH of 2.9. The glacial form is the pure form without water; it is 99.8% pure. Glacial acetic acid is a solid at normal temperatures. It is flammable, with a flash point of 110°F and a flammable range of 4 to 19.9%. The ignition temperature is 800°F. Acetic acid is "super-duper" polar and water soluble. It will require a polar solvent type foam to extinguish fires. The specific gravity is 1.05, which is slightly heavier than water, but being miscible it will mix rather than form layers. It is toxic by inhalation and ingestion, with a TLV of 10 ppm in air. Acetic acid is a strong irritant to skin and eyes. It is incompatible with nitric acid, peroxides, permanganates, ethylene glycol, hydroxyl compounds, perchloric acid, and chromic acid. The four-digit UN identification number is 2789. The NFPA 704 designation is health 3, flammability 2, and reactivity 0. It is a food additive at lower concentrations; it is used in the production of plastics, pharmaceuticals, dyes, insecticides, and photographic chemicals. The structure for acetic acid is shown in the Organic Acids section of this chapter.

**Caustic potash, KOH,** also known as potassium hydroxide and lye, is the thirty-ninth most produced industrial chemical, at 3.22 billion lbs. in 1995. Potassium hydroxide is a hydroxide salt. It is found as a white solid, in the form of pieces, lumps, sticks, pellets, or flakes. Potassium hydroxide may also be found as a liquid. It is water soluble and may absorb water and carbon dioxide from the air. The specific gravity is 2.04, which is heavier than water; however, it is miscible in water so it will mix rather than form layers. It is a strong base and is toxic by ingestion and inhalation. The TLV ceiling is 2 mg/m³ of air. The four-digit UN identification number is 1813 for the solid and 1814 for the solution. The NFPA 704 designation is health 3, flammability 0, and reactivity 1. It is used in soap manufacture, bleaching, as an electrolyte in alkaline storage batteries and some fuel cells, as an absorbent for carbon dioxide and hydrogen sulfide, and in fertilizers and herbicides.

**Adipic acid,** also known as hexanedioic acid, is an organic acid. It is the forty-eighth most produced industrial chemical, at 1.80 billion lbs. in 1995. It is a white, crystalline solid that is slightly soluble in water. In addition to being a corrosive, it is also flammable; however, it is a relatively stable compound. Adipic acid is used in the manufacture of nylon and polyurethane foams. It is also a food additive and adhesive. The structure and molecular formula are shown as follows; notice that the structure has two organic acid functional groups attached.

$$H-O-\overset{\overset{\displaystyle O}{\|}}{C}-\overset{\overset{\displaystyle H}{|}}{\underset{\underset{\displaystyle H}{|}}{C}}-\overset{\overset{\displaystyle H}{|}}{\underset{\underset{\displaystyle H}{|}}{C}}-\overset{\overset{\displaystyle H}{|}}{\underset{\underset{\displaystyle H}{|}}{C}}-\overset{\overset{\displaystyle H}{|}}{\underset{\underset{\displaystyle H}{|}}{C}}-\overset{\overset{\displaystyle O}{\|}}{C}-O-H$$

Adipic acid
COOH(CH₂)₄COOH

The naming of adipic acid does not follow any of the rules of naming organic acids under the trivial naming system. However, the formula and the structure indicate an organic acid and the name indicates acid. The hazards of the acids, except for flammability, are similiar. The fact that the name indicates acid should lead you to assume flammability and toxicity in addition to corrosiveness, until other information is known.

## DILUTION VS. NEUTRALIZATION

Dilution and neutralization are often tactics considered when dealing with spills of corrosive materials. Dilution involves placing water into the acid to reduce the pH level. The addition of water to a corrosive can create a very dangerous chemical reaction. Acids are very water reactive, creating vapors, heat, and splattering. With dilution you must consider the exponential values of the numbers on the pH scale. Just moving the pH from 1 to 2 on the scale will take an enormous amount of water. Dilution may not be a practical approach for large spills. For example, a 2000-gal spill of concentrated hydrochloric acid occurs. To have enough water to dilute the material to a pH of 6 would require the following efforts: one 1000-gpm pumper, pumping 24 hours a day, 7 days a week, 365 days a year, for 64 years! This would produce 1,440,000 gal of water per day! A large reservoir would be required to hold the water. As the process proceeds, it would become necessary to stir the mixture of water and acid to ensure uniformity in the dilution process. Dilution may work on small spills, but it will not work well on large spills.

Neutralization involves a chemical reaction that works very well under laboratory conditions, using small amounts of acids and bases. However, in the field, facing a large spill of a corrosive material, neutralization may not be feasible. The neutralization reaction will require a large amount of neutralizing agent. For the same spill of 2000 gallons of concentrated hydrochloric acid mentioned in the previous example, it would require 8.7 tons of sodium bicarbonate, 5.5 tons of sodium carbonate, or 4.15 tons of sodium hydroxide to neutralize the spill. The latter would not be recommended, because sodium hydroxide is a strong base and would be dangerous to work with by itself without trying to add it to a concentrated acid. There would also be a need for a method to apply the neutralizing agent. The reaction that occurs will be a violent one, producing heat, vapor, and splattering of product. Neutralization may not work well for emergency responders at the scene of an incident with a large spill. The method of choice may turn out to be one of cleaning up the product by a hazardous waste contractor. They may use vacuum trucks; absorbent, gelling materials; or neutralization to accomplish the task.

The main danger of corrosive materials to responders is the contact of these materials with the human body. Corrosive materials destroy living tissue. The destruction begins immediately upon contact with the corrosive material. Many of the strong acids and bases will cause severe damage upon contact with the skin. Weaker corrosives may not cause noticeable damage for several hours after exposure. A chemical burn is nine times more damaging than a thermal burn. There are four basic methods of reducing the chemical action of corrosives on the skin: physical removal, neutralization, dilution, and flushing. Flushing is the method of choice for corrosive materials. Removal of the material is difficult to accomplish and may leave a residue behind. Neutralization is a chemical reaction that may be violent and produce heat. This type of reaction on body tissues may cause more damage than it prevents. Neutralization should not be attempted on personnel wearing chemical suits, for the same reason as mentioned above. The layer of chemical protection is very thin and the heat from the neutralization may melt the suit and cause burns to the skin below the suit. Dilution takes a large amount of water to lower the pH to a neutral position. While dilution may be similar to flushing, the intended outcome is different. With dilution the goal is to reduce the pH number to as near neutral as possible. With flushing the goal is to remove as much of the material as possible with a large volume of water. Flushing is by far the method of choice and should be started as soon as possible to reduce the amount of chemical damage. Flushing should continue for a minimum of 15 min. This also applies to the eyes. Most corrosives are very water soluble. Contact lenses should not be worn by personnel at HazMat incident scenes. Contact with acids can "weld" the contact to the eye, which almost always produces blindness. The person being treated may be in a great deal of pain and may have to be restrained during the flushing operation. Treatment after flushing involves standard first-aid for burns.

Corrosives are transported in MC/DOT 312/412 tanker trucks. These trucks have a small-diameter tank with heavy reinforcing rings around the circumference of the tank. The tank diameter is small because most corrosives are very heavy.

MC/DOT 312/412 tanker for heavy corrosive materials.

No other type of hazardous material is carried in this type of tanker. The 312/412 is a corrosives tanker regardless of how it is placarded. The placard may indicate a poison, an oxidizer, or a flammable; but do not forget the "hidden hazard": the tank identifies corrosives. Lighter corrosives may also be found in MC/DOT 307/407 tankers and may be placarded corrosive, flammable, poison, and oxidizer. Corrosives may also be found in rail tank cars, intermodal containers, and varying sizes of portable containers. Portable containers may range from pint and gallon glass bottles to stainless steel carboys and 55-gal drums. Some are also shipped in plastic containers.

## CHEMICAL RELEASE STATISTICS

There are two corrosive materials on the EPA list of Top 15 chemical releases. Sulfuric and hydrochloric acids are numbers 3 and 5, respectively. There were over 2300 incidents of releases of sulfuric acid in 1994, involving over 39 million lbs. Hydrochloric acid was released during 1500 accidents, involving over 9 million lbs. Class 8 corrosives accounted for 20% of all tank car leaks in rail transportation. There are five corrosives in the railroad Top 25 commodities: sodium hydroxide (3) and sulfuric acid (4) are among the seven consistently highest-volume hazardous materials shipped by rail; the remaining three corrosives in the Top 25 include phosphoric acid (12), hydrochloric acid (16), and adipic acid (24). In 1994, there were 426 leaks from tank cars involving Class 8 corrosive liquids. Corrosives are the number one material involved in hazardous materials incidents (after flammable liquids and gases), according to the DOT. In 1995, over 4200 incidents involving 12 different corrosive materials occurred, resulting in 123 injuries. The corrosive materials involved in the incidents included corrosive liquids N.O.S.,* sodium hydroxide, hydrochloric acid, sulfuric acid, phosphoric acid, potassium hydroxide, compound cleaning liquid, ammonia solutions, caustic alkali liquid N.O.S., hypochlorite solution, and battery fluid acid.

## INCIDENTS

Emergency responders should have a thorough knowledge of corrosive materials. After flammable liquids and gases, corrosives are the most frequently encountered hazardous materials. Responders should have proper chemical protective equipment and self-contained breathing apparatus to deal safely with corrosive materials. Firefighter turnouts will not provide protection from corrosives. The most common exposures are contact with the hands and feet and inhalation of the vapors. Make sure that the chemical suits chosen for use are compatible with the corrosive material. No suit will protect you from chemicals indefinitely; they all have breakthrough times. Make sure personnel are rotated

---

* Not otherwise specified.

to avoid prolonged exposure and make sure they do not contact the material unless absolutely necessary. Safety should be your number one concern.

In California, an MC/DOT 312/412 tanker truck developed a leak along an interstate highway. On arrival, responders found a reddish-brown vapor cloud coming from the tank. The shipping papers indicated that the load was spent sulfuric acid; however, the color of the vapor coming from the trailer was in conflict with that information. As it turns out, the driver was hauling spent sulfuric acid but had room to pick up some nitric acid and put it in the same tank with the sulfuric acid. The nitric acid was not compatible with the tank and ate through it very quickly. The entire load of acid was spilled onto the highway when the tank failed. Certain hazardous materials have specific colors and responders should be aware of these colors.

A tank car placarded "empty", which contained an estimated 800 gal of anhydrous hydrogen fluoride, a corrosive liquid, was found leaking in a railyard. "Empty" or "residue" placarded tank cars, as they are now called, can have as much as 3000 gallons of product still in the tank if it has not been purged. Responders attempted to control the leak over a 4-hour period. In the meantime, a vapor cloud formed and traveled approximately 2.5 miles downwind. This forced the evacuation of 1500 people from a 1.1 sq. mile radius around the leaking tank car for 9 hours. Local hospitals treated approximately 75 people for minor skin and eye irritations.

## REVIEW QUESTIONS

1. Corrosive materials are actually two very different chemicals, known as _____ and _____.

2. There are two types of acids, _____ and _____.

3. The strength of an acid refers to the _____ of the solution resulting from a gas being passed through water.

4. The concentration refers to the amount of acid mixed with a certain amount of _____, and is usually expressed in terms of _____.

5. A hazardous material with a pH of 7.0 is considered _____.

6. A hazardous material with a pH of 2.3 is considered _____.

7. A hazardous material with a pH of 8.7 is considered _____.

8. Two methods of reducing the corrosive effects of an acid are _____ and _____.

9. Provide the names and structures for the following organic acids.

   CH$_3$COOH          C$_2$H$_5$COOH          C$_3$H$_7$COOH

10. List the formulas and names for the following organic acid structures.

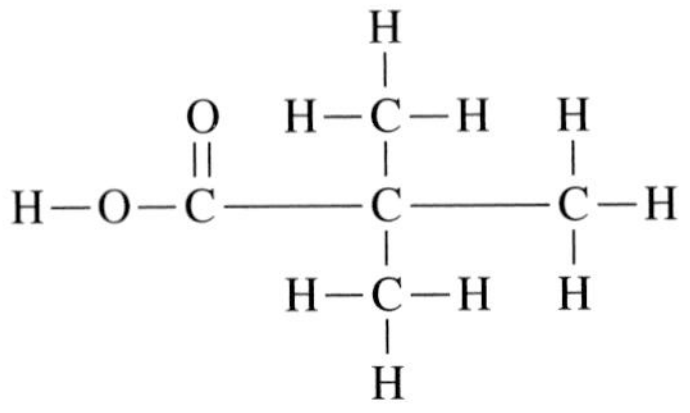

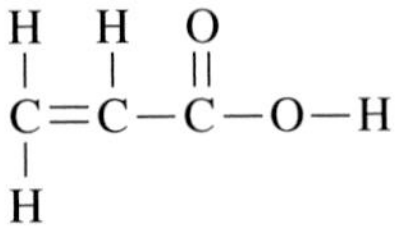

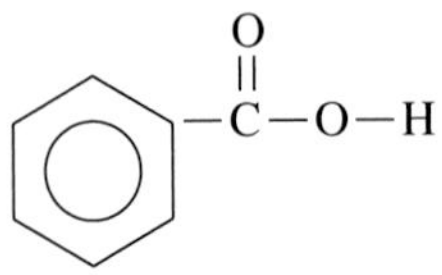

# 10      MISCELLANEOUS HAZARDOUS MATERIALS

Miscellaneous hazardous materials in DOT/UN Class 9 are defined as "a material which presents a hazard during transportation, but which does not meet the definition of any other hazard class." Other hazards might include anesthetics, noxious (harmful to health), elevated temperature, hazardous substance, hazardous waste, or marine pollutant. They may be encountered as solids of varying configurations, gases, and liquids. Examples are asbestos, dry ice, molten sulfur, and lithium batteries. These materials would be labeled and placarded with the Class 9 Miscellaneous Hazardous Materials placard, which is white with seven vertical black stripes on the top half.

Also included in the miscellaneous hazardous materials class are "Other Regulated Materials ORM-D, Consumer Commodities." They are "materials that present a limited hazard during transportation due to the form, quantity, and packaging." Some of these materials, if they were shipped in tank or box truck quantities, would fit into another hazard class. However, because the individual packaging quantities are so small, the hazard is considered limited by the DOT and they are labeled ORM-D. Generally, these ORM-D materials are destined for use in the home, industry, and institutions. The materials are in small containers, including aerosol cans, with a quantity that is usually a gallon or less. Caution should be observed if fire is involved in an incident, since many small containers can become projectiles as pressure builds up inside from the heat and the containers explode. Aerosol cans can be particularly dangerous because they are already pressurized and exposure to heat can cause them to explode and rocket from the pressure. Those materials used in industry and institutions are usually service products used in cleaning and maintenance rather than in industrial chemical processes. Examples of ORM-D materials include low concentration acids, charcoal lighter, spray paint, disinfectants, and cartridges for small firearms. Even though the container sizes may be small, the products inside can still cause contamination of responders or death and injury if not handled properly.

There is no one specific hazard that can be attributed to Class 9 materials. The hazards will vary and may include all of the other eight hazard classes. The

Intermodal container of miscellaneous hazardous materials.

physical and chemical characteristics mentioned in the first nine chapters of this book may be encountered with Class 9 materials. The difference is that the quantities may be very small or the materials may be classified as hazardous wastes, which can be almost any of the other hazard classes. With miscellaneous hazardous materials, it is very important to obtain more information about the shipment to determine the chemical names and the exact hazards of the materials involved.

Class 9 placards on transportation vehicles may include a four-digit UN identification number. The corresponding information in the *North American Emergency Response Guide (NAERG)* may not give detailed names of the materials. There may be generalizations such as "hazardous substance, n.o.s." When the material is not specifically identified in the *NAERG,* the shipping papers or other source will have to be consulted to determine the exact hazard of the shipment.

## ELEVATED TEMPERATURE MATERIALS

In addition to the Class 9 placard, a second placard may appear next to it with the word HOT. The word may also appear outside of a placard by itself. It indicates that the material inside has an elevated temperature that may be a hazard to anyone who comes in contact with it. An elevated temperature material is usually a solid that has been heated to the point that is melts and becomes a molten liquid. The change is in physical state only; the chemical characteristics of the material remain the same. There may, however, be vapors produced from molten materials that may not be present in the solid form. These vapors may be

Electrical transformers were once a primary source of PCBs, a miscellaneous hazardous material.

flammable or toxic. Water in contact with molten materials can cause a violent reaction and instantly turn to steam. If this happens inside a container, the pressure build-up from the steam can cause a boiler-type explosion that has nothing to do with the characteristics of the chemical inside. The steam, which is a gas, builds up pressure inside what is usually a nonpressure container. When the container can no longer withstand the pressure, it fails. The molten material inside may be splattered around by the explosion.

There are two molten materials specifically listed in the Hazardous Materials Tables in 49 CFR: molten aluminum and molten sulfur. **Molten aluminum** has a four-digit UN identification number of 9260. The *NAERG* refers to Guide 169 for hazards of the material. Molten aluminum is the only material that refers to this guide. The guide indicates that the material is above 1300°F and will react violently with water, which may cause an explosion and release a flammable gas. The molten material in contact with combustible materials may cause ignition, if the molten material is above the ignition temperature of the combustible material. For example, gasoline has an average ignition temperature of around 800°F. Diesel fuel has an average ignition temperature around 400°F, depending on the blend and additives. In an accident, gasoline or diesel fuel could be spilled. The molten material could be an ignition source for the gasoline or diesel fuel with which it came in contact. When contacting concrete on a roadway or at a fixed facility, molten materials could cause spalling and small pops. This could cause pieces of concrete to become projectiles. Contact with the skin would cause severe thermal burns. There is no personnel protective clothing that adequately protects responders from contact with molten materials.

Other molten materials are not as hot as molten aluminum. **Sulfur,** in the molten state, refers you to Guide 133 in the *NAERG*. Molten sulfur, four-digit UN identification number 2448, may ignite combustible materials that it comes in contact with, if it is above the ignition temperature of the material. Molten sulfur has a melting point of approximately 245°F. The molten sulfur in transportation would be above that temperature, but not as hot as molten aluminum. Contact would still cause severe thermal burns and the vapor is toxic. Hot asphalt in the liquid form can also cause combustion of combustible materials and severe thermal burns. Asphalt refers you to Guide 130 in the *NAERG* for hazard information. Asphalt has a boiling point of >700°F and a flash point of >400°F. The ignition temperature is 905°F. Fires involving asphalt should be fought with care. Water may cause frothing, as it does with all combustible liquids with flash points above 212°F. This does not mean that water should not be used, but be aware that the frothing may be violent and the water contacting the molten material may also cause a reaction. Asphalt has a four-digit UN identification number of 1999 for all forms. The NFPA 704 designation for asphalt is health 0, flammability 1, and reactivity 0.

## OTHER MISCELLANEOUS HAZARDOUS MATERIALS

White, gray, green, brown, and blue **asbestos** are impure magnesium silicate minerals that occur in fibrous form. Asbestos is noncombustible and was used extensively as a fire-retardant material until it was found to cause cancer. Asbestos is highly toxic by inhalation of dust particles. The four-digit UN identification number for white asbestos is 2590. The primary uses of asbestos are in fireproof fabrics, brake linings, gaskets, reinforcing agent in rubber and plastics, and as a cement reinforcement. Many uses of asbestos are being banned because of the cancer danger of the material.

**Ammonium nitrate fertilizers** that are not classified as oxidizers are classified as miscellaneous hazardous materials. This type of fertilizer has other materials in the mixture and there are controlled amounts of combustible materials. Mixtures of ammonium nitrate, nitrogen, and potash that are not more than 70% ammonium nitrate and do not have more than 0.4% combustible material are included as a miscellaneous hazardous material. Additionally, ammonium nitrate mixtures with nitrogen and potash, with not more than 45% ammonium nitrate, may have combustible material that is unrestricted in quantity. The four-digit UN identification number for these mixtures of ammonium nitrate fertilizer is 2071.

**Solid carbon dioxide,** also known as dry ice, presents a danger in transport because of the carbon dioxide gas produced as it warms. This warming is much like melting ice, although no liquid is formed in the case of dry ice. Dry ice sublimes and goes directly from a solid to a vapor or gas without becoming a liquid. While carbon dioxide gas is nontoxic, it is an asphyxiant and can displace oxygen in the air or in a confined space. It is nonflammable; in fact, carbon

Box truck with overpack drum underneath, usually hauling hazardous waste.

dioxide gas is used as a fire-extinguishing agent. The four-digit UN identification number for solid carbon dioxide is 1845.

**Solutions of formaldehyde,** 30 to 50%, such as those used in preservatives, are listed as miscellaneous hazardous materials. These solutions are nonflammable and the toxicity is below the requirements for a poison liquid. However, the material may still be carcinogenic. Formaldehyde solutions usually contain up to 15% methanol to retard polymerization. The four-digit UN identification number for nonflammable solutions is 2209.

**Polychlorinated biphenyls (PCBs)** are composed of two benzene rings attached together with at least two chlorine atoms in the compound. PCBs were widely used in industry since 1930 because of their stability; however, it was this same stability that led to their downfall. They are highly toxic, colorless liquids with a specific gravity of 1.4 to 1.5, which is heavier than water. They are known carcinogens. In the human body, they tend to settle in the liver and fat cells where they stay for a long period of time. They are not biodegraded and remain as an ecological hazard through water pollution. The only known way to remove PCBs from the environment is high temperature incineration (at least 2200°F) for a proper length of time. The manufacture was discontinued in the United States in 1976. The material that remains is considered hazardous waste and is shipped as a miscellaneous hazardous material. According to EPA statistics, from 1988 to 1992 PCBs accounted for 3586 accidental releases, resulting in 34 deaths. This was the number 1 chemical involved in accidental releases during that time period. The structure for PCB is shown in the following example:

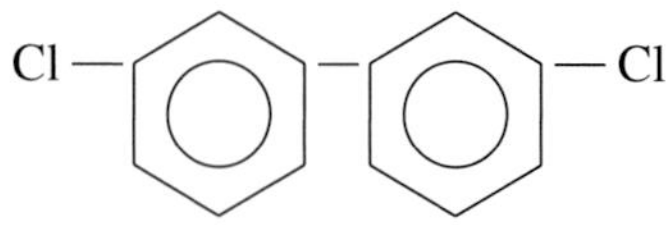

Cl —⬡—⬡— Cl

Polychlorinated biphenyl

**Batteries containing lithium** are listed as miscellaneous hazardous materials. The storage batteries are composed of lithium, sulfur, selenium, tellurium, and chlorine. These batteries have four-digit UN identification numbers assigned depending on the use and composition of the battery. Lithium batteries contained in some kind of equipment has the four-digit number 3091. Batteries with liquid or solid cathodes, not in any kind of equipment, are given the number 3090.

Other miscellaneous hazardous materials listed in CFR 49 Hazardous Material Tables include solid materials and fish meal or fish scrap that has been stabilized. Fish meal is subject to spontaneous heating. These materials are given the four-digit UN identification number of 2216. Castor beans, meal, or flakes may also undergo spontaneous heating. The four-digit UN identification number is 2969. Additional materials include cotton (wet) 1365, polystyrene beads 2211, life-saving appliances, self-inflating 2990, and not self-inflating 3072, environmentally hazardous liquids or substances 3077, hazardous waste liquid 3082, and hazardous waste solid 3077. Additionally, self-propelled vehicles, including internal combustion engines or other apparatus containing internal combustion engines, or electric storage batteries are regulated. Self-propelled vehicles include electric wheelchairs with spillable or nonspillable batteries.

Highway tanker of molten sulfur.

## TOP 50 INDUSTRIAL CHEMICALS

Several of the Top 50 industrial chemicals do not fit into the other eight hazard classes. The hazards may not be as significant as the other classes; however, because of the physical quantities, type of container, or physical form they may present responders with concern in an incident. Therefore, the remaining materials will be listed here to round out the Top 50.

**Titanium dioxide, $TiO_2$,** is the forty-second most produced industrial chemical, with 2.77 billion lbs. in 1995. It is a white powder and has the greatest hiding power of all white pigments. It is noncombustible; however, it is a powder and, when suspended in air, may cause a dust explosion if an ignition source is present. It is not listed in the DOT hazardous materials table and is not considered hazardous in transportation by the DOT. The primary uses are as white pigment in paints, paper, rubber, and plastics, in cosmetics, welding rods, and in radioactive decontamination of the skin.

**Sodium silicate, $2Na_2OSiO_2$,** also known as water glass, is the forty-sixth most produced industrial chemical, with 2.25 billion lbs. in 1995. It is the simplest form of glass. It is found as lumps of greenish glass soluble in steam under pressure, white powders of varying degrees of solubility, or liquids cloudy or clear. It is noncombustible; however, when the powdered form is suspended in air, it could cause a dust explosion if an ignition source is present. Breathing the dust may also cause health problems. The glass form could also create a hazard to responders in an accident. It is not listed as a hazardous material in the DOT Hazardous Materials Tables. The primary uses are as catalysts, soaps, adhesives, water treatment, bleaching, waterproofing, and as a flame retardant.

**Bisphenol A, $(CH_3)_2C(C_6H_4OH)_2$,** is the fiftieth most produced industrial chemical, with 1.62 billion lbs. in 1995. It is made up of white flakes that have a mild phenolic odor. It is insoluble in water. Bisphenol A is combustible, with a flash point of 175°F. It is not listed in the DOT Hazardous Materials Tables. It is used in the manufacture of epoxy, polycarbonate, polysulfone, and polyester resins, as a flame retardant, and as a fungicide. The structure is illustrated in the following example:

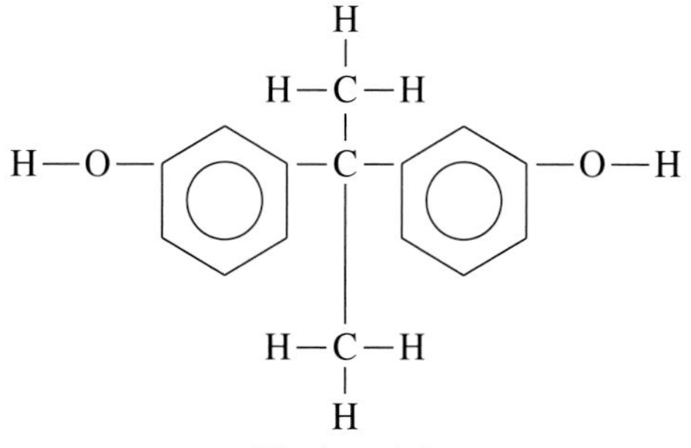

Bisphenol A
$(CH_3)_2C(C_6H_4OH)_2$

**Urea (carbamide), CO(NH$_2$)$_2$,** is the seventeenth most produced industrial chemical, with 15.59 billion lbs. in 1995. It is composed of white crystals or powder, almost odorless, with a saline taste. It is soluble in water and decomposes before reaching its boiling point. Urea is noncombustible. The primary uses of urea are in fertilizers, animal feed, plastics, cosmetics, flameproofing agents, pharmaceuticals, and as a stabilizer in explosives. Urea appears to be both a ketone and an amine by structure and molecular formula; however, it is neither, nor does it have any of the characteristics of either family. The structure is shown in the following illustration:

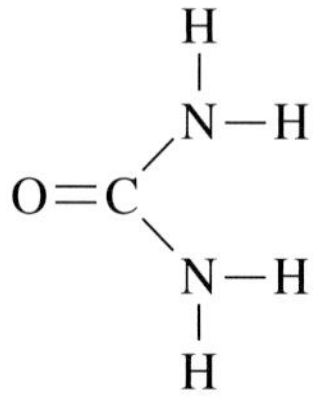

$$
\begin{array}{c}
\text{H} \\
| \\
\text{N}-\text{H} \\
\diagup \\
\text{O}=\text{C} \\
\diagdown \\
\text{N}-\text{H} \\
| \\
\text{H}
\end{array}
$$

Urea
CO(NH$_2$)$_2$

## CHEMICAL RELEASE STATISTICS

The number 1 released EPA Top 15 chemical is polychlorinated biphenyls (PCBs). There were 3586 incidents from 1988 to 1992, resulting in over 19 million lbs. being released into the environment. There are two Class 9 materials in the Top 25 rail commodities shipped: molten sulfur and elevated temperature materials. Molten sulfur is fourth, with over 63,000 tank carloads in 1994. Elevated temperature materials are seventh, with over 57,000 tank carloads. The DOT reports 676 incidents during 1995 involving hazardous wastes shipped under Hazard Class 9. Additionally, another 373 incidents occurred involving miscellaneous hazardous materials identified as formaldehyde solutions and environmentally hazardous solids and liquids. Ten injuries occurred with miscellaneous hazardous materials, two of them involving elevated temperature materials.

## INCIDENTS

Hot materials, such as asphalt, can cause serious thermal burns if contacted with parts of the body. The television show "Rescue 911" highlighted a rescue operation that involved hot asphalt. A dump truck being used to haul solid hot asphalt to patch holes in the road collided with a car. As a result of the collision, the load of hot asphalt was dumped into the car, covering the driver to the point he could not escape. Before rescuers could remove the driver, he suffered severe second- and third-degree thermal burns from the hot asphalt. This same hazard exists with all elevated temperature materials. Responders should work carefully

Railcar with molten sulfur, a Class 9 miscellaneous hazardous material.

Highway tanker used to haul molten asphalt with "HOT" marking.

around transportation vehicles that have the HOT placard or the word "HOT" on the container. Miscellaneous hazardous materials can expose emergency responders to a wide variety of hazards. The placard itself does not indicate what those hazards may be. Do not treat this hazard class lightly; as there are some materials that can cause injury or death to responders if not handled properly.

In Benicia, CA, a truck pulling two tank trailers of molten sulfur was involved in a collision on the Benicia–Martinez Bridge. One of the tanks ruptured, spilling the molten sulfur onto two other vehicles. The truck driver died along with a passenger in one of the cars; another passenger was severely burned. At the time of the accident, molten sulfur was not regulated as a hazardous material. The molten sulfur produced sulfur dioxide vapors, which hampered visibility along with fog.

Hazardous materials, regardless of class, almost always have multiple hazards. It is important for emergency responders to recognize that the hazard classes indicate only the most severe hazard of the materials as determined by the DOT. Research has to be conducted with all hazardous materials, even if the hazard class is known. The correct chemical name must be identified and all of the associated hazards evaluated before tactics are determined. Responders should have a thorough understanding of the physical and chemical characteristics of hazardous materials, including parameters of combustion, water and air reactivity, incompatibilities with other materials, and the effects of temperature and pressure on hazardous materials. Emergency responders should have the same level of understanding of hazardous materials as they do for firefighting, EMS protocols, and law enforcement procedures respectively. All emergency response incidents have the potential to involve hazardous materials; your knowledge of the physical and chemical characteristics of hazardous materials will help ensure a safe outcome to incidents.

## REVIEW QUESTIONS

1. Elevated temperature materials may have the word _____ on a placard or on the tank itself.

2. ORM-D materials are also known as _____.

3. A Class 9 hazardous material is one that presents a hazard during transportation but does not fit into any of the _____.

# GLOSSARY

**Absorption:** A route of exposure. It occurs when a toxic material contacts the skin and enters the bloodstream by passing through the skin.

**Accidental explosion:** An unplanned or premature detonation/ignition of explosive/incendiary material or material possessing explosive properties. The activity leading to the detonation/ignition had no criminal intent. Primarily associated with legal, industrial, or commercial activities.

**Acid:** 1. Any of a class of chemical compounds whose aqueous solutions turn litmus paper red (have a pH less than 7) or react with and dissolve certain metals or react with bases to form salts. 2. A compound capable of transferring a hydrogen ion in solution. 3. A molecule or ion that combines with another molecule or ion by forming a covalent bond with two electrons from other species.

**Acid, corrosive:** A material that usually contains an $H^+$ ion and is capable of dehydrating other materials.

**Acute exposure:** The adverse effects resulting from a single dose or exposure to a material. Ordinarily used to denote effects observed in experimental animals.

**Acute toxicity:** Any harmful effect produced by a single short-term exposure that may result in severe biological harm or death.

**Aerosol:** The dispersion of very fine particles of a solid or liquid in a gas, fog, foam, or mist.

**Alcohol foam:** A type of foam developed to suppress ignitable vapors on polar solvents (those miscible in water). Examples of polar flammable liquids are alcohols and ketones.

**Alkaline:** Any compound having the qualities of a base. Simplified, a substance that readily ionizes in aqueous solution to yield hydroxyl ($OH^-$) anions. Alkalis have a pH greater than 7 and turn litmus paper blue.

**Alpha particle:** A form of ionizing radiation that consists of two protons and neutrons.

**Ambient temperature:** The normal temperature of the environment.

**ANFO:** An ammonium nitrate and fuel oil mixture, commonly used as a blasting agent. The proportions are determined by the manufacturer or user. It is commonly mixed with the addition of an "enhancer", such as magnesium or aluminum, to increase the rate of burn.

**Anhydrous:** Describes a material that contains no water (water-free).

**Anion:** A negatively charged ion that moves toward the anode (+ terminal) during electrolysis. Oxidation occurs at the anode.

**Antidote:** A material administered to an individual who has been exposed to a poison in order to counteract its toxic effects.

**Asphyxia:** Lack of oxygen and interference with oxygenation of the blood. Can lead to unconsciousness.

**Asphyxiant:** A vapor or gas that can cause unconsciousness or death by suffocation (lack of oxygen). Most simple asphyxiants are harmful to the body when they become so concentrated that they reduce (displace) the available oxygen in air (normally about 21%) to dangerous levels (18% or lower). Chemical asphyxiants, like carbon monoxide (CO), reduce the blood's ability to carry oxygen or, like cyanide, interfere with the body's utilization of oxygen.

**Asphyxiation:** Asphyxia or suffocation. Asphyxiation is one of the principal potential hazards of working in confined spaces.

**Atmospheric container:** A type of container that holds products at atmospheric pressure (760 mm).

**Atom:** The smallest unit into which a material may be broken by chemical means. In order to be broken into any smaller units, a material must be subjected to a nuclear reaction.

**Atomic weight (at. wt.):** The relative mass of an atom. Basically, it equals the number of protons plus neutrons.

**Autoignition:** A process in which a material ignites without any apparent outside ignition source. In the process, the temperature of the material is raised to its ignition temperature by heat transferred by radiation, convection, combustion, or some combination of all three.

**Autoignition temperature:** *See* Ignition temperature.

**Base:** A chemical compound that reacts with an acid to form a salt. The term is applied to the hydroxides of the metals, to certain metallic oxides, and to groups of atoms containing one or more hydroxyl groups (OH⁻) in which hydrogen is replaceable by an acid radical. *See* Alkaline.

**Beta particle:** A form of ionizing radiation that consists of either electrons or positrons.

**Biohazard:** Those organisms that have a pathogenic effect on life and the environment and can exist in normal ambient environments. These hazards can represent themselves as disease germs and viruses.

**Blasting agent:** A material designed for blasting that has been tested in accordance with Sec. 173.114a (49 CFR). It must be so insensitive that there is very little probability of accidental explosion or going from burning to detonation.

**BLEVE:** See Boiling liquid, expanding vapor, explosion.

**Blood asphyxiant:** A chemical that is absorbed by the blood and changes or prevents the blood from flowing or carrying oxygen to cells. An example is carbon monoxide poisoning.

**Boiling liquid, expanding vapor, explosion (BLEVE):** The explosion and rupture of a container caused by the expanding vapor pressure as liquids in the container become overheated.

**Boiling point:** At this temperature, vapor pressure of a liquid now equals the surrounding atmospheric pressure (14.7 psi at sea level).

**BTU:** British Thermal Unit. Amount of heat required to raise 1 lb. of $H_2O$ –1°F at sea level.

**Cation:** A positively charged ion that moves toward the cathode during eletrolysis. Reduction occurs at the cathode.

**Carcinogen:** A material that either causes cancer in humans or, because it causes cancer in animals, is considered capable of causing cancer in humans.

**Caustic:** 1. Burning or corrosive. 2. A hydroxide of a light metal. Broadly, any compound having highly basic properties. A compound that readily ionizes in aqueous solution to yields OH⁻ anions, with a pH above 7, and turns litmus paper blue. *See* Alkaline; Base.

**Cellular asphyxiant:** A material that upon entering the body inhibits the normal function of cells. Examples are CO, hydrogen cyanide, or hydrogen sulfide poisoning.

**Central nervous system (CNS):** In humans, the brain and spinal cord, as opposed to the peripheral nerves found in the fingers, etc.

**Chemical burn:** A burn that occurs when the skin comes into contact with strong acids, strong alkalis, or other corrosive materials. These agents literally eat through the skin and in many cases continue to do damage as long as they remain in contact with the skin.

**Chemical properties:** A property of matter that describes how it reacts with other substances.

**Chemical reaction:** A process that involves the bonding, unbonding, or rebonding of atoms. A chemical change takes place that actually changes substances into other substances.

**Chemical reactivity:** The process whereby substances are changed into other substances by the rearrangement, or recombination, of atoms.

**CHEMTREC:** Chemical Transportation Emergency Center operated by the Chemical Manufacturers Association. Provides information and/or assistance to emergency responders. Chemtrec contacts the shipper or producer of the material for more detailed information, including on-scene assistance when feasible. Can be reached 24 hours a day by calling 800-424-9300.

**CHLOREP:** Chlorine Emergency Plan operated by the Chlorine Institute. A 24-hour mutual aid program. Response is activated by a CHEMTREC call to the designated CHLOREP's geographical sector assignments for teams.

**Chronic:** Applies to long periods of action such as weeks, months, or years.

**Chronic effects:** An adverse health effect on a human or animal body with symptoms that develop slowly or that recur frequently due to the exposure of hazardous chemicals.

**Chronic exposure:** Repeated doses or exposure to a material over a relatively prolonged period of time.

**Closed-cup tester:** A device for determining flash points of flammable and combustible liquids, utilizing an enclosed cup, or container, for the liquid. Recognized types are the Tagliabue (Tag) Closed Tester, the Pensky-Martens Closed Tester, and the Setaflash Closed Cup Tester.

**CNS:** *See* Central nervous system.

**Combustibility:** The ability of a substance to undergo rapid chemical combination with oxygen, with the evolution of heat.

**Combustible dust:** Particulate material that when mixed in air, will burn or explode.

**Combustible liquid:** Term commonly used for liquids that emit burnable vapors or mists. Technically, a liquid whose vapors will ignite at a temperature of 100°F or above.

**Compound:** A substance composed of two or more elements that have chemically reacted. The compound that results from the chemical reaction is unique in its chemical and physical properties.

**Compressed gas:** Any material or mixture having in the container an absolute pressure exceeding 40 psi at 70°F or, regardless of the pressure at 70°F, having an absolute pressure exceeding 104 psi at 130°F; or any liquid flammable material having a vapor pressure exceeding 40 psi absolute at 100°F as determined by testing. Also includes cryogenic or "refrigerated liquids" (DOT) with boiling points lower than –130°F at 1 atmosphere.

**Concentration:** The amount of a material that is mixed with another material.

**Concentration (corrosives):** In corrosives the amount of acid or base compared to the amount of water present. Corrosives have "strength" and "concentration." *See* Strength.

**Contaminant:** 1. A toxic substance that is potentially harmful to people, animals, and the environment. 2. A substance not in pure form.

**Corrosive:** A chemical that causes visible destruction of or irreversible alterations in living tissue by chemical action at the site of contact; a liquid that causes a severe corrosion rate in steel. A corrosive is either an acid or a caustic (a material that reads at either end of the pH scale.)

**Corrosive material (DOT):** A material that causes the destruction of living tissue and metals.

**Covalent bond:** A chemical bond in which atoms share electrons in order to form a molecule.

**Critical pressure:** The pressure required to liquefy a gas at its critical temperature.

**Critical temperature:** The temperature above which a gas cannot be liquefied by pressure.

**Cryogenic burn:** Frostbite; damage to tissues as the result of exposure to low temperatures. It may involve only the skin, extend to the tissue immediately beneath it, or lead to gangrene and loss of affected parts.

**Cryogenic cylinder:** An insulated metal cylinder contained within an outer protective metal jacket. The area between the cylinder and the jacket is normally under vacuum. The cylinders range in size from a Dewier (similar to a small thermos) up to 24 in. in diameter and 5 ft in length. Examples of materials found in these types of cylinders are argon, helium, nitrogen, and oxygen.

**Cryogenic liquid:** A liquid with a boiling point below –130°F.

**Cylinder:** A container for liquids, gases, or solids under pressure. Ranges in size from aerosol containers found at home, such as spray deodorant, to the cryogenic (insulated) cylinders for nitrogen that can be approximately 24 in. in diameter and 5 ft in length. Pressure ranges from a few pounds to 6000 pounds per square inch.

**Dangerous when wet:** Materials that when exposed to water allow a chemical reaction to take place and often produce flammable or poisonous gases, heat, and a caustic solution. An example is sodium.

**Decomposition:** Separation of larger molecules into separate constituent and smaller parts.

**Decomposition (chemical):** A reaction in which the molecules of a chemical break down to its basic elements, such as carbon, hydrogen, or nitrogen, or to more simple compounds. This often occurs spontaneously, liberating considerable heat and often large volumes of gas.

**Decontamination:** The physical and/or chemical process of reducing and preventing the spread of contamination from persons and equipment used at a hazardous materials incident.

**Deflagration:** Explosion, very rapid combustion, up to 1250 feet per second.

**Detonating cord:** A flexible cord containing a center cord of high explosives used to detonate other explosives with which it comes in contact.

**Detonation:** An explosion at speeds above 1250 feet per second and many times over 3300 feet per second.

**Detonator:** Any device containing a detonating charge that is used for initiating detonation in an explosive. This term includes but is not limited to electric and nonelectric detonators (either instantaneous or delayed) and detonating connectors.

**Dewier container:** Small (less than 25 gallons) used for temporary storage or handling of cryogenic liquids.

**Dilution:** The application of water to water-miscible hazardous materials. The goal is to reduce the hazard of a material to safe levels by reducing its concentration.

**Dose:** The accumulated amount of a chemical to which a person is exposed.

**DOT:** U.S. Department of Transportation. Regulates transportation of materials to protect the public as well as fire, law, and other emergency-response personnel.

**Dry bulk:** A type of container used to carry large amounts of solid materials (more than 882 lbs., or 400 kg). It can either be placed on or in a transport vehicle or vessel constructed as an integral part of the transport vehicle.

**Element:** A substance that cannot be broken down into any other substance by chemical means.

**Empirical formula:** Describes the *ratio* of the number of each element in the molecule, but not the exact number of atoms in the molecule.

**Emulsification:** The process of dispersing one liquid in a second immiscible liquid. The largest group of emulsifying agents are soaps, detergents, and other compounds whose basic structure is a paraffin chain terminating in a polar group.

**Endothermic:** A process or chemical reaction that is accompanied by absorption of heat.

**Etiologic agent:** Those living organisms or their toxins that contribute to the cause of infection, disease, or other abnormal condition.

**Evaporation:** The process in which liquid becomes vapor as more molecules leave the vapor than return.

**Exothermic reaction:** A chemical reaction that liberates heat during the reaction.

**Expansion ratio:** The amount of gas produced from a given volume of liquid escaping from a container at a given temperature.

**Explosion:** The sudden and rapid production of gas, heat, and noise and many times a shock wave, within a confined space.

**Explosive (DOT):** Any chemical compound or mixture whose primary function is to produce an explosion.

**Explosives, high:** Explosive materials that can be used to detonate by means of a detonator when unconfined (e.g., dynamite).

**Explosives, low:** Explosive materials that deflagrate rather than detonate (e.g., black powder, safety fuses, and "special fireworks" as defined by Class 1.3 explosives).

**Explosive limits:** *See* Flammable limits.

**Fire point:** The lowest temperature at which the vapor above the liquid will ignite and continue to burn, usually a few degrees above the flash point.

**Flammable gas:** A gas that at ambient temperature and pressure forms a flammable mixture with air at a concentration of 13% by volume or less; or a gas that at ambient temperature and pressure forms a range of flammable mixtures with air greater than 12% by volume, regardless of the lower explosive limit.

**Flammable limits:** The range of the percentages of vapor mixed with air that are capable of being ignited, as opposed to those mixtures that have too much or too little vapor to be ignited. Also called explosive limits.

**Flammable liquid:** A liquid that gives off readily ignitable vapors. Defined by the NFPA and DOT as a liquid with a flashpoint below 100°F (38°C).

**Flammable range:** The percentage of fuel vapors in air where ignition can occur. Flammable range has an upper and lower limit.

**Flammable solid:** A solid (other than an explosive) that ignites readily and continues to burn. It is liable to cause fires under ordinary conditions or during transportation through friction or retained heat from manufacturing or processing. It burns so vigorously and persistently as to create a serious transportation hazard. Included in this class are spontaneously combustible and water reactive materials. Example: white phosphorus.

**Flash back:** The ignition of vapors and the travel of the flame back to the liquid/vapor release source.

**Flash point:** The minimum temperature at which a liquid gives off vapor within a test vessel in sufficient concentration to form an ignitable mixture with air near the surface of the liquid.

**Foam:** Firefighting material consisting of small bubbles of air, water, and concentrating agents. Chemically, the air in the bubbles is suspended in the fluid. The foam clings to vertical and horizontal surfaces and flows freely over burning or vaporizing materials. Foam puts out a fire by blanketing it, excluding air, and blocking the escape of volatile vapor. Its flowing properties resist mechanical interruption and reseal the burning material.

**Formula:** A combination of the symbols for atoms or ions that are held together chemically.

**Freezing point:** The temperature at which a material changes its physical state from a liquid to a solid.

**Frothing:** A foaming action caused when water, turning to steam in contact with a liquid at a temperature higher than the its boiling point (212°F), picks up a part of a viscous liquid.

**Gas:** A formless fluid that occupies the space of its enclosure. It can settle to the bottom or top of an enclosure when mixed with other chemicals. It can be changed to its liquid or solid state only by increased pressure and decreased temperature.

**Halogens:** A chemical family that includes fluorine, chlorine, bromine, and iodine.

**Halon:** Halogenated hydrocarbons (contain the elements F, Cl, Br, or I) used to suppress or prevent combustion.

**Hazard class:** One of nine classes of hazardous materials as categorized and defined by the DOT in 49 CFR.

**HazMat foam:** A special vapor suppressing mix that can be applied to liquids or solids to prevent off-gassing.

**Hepatoxin:** A chemical that is injurious to the liver.

**High-expansion foam:** A detergent-based foam (low water content) that expands at ratios of 1000 to 1.

**High-order explosion:** Materials that require moderate heat and reducing agents to initiate combustion.

**Hypergolic materials:** Materials that ignite on contact with one another.

**Hypergolic reaction:** The immediate spontaneous ignition when two or more materials are mixed.

**IDLH:** Abbreviation for Immediately Dangerous to Life and Health. The maximum level to which a healthy worker can be exposed for 30 min to a chemical and escape without suffering irreversible health effects or escape impairing symptoms.

**Ignition temperature:** The minimum temperature at which a material will ignite without a spark or flame being present. This is also the temperature the ignition source must be.

**Immiscible:** Matter that cannot be mixed. For example, water and gasoline are immiscible.

**Incompatibility:** The inability to function or exist in the presence of something else such as when a chemical will destroy the container.

**Inert:** A material that under normal temperatures and pressures does not react with other materials.

**Inhibited:** A substance that has had another substance added to prevent or deter its reaction either with other materials or itself (polymerization). Usually used to deter polymerization.

**Inhibitor:** A substance that is capable of stopping or retarding a chemical reaction. To be technically useful, it must be effective in low concentration (i.e., to stop polymerization).

**Initiator:** The substance or molecule (other than reactant) that initiates a chain reaction, as in polymerization.

**Inorganic:** Pertaining to or composed of chemical compounds that do not contain carbon as the principal element (except carbonates, cyanides, and cyanates). Matter other than plant or animal.

**Inorganic peroxides:** Inorganic compounds containing an element at its highest state of oxidation (such as sodium peroxide), or having the peroxy group $-O-O-$ (such as perchloric acid).

**Ion:** An atom that possesses an electrical charge, either (+) positive or (−) negative.

**Ionic bond:** A chemical bond in which atoms of different elements transfer (exchange) electrons. As the electrons are exchanged, charged particles known as ions are formed.

**Ionizing radiation:** High-energy radiation, such as an X-ray, that causes the formation of ions in substances through which it passes (gamma rays). Excessive amounts of ionizing radiation will cause permanent genetic or bodily damage.

**Irritant:** A noncorrosive material that causes a reversible inflammatory effect on living tissue by chemical action at the site of contact.

**Lacrimation:** Secretion and discharge of tears.

**LC$_{50}$:** Lethal concentration 50, median lethal concentration. The concentration of a material in air that on the basis of laboratory tests (respiratory route) is expected to kill 50% of a group of test animals when administered as a single exposure in a specific time period.

**LD$_{50}$:** Lethal dose 50. The single dose of a substance that causes death of 50% of an animal population from exposure to the substance by any route other than inhalation.

**Liquid:** A substance that is neither a solid nor a gas; a substance that flows freely like water.

**Low-order explosion:** Materials that require excessive heat and reducing agents to initiate combustion.

**Low-pressure container:** A container designed to withstand pressures from 5 to 100 psi.

**Lower explosive limit:** The lowest concentration of gas or vapor (% by volume in air) that burns or explodes if an ignition source is present at ambient temperatures.

**LOX:** Liquid oxygen.

**Mechanical foam:** A substance introduced into the water line by various means at a 6% concentration. Air is then introduced to yield a foam consisting generally of 90 volumes air, 9.4 volumes water, and 0.6 volumes foam liquid. It uses hydrolyzed soybean, fish scales, hoof and horn meal, and peanut or corn protein as a base.

**Melting point:** The degree of temperature at which a solid substance becomes a liquid, especially under a pressure of one atmosphere.

**Miscible:** Mixable in any and all proportions to form a uniform mixture. Water and alcohol are miscible; water and oil are not.

**Molecular formula:** Shows the *exact number* of each atom in the molecule.

**Molecular weight:** The sum of the atomic weights of all the atoms in a molecule.

**Molecule:** The smallest possible particle of a chemical compound that can exist in the free state and still retain the characteristics of the substance. Molecules are made up of atoms of various elements that form the compound.

**Monomer:** A simple molecule capable of combining with a number of like or unlike molecules to form a polymer. It is a repeating structure unit within a polymer.

**Mutagen:** A material that induces genetic changes (mutations) in the DNA of chromosomes. Chromosomes are the "blueprints" of life within individual cells.

**Necrosis:** Cell or tissue death due to disease or injury.

**Neutralization:** A chemical reaction used to remove $H^+$ ions from acidic solutions and $OH^-$ ions from basic solutions. The reaction can be violent and usually produce water, a salt, heat, and many times a gas.

**Normal:** A solution that contains 1 equivalent of solute per liter of solution.

**NRC:** National Response Center, a communications center for activities related to response actions, is located at Coast Guard headquarters in Washington, D.C. The toll-free number, (800)424-8802, can be reached 24 hours a day for reporting actual or potential pollution incidents.

**NRT:** National Response Team, consisting of representatives of 14 government agencies (DOD, DOI, DOT/RSPA, DOT/USCG, EPA, DOC, FEMA, DOS, USDA, DOJ, HHS, DOL, NRC, and DOE), is the principal organization for implementing the NCP. When the NRT is not activated for a response action, it serves as a standing committee to develop and maintain preparedness, to evaluate methods of responding to discharges or releases, to recommend needed changes in the response organization, and to recommend revisions to the NCP.

**Odor:** A quality of something that affects the sense of smell; fragrance.

**Odor threshold:** The greatest dilution of a sample with odor-free water to yield the least definitely perceptible odor.

**OHMTADS:** Oil and Hazardous Materials Technical Assistance Data System, a computerized data base containing chemical, biological, and toxicological information about hazardous substances. OSCs use OHMTADS to identify unknown chemicals and to learn how to best handle known chemicals.

**Open-cup tester:** A device for determining flash points of flammable and combustible liquids, utilizing an open cup, or container, for the liquid. Recognized types are the Tagliabue (Tag) Open Cup Apparatus and the Cleveland Open Cup Apparatus.

**Organic:** A material that comes from living plants or animals, such as waste or decay products. Distinguished from mineral matter. Organic chemistry deals with materials that contain the element carbon (C).

**Organic peroxides:** Any organic compound containing oxygen (O) in the bivalent –O–O– structure and that may be considered a derivative of hydrogen peroxide, where one or more of the hydrogen atoms have been replaced by organic radicals.

**Oxidizer:** A material that gives up oxygen easily, removes hydrogen from another compound, or attracts negative electrons (such as chlorine or fluorine), enhancing the combustion of other materials.

**Oxidizing agent:** A material that gains electrons during combustion. The electrons are gained from the fuel.

**Oxygen deficient:** Defined by OSHA as ambient air containing less than 19.5% oxygen concentration.

**Oxygen enriched:** Defined by OSHA as ambient air containing above 24% oxygen concentration.

**PEL (permissible exposure limit):** Term used by OSHA for its health standards covering exposures to hazardous chemicals. PEL generally relates to legally enforceable TLV limits.

**pH:** The "power of hydrogen." A measure of the acidity or basicity of a solution, that is, of the concentration of $H^+$ or $OH^-$ ions in solution. Scale ranges from 0 to 14, where a reading of 7 is neutral.

**Physical properties:** A property of matter that describes only its condition, not the way it reacts with other substances. Examples are size, density, color, and electrical conductivity.

**Poison:** Any substance (solid, liquid, or gas) that by reason of an inherent deleterious property tends to destroy life or impair health.

**Polar solvent liquids:** Those liquids that mix (are miscible with water).

**Polymer:** A giant, long chain of molecules having extremely high molecular weights made up of many repeating smaller units called monomers or comonomers.

**Polymerization:** A chemical reaction in which small molecules combine to form larger molecules. A hazardous polymerization is a reaction that takes place at a rate that releases large amounts of energy that can cause fires or explosions or burst containers. Materials that can polymerize usually contain inhibitors that can delay the reaction.

**Powder:** A solid reduced to dust by pounding, crushing, or grinding.

**ppm (parts per million):** Parts of vapor or gas per million parts of contaminated air by volume at 25°C and 1 torr pressure.

**Pressure vessel:** A tank or other container constructed so as to withstand interior pressure greater than that of the atmosphere.

**psi:** Pounds per square inch.

**Pyrolysis:** A chemical decomposition or breaking apart of molecules produced by heating in the absence of air.

**Pyrophoric:** Material that ignites spontaneously in air below 130°F (54°C). Occasionally caused by friction.

**Pyrophoric gas:** Gaseous materials that spontaneously ignite when exposed to air under ambient conditions. Example: trimethyl aluminum.

**Pyrophoric liquid:** Liquid materials that spontaneously ignite when exposed to air under ambient conditions.

**Pyrophoric solid:** Solid materials that spontaneously ignite when exposed to air under ambient conditions. An example is phosphorus.

**RAD:** Radiation absorbed dose.

**Radiation:** Ionizing energy, either particulate or wave, that is spontaneously emitted by a material or combination of materials.

**Radioactive material (DOT):** Materials that emit ionizing radiation.

**Radioactivity:** Any process by which unstable nuclei increase their stability by emitting particles (alpha or beta) or gamma rays.

**Rate of explosion:** Rate of decomposition measured in feet per second in relation to the speed of sound. If subsonic, the rate is described as a deflagration. If supersonic, the rate of decomposition is defined as a detonation.

**Reducing agent:** A substance that gives electrons to (and thereby reduces) another substance.

**Respiratory asphyxiant:** A material that prevents or reduces the available oxygen necessary for normal breathing. Divided into simple and chemical asphyxiants.

**RCRA:** Resource Conservation and Recovery Act (of 1976) established a framework for the proper management and disposal of all wastes.

**Roentgen:** The amount of ionization that occurs per cubic centimeter of air.

**Routes of exposure:** Ways in which chemicals get in contact with or enter the body. These are inhalation, absorption, ingestion, or injection.

**RRT:** Regional response teams composed of representatives of Federal agencies and a representative from each state in the Federal region.

**SADT:** See self-accelerating decomposition temperature.

**Safety fuse:** A flexible cord containing an internal burning medium by which fire or flame is conveyed at a uniform rate from point of ignition to point of use, usually a detonator.

**Safety relief valve:** A safety relief device containing an operating part that is held normally in a position closing a relief channel by spring force and is intended to open and close at a predetermined pressure.

**SARA:** The Superfund Amendments and Reauthorization Act of 1986. Title III of SARA includes detailed provisions for community planning.

**Self-accelerating decomposition temperature (SADT):** Organic peroxides or other synthetic chemicals that decompose at ambient temperature, or react to light or heat, resulting in a chemical breakdown. This releases oxygen, energy, and fuel in the form of rapid fire or explosion. To ensure stabilization, these materials must be kept in a dark and/or refrigerated environment.

**Sensitizer:** A substance that on first exposure causes little or no reaction in humans or test animals, but that on repeated exposure may cause a marked response not necessarily limited to the contact site.

**Simple asphyxiant:** A material that replaces the amount of oxygen admitted into the body without further damage to tissue or poisoning. Examples are nitrogen and carbon dioxide.

**Slurry:** A pourable mixture of solid and liquid.

**Solid:** The state of matter having definite volume and rigid shapes. Its atoms or molecules are restricted to vibration only.

**Solubility:** The ability of a substance to form a solution with another substance.

**Solution:** The even dispersion (mixing) of molecules of two or more substances. The most commonly encountered solutions involve mixing of liquids and liquids or solids and liquids.

**Solvent:** A substance, usually a liquid, capable of absorbing another liquid, gas, or solid to form a homogeneous mixture.

**Specific gravity:** The weight of a solid or liquid substance as compared to the weight of an equal volume of water; specific gravity of water equals 1.

**Spontaneous combustion:** A process by which heat is generated within material by either a slow oxidation reaction or by microorganisms.

**Spontaneous ignition:** Ignition that can occur when certain materials, such as tung oil, are stored in bulk, resulting from the generation of heat, which cannot be readily dissipated; often heat is generated by microbial action.

**Spontaneous ignition temperature:** *See* Ignition temperature.

**States of matter:** Any of three physical forms of matter: solid, liquid, or gas.

**Strength (acid/base):** The amount of ionization that occurs when an acid or a base is dissolved in a liquid.

**Subacute exposure:** 1. Less than acute. 2. Of or pertaining to a disease or other abnormal condition present in a person who appears to be clinically well. The condition may be identified or discovered by means of a laboratory test or by radiologic examination.

**Sublimation:** The direct change of state from solid to vapor.

**Subscripts:** Identify the number of atomic weights of the element present in the molecule.

**Symbol:** Letters used to identify each element. The symbol for an element represents a definite weight (1 atomic weight) of that element.

**Target organ:** The primary organ to which specific chemicals cause harm. Examples are the lungs, liver, or kidneys.

**Temperature:** Measure of the vibratory rate of a molecule.

**Teratogen:** Material that affects the offspring when a developing embryo or fetus is exposed to that material.

**Thermal burn:** Pertaining to or characterized by heat.

**TLV:** Threshold limit value, *estimated* exposure value, below which no ill health effects *should* occur to the individual.

**Toxicity:** The ability of a substance to cause damage to living tissue, impairment of the central nervous system, severe illness, or death when ingested, inhaled, or absorbed by the skin.

**TTL:** Threshold toxic limit, *estimated* exposure value, below which no ill health effects *should* occur to the individual.

**Upper explosive limit:** The maximum fuel-to-air mixture in which combustion can occur.

**Vapor density:** The weight of a vapor or gas compared to the weight of an equal volume of air; an expression of the density of the vapor or gas calculated as the ratio of the molecule weight of the gas to the average molecule weight of air, which is 29. Materials lighter than air have vapor densities less than 1.

**Vapor pressure:** The pressure exerted by a saturated vapor above its own liquid in a closed container. Vapor pressure reported on MSDS are in millimeters of mercury at 68°F (20°C), unless stated otherwise.

**Vapors:** Molecules of liquid in air; moisture such as steam, fog, mist, etc. often forming a cloud suspended or floating in the air, usually due to the effect of heat upon a liquid.

**Violent reaction:** The action by which a chemical changes its composition near or exceeding the speed of sound, often releasing heat and gases.

**Viscosity:** The measurement of the flow properties of a material expressed as its resistance to flow. Unit of measurement and temperature are included.

**Water reactive material:** A material that will decompose or react when exposed to moisture or water.

**Water solubility:** The ability of a substance to mix with water.

# APPENDIX

## APPENDIX A

*This Appendix is not a part of the requirements of this NFPA document, but is included for information purposes only.*

**A-1** This is a system for the identification of hazards to life and health of people in the prevention and control of fires and explosions in the manufacture and storage of materials. The bases for identification are the physical properties and characteristics of materials that are known or can be determined by standard methods. Technical terms, expressions, trade names, etc., are purposely avoided as this system is concerned only with the identification of the involved hazard from a standpoint of safety.

The explanatory material in this Appendix is to assist users of this standard, particularly the person who assigns the degree of hazard in each category.

| Identification of Health Hazard Color Code: BLUE | | Identification of Flammability Color Code: RED | | Identification of Reactivity (Stability) Color Code: YELLOW | |
|---|---|---|---|---|---|
| Signal | Type of Possible Injury | Signal | Susceptibility of Materials to Burning | Signal | Susceptibility to Release of Energy |
| **4** | Materials that on very short exposure could cause death or major residual injury. | **4** | Materials that will rapidly or completely vaporize at atmospheric pressure and normal ambient temperature, or that are readily dispersed in air and that will burn readily. | **4** | Materials that in themselves are readily capable of detonation or of explosive decomposition or reaction at normal temperatures and pressures. |
| **3** | Materials that on short exposure could cause serious temporary or residual injury. | **3** | Liquids and solids that can be ignited under almost all ambient temperature conditions. | **3** | Materials that in themselves are capable of detonation or explosive decomposition or |

Reprinted with permission from NFPA 704-1990, *Identification of the Fire Hazards of Materials,* Copyright © 1990, National Fire Protection Association, Quincy, MA 02269. This reprinted material is not the complete and official position of the National Fire Protection Association, on the referenced subject which is represented only by the standard in its entirety.

| Identification of Health Hazard Color Code: BLUE | | Identification of Flammability Color Code: RED | | Identification of Reactivity (Stability) Color Code: YELLOW | |
|---|---|---|---|---|---|
| Signal | Type of Possible Injury | Signal | Susceptibility of Materials to Burning | Signal | Susceptibility to Release of Energy |
| | | | | | reaction but require a strong initiating source or which must be heated under confinement before initiation or which react explosively with water. |
| **2** | Materials that on intense or continued but not chronic exposure could cause temporary incapacitation or possible residual injury. | **2** | Materials that must be moderately heated or exposed to relatively high ambient temperatures before ignition can occur. | **2** | Materials that readily undergo violent chemical change at elevated temperatures and pressures or which react violently with water or which may form explosive mixtures with water. |
| **1** | Materials that on exposure would cause irritation but only minor residual injury. | **1** | Materials that must be preheated before ignition can occur. | **1** | Materials that in themselves are normally stable, but which can become unstable at elevated temperatures and pressures. |
| **0** | Materials that on exposure under fire conditions would offer no hazard beyond that of ordinary combustible material. | **0** | Materials that will not burn | **0** | Materials that in themselves are normally stable, even under fire exposure conditions, and which are not reactive with water. |

**A-2**   In developing this edition of NFPA 704, the Committee on Fire Hazards of Materials determined that the standard should provide quantitative guidelines for determining the numerical health hazard rating of a material. In addition, the Committee agreed that a "4" or a "3" health hazard rating should be assigned to any material classified as a "Poison-Inhalation Hazard" by the U.S. Department of Transportation (DOT). This classification, "Poison-Inhalation Hazard", was added by DOT from the United Nations (UN) criteria detailed in the UN publication, *Recommendations on the Transport of Dangerous Goods,* 4th Edition — Revised, 1986. (See also Notice of Proposed Rulemaking, *Federal Register,* Vol. 50, p. 5270 et seq., February 7, 1985, and Notice of Final Rule, *Federal Register,* Vol. 50, p. 41092 et seq., October 8, 1985.)

The UN criteria for inhalation toxicity is based upon the $LC_{50}$ and saturated vapor concentration of the material. Furthermore, in addition to inhalation toxicity, the UN has established criteria for oral and dermal toxicity, as well as corrosivity. Based upon these criteria, the UN assigns a given material to categories called Packing Groups I, II, or III. Packing Group I materials represent a severe hazard in transport, Group II materials a serious hazard, and Group III materials a low hazard.

The Committee decided to adopt the UN criteria for toxicity and corrosivity, and correlate Packing Groups I, II, and III with the health hazard ratings "4," "3," and "2," respectively. Adoption of the UN system has several advantages. First, it addresses hazards in transportation, which are similar to the type of emergencies likely to be encountered by fire fighting personnel and emergency responders. Most other hazard ranking systems have been developed for occupational exposures. Secondly, the UN system is well established, and it is presumed that a large number of chemical manufacturers have already classified (or can easily classify materials into the appropriate Packing Groups). Finally, users of chemicals can assign "4," "3," or "2" health hazard ratings by establishing if chemicals have been assigned to UN Packing Groups due to toxicity or corrosivity.

In order to establish "1" and "0" health hazard rankings, the Committee utilized criteria for the "1" and "0" ratings contained in the Hazardous Materials Identification System (HMIS™) developed by the National Paint & Coatings Association (NPCA) (Hazardous Materials Identification System Revised, Implementation Manual, 1981). Although the NPCA criteria were developed for occupational exposure, the "1" and "0" criteria are on the low end of the hazard spectrum and are fairly consistent with, and complementary to, the "4," "3," and "2" ratings based upon the UN criteria. No UN criteria was established for eye irritation, and the Committee adopted NPCA "3," "2," and "1," and "0" criteria as health hazard ratings for eye irritation.

The Committee made a number of revisions to the proposed hazard rating system to provide conformity with existing industrial practice and to recognize limitations and availability of corrosivity and eye irritation into a single "skin/eye contact" category and utilize descriptive terms for the health hazard ratings. Minor changes were made to the "2," "1," and "0" criteria for oral toxicity and to the "1" and "0" criteria for dermal toxicity. Specifically, the distinction between solids and liquids in the oral toxicity criteria was eliminated, and the cutoff between "1" and "0" rankings for oral and dermal toxicity was lowered from 5000 to 2000 mg/kg.

In summary, the "4," "3," and "2" health hazard rankings for oral, dermal, and inhalation toxicity are based primarily on UN criteria. The "1" and "0" health hazard rankings for oral, dermal, and inhalation toxicity, and all of the "skin/eye contact" rankings are based primarily on NPCA criteria.

For the assistance of the user of this standard, the following definitions are quoted from Section 6.5 of Recommendations on the Transport of Dangerous Goods, Fourth Revised Edition, 1986, published by the United Nations, New York, NY.

$LD_{50}$ for acute oral toxicity:

"That dose of the substance administered which is most likely to cause death within 14 days in one half of both male and female young adult albino rats. The number of animals tested shall be sufficient to give a statistically significant result and be in conformity with good pharmacological practice. The result is expressed in milligrams per kilogram of body weight."

$LD_{50}$ for acute dermal toxicity:

"That dose of the substance which, administered by continuous contact for 24 hours with the bare skin of albino rabbits, is most likely to cause death within 14 days in one half of the animals tested. The number of animals tested shall be sufficient to give a statistically significant result and be in conformity with good pharmacological practice. The result is expressed in milligrams per kilogram of body weight."

$LC_{50}$ for acute toxicity on inhalation:

"That concentration of vapor, mist or dust which, administered by continuous inhalation to both male and female young adult albino rats for one hour, is most likely to cause death within 14 days in one half of the animals tested. If the substance is administered to the animals as dust or mist, more than 90 percent of the particles available for inhalation in the test must have a diameter of 10 microns or less, provided that it is reasonably forseeable that such concentrations could be encountered by man during transport. The result is expressed in milligrams per liter of air for dusts and mists or in milliliters per cubic meter of air (parts per million) for vapors."

The following information quoted from Section 6.4 of the above-cited Recommendations also applies:

"The criteria for inhalation toxicity of dusts and mists are based on $LC_{50}$ data relating to 1 hour exposures and where such information is available it should be used. However, where only $LC_{50}$ data relating to 4 hour exposures to dusts and mists are available, such figures can be multiplied by four and the product substituted in the above criteria, i.e. $LC_{50}$ (4 hour) $\times$ 4 is considered equivalent of $LC_{50}$ (1 hour)."

"The criteria for inhalation toxicity of vapors are based on $LC_{50}$ data relating to 1 hour exposures, and where such information is available it should be used. However, where only $LC_{50}$ data relating to 4 hour exposures to dusts and mists are available, such figures can be multiplied by two and the product substituted in the above criteria, i.e. $LC_{50}$ (4 hour) $\times$ 2 is considered equivalent of $LC_{50}$ (1 hour)."

## APPENDIX B

*This Appendix is not a part of the requirements of this NFPA document, but is included for information purposes only.*

The information contained within Appendix B is derived from introductory explanatory material on the 704 system contained within NFPA 49, *Hazardous Chemicals Data*; and NFPA 325M, *Figure Hazard Properties of Flammable Liquids, Gases, and Volatile Solids.* The following paragraphs summarize the meanings of the numbers in each hazard category and explain what a number should tell fire fighting personnel about protecting themselves and how to fight fires where the hazard exists.

## Health

In general, health hazard in fire fighting is that of a single exposure that may vary from a few seconds up to an hour. The physical exertion demanded in fire fighting or other emergency conditions may be expected to intensify the effects of any exposure. Only hazards arising out of an inherent property of the material are considered. The following explanation is based upon protective equipment normally used by fire fighters.

**4** Materials too dangerous to health to expose fire fighters. A few whiffs of the vapor could cause death, or the vapor or liquid could be fatal on penetrating the fire fighter's normal full protective clothing. The normal full protective clothing and breathing apparatus available to the average fire department will not provide adequate protection against inhalation or skin contact with these materials.

**3** Materials extremely hazardous to health but areas may be entered with extreme care. Full protective clothing, including self-contained breathing apparatus, coat, pants, gloves, boots, and bands around legs, arms, and waist should be provided. No skin surface should be exposed.

**2** Materials hazardous to health, but areas may be entered freely with full-faced mask self-contained breathing apparatus that provides eye protection.

**1** Materials only slightly hazardous to health. It may be desirable to wear self-contained breathing apparatus.

**0** Materials that on exposure under fire conditions would offer no hazard beyond that of ordinary combustible material.

## Flammability

Susceptibility to burning is the basis for assigning degrees within this category. The method of attacking the fire is influenced by this susceptibility factor.

**4** Very flammable gases or very volatile flammable liquids. Shut off flow and keep cooling water streams on exposed tanks or containers.

**3** Materials that can be ignited under almost all normal temperature conditions. Water may be ineffective because of the low flash point.

**2** Materials that must be moderately heated before ignition will occur. Water spray may be used to extinguish the fire because the material can be cooled below its flash point.

**1** Materials that must be preheated before ignition can occur. Water may cause frothing if it gets below the surface of the liquid and turns to steam. However, water fog gently applied to the surface will cause a frothing that will extinguish the fire.

**0** Materials that will not burn.

## Reactivity (Stability)

The assignment of degrees in the reactivity category is based upon the susceptibility of materials to release energy either by themselves or in combination with water. Fire exposure is one of the factors considered along with conditions of shock and pressure.

**4** Materials that (in themselves) are readily capable of detonation or of explosive decomposition or explosive reaction at normal temperatures and pressures. Includes materials that are sensitive to mechanical or localized thermal shock. If a chemical with this hazard rating is in an advanced or massive fire, the area should be evacuated.

**3** Materials that (in themselves) are capable of detonation or of explosive decomposition or of explosive reaction but which require a strong initiating source or which must be heated under confinement before initiation. Includes materials that are sensitive to thermal or mechanical shock at elevated temperatures and pressures or that react explosively with water without requiring heat or confinement. Fire fighting should be done from a location protected from the effects of an explosion.

**2** Materials that (in themselves) are normally unstable and readily undergo violent chemical change but do not detonate. Includes materials that can undergo chemical change with rapid release of energy at normal temperatures and pressures or that can undergo violent chemical change at elevated temperatures and pressures. Also includes those materials that may react violently with water or that may form potentially explosive mixtures with water. In advanced or massive fires, fire fighting should be done from a safe distance or from a protected location.

**1** Materials that (in themselves) are normally stable but which may become unstable at elevated temperatures and pressures or which may react with water with some release of energy but not violently. Caution must be used in approaching the fire and applying water.

**0** Materials that (in themselves) are normally stable even under fire exposure conditions and that are not reactive with water. Normal fire fighting procedures may be used.

## APPENDIX C: FLAMMABILITY

*This Appendix is not a part of the requirements (recommendations) of this NFPA document, but is included for information purposes only.*

The selection of the flash point breaks for the assigning of degrees within the Flammability category has been based upon the recommendations of the Technical Committee on Classification and Properties of Flammable Liquids of the NFPA Committee on Flammable Liquids. This Technical Committee initiated the study that led to the development of this standard. Close cooperation between the Technical Committee and the Committee on Fire Hazards of Materials has continued.

Flash point tells several things. One, if the liquid has no flash point, it is not a flammable liquid. Two, if it has a flash point, it must be considered flammable or combustible. Three, the flash point is normally an indication of susceptibility to ignition.

The flash point test may give results that would indicate that the liquid is nonflammable or that it comes under degree 1 or 2 when it is a mixture containing, for example, carbon tetrachloride. As a specific example, sufficient carbon tetrachloride can be added to gasoline so that the mixture has no flash point. However, on standing in an open container, the carbon tetrachloride will evaporate more rapidly than the gasoline. Over a period of time, therefore, the residual liquid will first show a high flash point, then a progressively lower one until the flash point of the final 10 percent of the original sample will approximate that of the heavier fractions of the gasoline. In order to evaluate the fire hazard of such liquid mixtures, fractional evaporation tests can be conducted at room temperature in open vessels. After evaporation of appropriate fractions, such as 10, 20, 40, 60, and 90 percent of the original sample, flash point tests can be conducted on the residue. The results of such tests indicate the grouping into which the liquid should be placed if the conditions of use are such as to make it likely that appreciable evaporation will take place. For open system conditions, such as in open dip tanks, the open-cup test method may give a more reliable indication of the flammability hazard.

In the interest of reproducibility of results, it is recommended that:

The flash point of liquids having a viscosity less than 45 SUS (Saybolt Universal Seconds) at 100°F (37.8°C) and a flash point below 200°F (93.4°C) may be determined in accordance with ASTM D-56-79, *Standard Method of Test for Flash Point by the Tag Closed Tester.* (In those countries that use the Abel or Abel-Pensky closed cup tests as an official standard, these tests will be equally acceptable to the Tag Closed Cup Method.)

The flash point of aviation turbine fuels may be determined in accordance with ASTM D3828-81, *Test Method for Flash Point by Setaflash Closed Tester.*

For liquids having flash points in the range of 32°F (0°C) to 230°F (110°C) the determination may be made in accordance with ASTM D3278-82, *Flash Point of Liquids by Setaflash Closed Tester.*

For viscous and solid chemicals the determination may be made in accordance with ASTM E502-74, *Flash Point of Chemicals by Closed Cup Methods.*

The flash point of liquids having a viscosity of 45 SUS (Saybolt Universal Seconds) or more at 100°F (37.8°C) or a flash point of 200°F (93.4°C) or higher may be determined in accordance with ASTM D-93-79, *Standard Method of Test for Flash Point by the Pensky-Martens Closed Tester.*

## APPENDIX D: REACTIVITY, DIFFERENTIAL SCANNING CALORIMETRY (DSC)

*This Appendix is not a part of the requirements (recommendations) of this NFPA document, but is included for information purposes only.*

Differential Scanning Calorimetry (DSC) is the primary screening test for assessing reactivity hazard. It indicates whether a material undergoes an exothermic or endothermic reaction, and a general temperature range in which the reaction occurs.

This test is routinely run before other more sophisticated tests are run, such as an Accelerating Rate Calorimetry (ARC) test or drop weight testing. Heats of reaction, heats of decomposition, and heats of fusion, as well as kinetic information, can be determined by DSC for homogenous solids and liquids, as well as heterogenous systems.

DSC data should be used with caution, avoiding any inference that the test conditions duplicate those that the material will experience in a foreign environment.

A DSC test consists of heating a small quantity of material (typically 1–10 milligrams) held in a sample container from room temperature to approximately 500°C. Exotherms are usually detected by the DSC test at temperatures higher than that expected in systems that are more adiabatic (insulated tanks, large masses of material, etc.)

Small changes in the composition of a material can have a significant effect on its thermal behavior. For example, a material may not decompose in the container in which DSC is done, but it may be catalytically decomposed by the material of construction of the container used in service.

The DSC is a screening test that is used primarily to determine if further testing is required.

## APPENDIX E: REFERENCED PUBLICATIONS

**E-1**   The following documents or portions thereof are referenced within this standard for informational purposes only and thus are not considered part of the requirements of this document. The edition indicated for each reference is the current edition as of the date of the NFPA issuance of this document.

**E-1.1 NFPA Publications.**   National Fire Protection Association, 1 Batterymarch Park, P.O. Box 9101, Quincy MA 02269-9101.

NFPA 49-1975, *Hazardous Chemicals Data.*

NFPA 325M-1984, *Fire Hazard Properties of Flammable Liquids, Gases, and Volatile Solids.*

**E-1.2 ASTM Publications.**   American Society for Testing and Materials, 1916 Race Street, Philadelphia, PA 19103.

ASTM D-56-87, *Standard Method for Test for Flash Point by the Tag Closed Tester.*

ASTM D-3828-87, *Test Method for Flash Point by Setaflash Closed Tester.*
ASTM D-3278-82, *Flash Point of Liquids by Setaflash Closed Tester.*
ASTM D-93-85, *Test Methods for Flash Point by the Pensky-Martens Closed Tester.*
ASTM E-502-84, *Flash Point of Chemicals by Closed Cup Methods.*

### E-1.3 Other Publications.

Tou, J.C. and Whiting, L.F.; "A Cradle-Glass Ampoule Sample Container for Differential Scanning Calorimetric Analysis"; Thermochimica Acta; Vol. 42; Elsevier Scientific Publishing Co.; Amsterdam; 1980.

Whiting, L.F., LaBean, M.S., and Eadie, S.S.; "Evaluation of a Capillary Tube Sample Container for Differential Scanning Calorimetry"; Thermochimica Acta; Vol. 136; Elsevier Scientific Publishing Co.; Amsterdam, 1988.


## LIST OF ACRONYMS AND RECOGNIZED ABBREVIATIONS

**AAR/BOE:** Association of American Railroads/Bureau of Explosives
**AIChE:** American Institute of Chemical Engineers
**ASCS:** Agricultural Stabilization and Conservation Service
**ASME:** American Society of Mechanical Engineers
**ASSE:** American Society of Safety Engineers
**ATSDR:** Agency for Toxic Substances and Disease Registry (HHS)
**CAER:** Community Awareness and Emergency Response (CMA)
**CDC:** Centers for Disease Control (HHS)
**CEPP:** Chemical Emergency Preparedness Program
**CERCLA:** Comprehensive Environmental Response, Compensation, and Liability Act of 1980 (PL 96-510)
**CFR:** Code of Federal Regulations
**Chemnet:** A mutual aid network of chemical shippers and contractors
**CHEMTREC:** Chemical Transportation Emergency Center
**CHLOREP:** A mutual aid group comprised of shippers and carriers of chlorine
**CHRIS/HACS:** Chemical Hazards Response Information System/Hazard Assessment Computer System
**CMA:** Chemical Manufacturers Association
**CSEPP:** Chemical Stockpile Emergency Preparedness Program
**CWA:** Clean Water Act
**DOC:** U.S. Department of Commerce
**DOD:** U.S. Department of Defense
**DOE:** U.S. Department of Energy
**DOI:** U.S. Department of Interior
**DOJ:** U.S. Department of Justice
**DOL:** U.S. Department of Labor
**DOS:** U.S. Department of State
**DOT:** U.S. Department of Transportation

**EENET:** Emergency Education Network (FEMA)
**EMA:** Emergency Management Agency
**EMI:** Emergency Management Institute
**EOC:** Emergency Operating Center
**EOP:** Emergency Operations Plan
**EPA:** U.S. Environmental Protection Agency
**ERD:** Emergency Response Division (EPA)
**FEMA:** Federal Emergency Management Agency
**FWPCA:** Federal Water Pollution Control Act
**HazMat:** Hazardous materials
**HazOp:** Hazard and Operability Study
**HHS:** U.S. Department of Health and Human Services
**ICS:** incident command system
**IEMS:** integrated emergency management system
**LEPC:** Local Emergency Planning Committee
**MSDS:** material safety data sheets
**NACA:** National Agricultural Chemical Association
**NCP:** national contingency plan
**NCRIC:** National Chemical Response and Information Center (CMA)
**NETC:** National Emergency Training Center
**NFA:** National Fire Academy
**NFPA:** National Fire Protection Association
**NIOSH:** National Institute of Occupational Safety and Health
**NOAA:** National Oceanic and Atmospheric Administration
**NRC:** National Response Center
**NRT:** National Response Team
**OHMTADS:** Oil and Hazardous Materials Technical Assistance Data System
**OSC:** on-scene coordinator
**OSHA:** Occupational Safety and Health Administration (DOL)
**PSTM:** Pesticide Safety Team Network
**RCRA:** Resource Conservation and Recovery Act
**RQs:** reportable quantities
**RRT:** Regional Response Team
**RSPA:** Research and Special Programs Administration (DOT)
**SARA:** Superfund Amendments and Reauthorization Act of 1986 (PL 99-499)
**SCBA:** self-contained breathing apparatus
**SERC:** State Emergency Response Commission
**SPCC:** spill prevention control and countermeasures
**TSD:** treatment, storage, and disposal facilities
**USCG:** U.S. Coast Guard (DOT)
**USDA:** U.S. Department of Agriculture
**USGS:** U.S. Geological Survey
**USNRC:** U.S. Nuclear Regulatory Commission

## NUMBERS TO REMEMBER

| | |
|---|---|
| Agency For Toxic Substance & Disease Registry | (404) 639-0615 |
| American Petroleum Institute | (202) 682-8000 |
| American Trucking Association | (800) 282-5463 |
| Ashland Chemical Company | (614) 889-3333 |
| Association of American Railroads | (202) 639-2100 |
| Bureau of Explosives | (202) 639-2222 |
| CAS Registry Service | (800) 848-6538 |
| Centers for Disease Control (CDC) | (404) 633-5313 |
| CHEMTREC | (800) 424-9300 |
| Non-Emergency | (800) 262-8200 |
| Chlorine Institute | (202) 775-2790 |
| CMA Main Phone | (703) 741-5000 |
| CHEMTREC International Emergency | (202) 483-7616 |
| Compressed Gas Association | (703) 412-0900 |
| Department of Defense Nuclear Accident Center | (703) 325-2102 |
| DOW Chemical Emergency | (517) 636-4400 |
| Du Pont Chemical Emergency | (302) 774-7500 |
| FEMA 24-hour Emergency | (202) 646-2400 |
| Fertilizer Institute | (202) 675-8250 |
| Kerr-McGee Chemical Company | (405) 270-1313 |
| Malline Krodf Chemical Company | (314) 530-2000 |
| Monsanto Chemical Company | (314) 694-1000 |
| National Fire Protection Association | (617) 482-8755 |
| National Foam "Red Alert" 24-hour | (610) 363-1400 |
| National Pesticide Telecommunications Network | (800) 858-7378 |
| National Propane Gas Association | (703) 979-3563 |
| National Response Center | (800) 424-8802 |
| National Safety Council | (202) 293-2270 |
| National Transportation Safety Board (NTSB) | (202) 382-6600 |
| Nuclear Regulatory Commission | (301) 492-7000 |
| Poison Control Center | (800) 922-1117 |
| US DOT Office Of Hazardous Materials | (800) 467-4922 |
| US DOT Information | (202) 366-4000 |
| US DOT Hazardous Materials Registration | (202) 366-4109 |
| US Army Explosives Disposal | (301) 677-9770 |
| Toxic Substance Control Act Hotline | (202) 554-1404 |

## RESOURCE GUIDE

*Chemical Hazard Information Response System* . . . . . . . . . . . . . . . . . . . .CHRIS
*Condensed Chemical Dictionary* . . . . . . . . . . . . . . . . . . . . . . . . . . . . . . CCD
*Cross-handling Guide For Potentially Hazardous Materials* . . . . . . . . . CROSS
*Dangerous Properties Of Hazardous Materials*. . . . . . . . . . . . . . . . . . . . .SAX
*North American Emergency Response Guidebook*. . . . . . . . . . . . . . . . . .NAERG

*Emergency Handling of Hazardous Materials in Surface*
  *Transportation* . . . . . . . . . . . . . . . . . . . . . . . . . . . . . . . . . . . . . . . . .EHHM
*NFPA Fire Protection Guide On Hazardous Materials* . . . . . . . . . . . . . . . .FPG
*NIOSH/OSHA Pocket Guide To Chemical Hazards.* . . . . . . . . . . . . . . . . NIOSH
*TLV Guide* (American Conference of Governmental Industrial
  Hygienists). . . . . . . . . . . . . . . . . . . . . . . . . . . . . . . . . . . . . . . . . . . . . .TLV

## Sources of Specific Information

Chemical name to four-digit UN number. . . . . . . . . . . . . . . . . . . . . . . .NAERG
                                                       EHHM
Four-digit UN number to chemical name. . . . . . . . . . . . . . . . . . . . . . . .NAERG
                                                       EHHM
Chemical name to STCC . . . . . . . . . . . . . . . . . . . . . . . . . . . . . . . . . . . . .EHHM
STCC to chemical name . . . . . . . . . . . . . . . . . . . . . . . . . . . . . . . . . . . . . .EHHM
STCC to four-digit UN number . . . . . . . . . . . . . . . . . . . . . . . . . . . . . . . .EHHM
Four-digit UN number to STCC . . . . . . . . . . . . . . . . . . . . . . . . . . . . . . . .EHHM
Chemical name to synonym . . . . . . . . . . . . . . . . . . . . . . . . . . CHRIS (MAN 2)
                                                       NIOSH
                                                       CROSS
                                                 CCD (Limited)
                                                 Sax (Limited)
Synonym to chemical name . . . . . . . . . . . . . . . . . . . . . . . . . . . . CHRIS(MAN 2)
                                                        CCD
                                                       Sax

MNFC and address
  (From trade name). . . . . . . . . . . . . . . . . . . . .FPG (Flash Point Index) CCD
  (From chemical name) . . . . . . . . . . . . . . . . . . . . . . . . . . . . CHRIS(MAN 2)

## Product Uses

(From trade name) (flash point index) . . . . . . . . . . . . . . . . . . . . . . . . . . . .FPG
                                                        CCD
(From chemical name) . . . . . . . . . . . . . . . . . . . . . . . . . . . . . . . . . . . . . . . CCD
                                                      CROSS
Product trade name to chemical composition . . . . . . . . . (contact manufacturer)
NFPA 704 designation . . . . . . . . . . . . . . . . . . . . . . . . . . . . . . . . . . . . . . . .FPG
                                          (Sec. 325M, Sec. 49)
Chemical formula . . . . . . . . . . . . . . . . . . . . . . . . . . . . . . . . . . . . . . . . NIOSH
                                                       CCD
                                             CHRIS(MAN 2)
                                    FPG (Sec. 325M, Sec. 49)
                                                    Sax
Reactions . . . . . . . . . . . . . . . . . . . . . . . . . . . . . . . . . . . . . . . . FPG (Sec. 491M)
                                                   NIOSH
                                                  CHRIS

## SELECTED TECHNICAL REFERENCES

*Computer-Aided Management of Emergency Operations (CAMEO),* National Safety Council, 1019 19th Street N.W., Suite 401, Washington, D.C. 20036.

*Chemical Hazard Response Information System (CHRIS),* Superintendent of Documents, U.S. Government Printing Office, Washington, D.C. 20402-9328.

*Condensed Chemical Dictionary,* Van Nostrand and Reinhold Co. Inc., 115 Fifth Avenue, New York, 10003.

*Dangerous Properties of Industrial Chemicals (Sax),* Van Nostrand and Reinhold Co. Inc., 115 Fifth Avenue, New York, 10003.

*Dictionary of Chemical Names and Synonyms,* CRC/Lewis Publishers, 2000 Corporate Blvd. N.W., Boca Raton, FL 33431.

*North American Emergency Response Guide Book (NAERG),* U.S. Department of Transportation, RSPA, 400 Seventh Street, S.W., Washington, D.C. 20590-0001.

*Emergency Action Guides,* Bureau of Explosives, Association of American Railroads, 50 "F" Street, N.W., Washington, D.C., 20001.

*Emergency Handling of Hazardous Materials in Surface Transportation,* Bureau of Explosives, Association of American Railroads, 50 "F" Street, N.W., Washington, D.C. 20001.

*Farm Chemical Handbook,* Meister Publishing Co., 37841 Euclid Ave., Willoughby, OH, 44094.

*Fire Protection Guide on Hazardous Materials,* National Fire Protection Association (NFPA), One Batterymarch Park, Quincy, MA, 02110.

*Handbook of Chemistry and Physics,* CRC/Lewis Publishers, 2000 Corporate Blvd. N.W., Boca Raton, FL 33431.

*Merck Index,* Merck and Company Inc., Rahway, NJ, 07065

*NIOSH Pocket Guide to Chemical Hazards,* Superintendent of Documents, U.S. Government Printing Office, Washington, D.C. 20402-9328.

## IUPAC RULES OF NOMENCLATURE

### Hydrocarbons With More Than 10 Carbons

| | | | |
|---|---|---|---|
| $C_{11}H_{24}$ | Undecane | $C_{20}H_{42}$ | Eicosane |
| $C_{12}H_{26}$ | Dodecane | $C_{21}H_{42}$ | Heneicosane |
| $C_{13}H_{28}$ | Tridecane | $C_{22}H_{46}$ | Docosane |
| $C_{14}H_{30}$ | Tetradecane | $C_{23}H_{48}$ | Tricosane |
| $C_{15}H_{32}$ | Pentadecane | $C_{26}H_{54}$ | Hexacosane |
| $C_{16}H_{34}$ | Hexadecane | $C_{30}H_{62}$ | Triacontane |
| $C_{17}H_{36}$ | Heptadecane | $C_{31}H_{64}$ | Hentriacontane |
| $C_{18}H_{38}$ | Octadecane | $C_{32}H_{66}$ | Dotriacontane |
| $C_{19}H_{40}$ | Nonadecane | $C_{33}H_{68}$ | Tritriacontane |

| | |
|---|---|
| $C_{40}H_{82}$ | Tetracontane |
| $C_{49}H_{100}$ | Nonatetracontane |
| $C_{50}H_{102}$ | Pentacontane |
| $C_{60}H_{122}$ | Hexacontane |
| $C_{70}H_{142}$ | Heptacontane |
| $C_{80}H_{162}$ | Octacontane |
| $C_{90}H$ | Nonacontane |
| $C_{100}H_{202}$ | Hectane |
| $C_{132}H_{266}$ | Dotriacontahectane |

There are four types of structures:

*Molecular*    $C_2H_{10}$

*Structural*

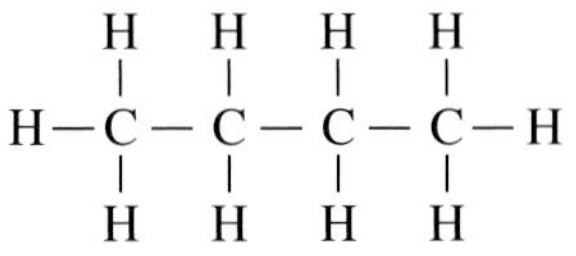

```
      H    H    H    H
      |    |    |    |
  H — C — C — C — C — H
      |    |    |    |
      H    H    H    H
```

*Condensed structural*      $CH_3CH_2CH_2CH_3$

*Skeleton*      C–C–C–C

The naming of all the alkanes is based upon the number of carbon atoms in the longest continuous chain of carbon atoms. If, for example, the longest chain contains four carbon atoms, the compound would be called butane. If it has five carbon atoms, it is pentane, and so on.

Names of branched-chain hydrocarbons and hydrocarbon derivatives using the IUPAC system are based on the name of the longest continuous carbon chain in the molecule with a number indicating the location of a branch or substituent.

In order to locate the position of a branch or substituent, the carbon chain is numbered consecutively from one end to the other, starting at that end which will give the lowest number(s) to the substituent(s). Examples:

```
   1      2      3      4      5
   H      H      H      H      H
   |      |      |      |      |
H—C  —  C  —  C  —  C  —  C—H    is 2-Methylpentane
   |      |      |      |      |
   H      |      H      H      H
          |
       H—C—H
          |
          H
```

```
   1      2      3      4
   H      Cl     H      H
   |      |      |      |
H—C  —  C  —  C  —  C—H    is 2,2-Dichlorobutane
   |      |      |      |
   H      Cl     H      H
```

The prefixes "di", "tri", "tetra", "penta", "hexa", etc. indicate how many of each substituent are in the molecule. A cyclic (ring) hydrocarbon is designated by the prefix **cyclo.**

Double bonds in hydrocarbons are indicated by changing the suffix "ane" to "ene" and triple bonds by changing to "yne". The position of the multiple bond within the structure is indicated by the number of the first or lowest-numbered carbon atom attached to the multiple bond. Examples:

```
      H    H    H    H    H
      |    |    |    |    |
   H—C  — C  — C = C — C—H        is 2-Pentene
      |    |              |
      H    H              H
```

```
    5   4   3   2   1
    H   H   H
    |   |   |
H—C — C — C — C ≡ C — H      is 1-Pentyne
    |   |   |
    H   H   H
```

```
    4   3   2   1
    H   H   H   H
    |   |   |   |
H—C = C — C = C — H          is 1,3,-Butadiene
    |   |   |   |
    H   H   H   H
```

Most of the hydrocarbon derivative functional groups in organic compounds are designated by either a suffix or a prefix as shown in Table B. Rules regarding whether a prefix or a suffix designation is used are as follows:

1. When one such group is present, the suffix will be used.
2. When more than one such group is present, only one will be designated by a suffix, the other or others by prefix or prefixes.
3. The order of precedence for deciding which group takes the suffix designation is the same as the order in the Table B.

Examples:

```
    H   H
    |   |
H—C — C — O — H          is Ethanol, not Hydroxyethane
    |   |
    H   H
```

```
        3   2   1
        H   H   O
        |   |   ||
H—O — C — C — C — O — H   is 3-Hydroxypropanoic acid
        |   |
        H   H
```

In numbering the carbon chain, the lowest numbers will be given preference to:

1. Groups in Table B named by suffixes, then
2. Double bonds, then
3. Triple bonds, then
4. Groups named by prefixes (groups named by prefixes are listed in alphabetical order).

## Nomenclature of Aromatic Compounds

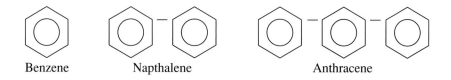

Benzene     Napthalene     Anthracene

## Monosubstituted Compounds

Names are derived using prefixes from Table B and C followed by the name benzene.

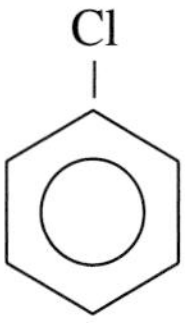

Chloro
Benzene

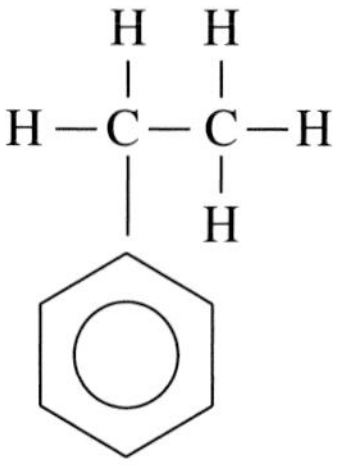

Ethyl
Benzene

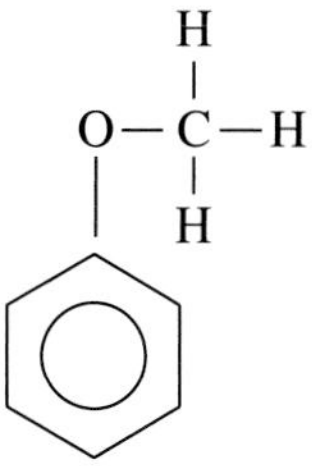

Methyloxy
Benzene

Name can be indicated by commonly accepted names:

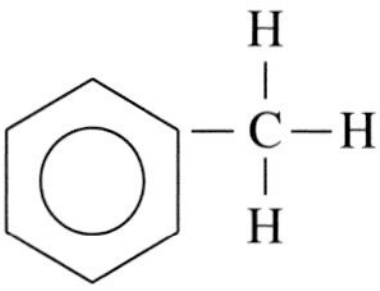

Toluene

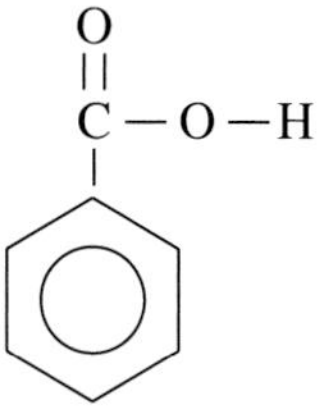

Benzoic
acid

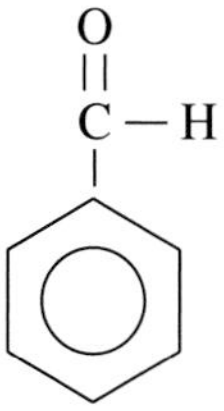

Benzaldehyde

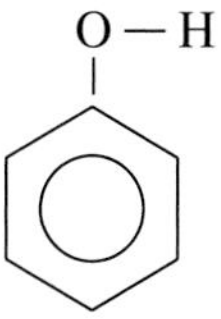

Phenol

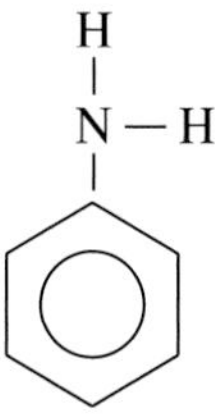

Analine

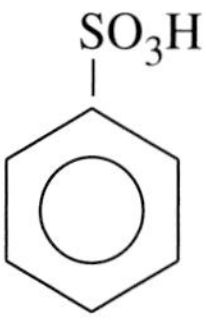

Benzene
sulfonic acid

## Disubstituted Compounds

Names derived using prefixes (including commonly accepted names) and numbers or words:

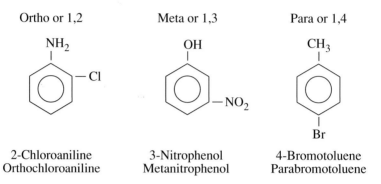

Ortho or 1,2

2-Chloroaniline
Orthochloroaniline

Meta or 1,3

3-Nitrophenol
Metanitrophenol

Para or 1,4

4-Bromotoluene
Parabromotoluene

## TABLES

| Class | Functional group | Prefix | Suffix |
|---|---|---|---|
| **A. Groups indicated by suffix only** | | | |
| Alkanes | C–C | | -ane |
| Alkenes | C=C | | -ene |
| Alkynes | C≡C | | -yne |
| **B. Groups indicated by prefix or suffix** | | | |
| Carboxylic acids | –COOH(App. 17)–CO$_2$H | Carboxy- | -oic acid |
| Aldehydes | –CHO(App. 18) | Carboxaldo- | -al |
| Ketones | –CO(App. 19) | Keto- | -one |
| Alcohols | –OH | Hydroxy- | -ol |
| Amines | –NH$_2$ | Amino- | -amine |
| **C. Groups indicated by prefix only** | | | |
| Halogenated compounds | –F | Fluoro- | |
| | –Cl | Chloro- | |
| | –Br | Bromo- | |
| | –I | Iodo- | |
| Nitrated compounds | –NO$_2$ | Nitro- | |
| Alkylated compounds | –R | Alkyl- | |
| Ethers | –OR | Alkoxy- | |

## ANSWERS TO REVIEW QUESTIONS

**Chapter 1**
**1**-A
**2**-C
**3**-D, C, E, A, B
**4**-Ionic
**5**-Covalent
**6**-Protons, neutrons, electrons
**7**-2 or 8
**8**-Duet and octet
**9**-NaCl　Sodium chloride　Binary salt　Varying
　$Ca_3(PO_4)_2$　Calcium phosphate　Oxysalt　Oxidizer
　$Al_2(O_2)_3$　Aluminum peroxide　Peroxide　RH, CL, RO
　$CuBr_2$　Copper II bromide　Binary salt　Varying
　KOH　Potassium hydroxide　Hydroxide　RH, CL
　$Li_2O$　Lithium oxide　Metal oxide　RH, CL
　$Mg(ClO)_2$　Magnesium hypochlorite　Oxysalt　Oxidizer
　$HgO_2$　Mercury II peroxide　Peroxide　RH, CL, RO
　NaF　Sodium fluoride　Binary salt　Varying
　$FeCO_3$　Iron II carbonate　Oxysalt　Oxidizer
**10**-$Ca(ClO)_2$　Binary salt　Varying
　$AlCl_3$　Binary salt　Varying
　LiOH　Hydroxide　RH, CL
　$CuO_2$　Peroxide　RH, CL, RO
　$Na_2O$　Metal oxide　RH, CL
　KI　Binary salt　Varying
　$Mg_3P_2$　Binary salt　Varying
　$HgClO_4$　Oxysalt　Oxidizer
　$Fe(FO_3)_3$　Oxysalt　Oxidizer

**Chapter 2**
**1**-Positive and negative
**2**-Mechanical, mechanical chemical, chemical, dust, nuclear
**3**-Detonation and deflagration
**4**-(App. 20)
**6**-B
**7**-Faster, slower

**Chapter 3**
**1**-E
**2**-A
**3**-Upper and lower explosive limits
**4**-Heavier

**5**-Alkane, alkyne, alkene, alkane, alkyne
**6**-Isobutane i-$C_4H_{10}$, ethane $C_2H_6$, methane $CH_4$, ethene $C_2H_4$
**7**-yne, ane, one, ene

## Chapter 4
**1**-Vapor pressure of the liquid equals atmospheric pressure
**2**-Liquid
**3**-High vapor pressure and high vapor content
**4**-Low
**5**-Weight, polarity, branching
**6**-B, C, B, A, D, A
**7**-Monomers
**8**-Alkane, aromatic, alkene, aromatic, alkene
**9**-(App. 21)
**10**-Acetaldehyde $CH_3CHO$ WFR/toxic, butadiene $C_4H_6$ flammable, isopropyl
  alcohol $C_3H_7OH$ WFR/toxic, methyl amine $CH_3NH_2$ toxic/flammable

## Chapter 5
**1**-Flammable solid, spontaneously combustible, dangerous when wet
**2**-Solid, gas, liquid
**3**-Spontaneously combust, air
**4**-E
**5**-A, C, D
**6**-D
**7**-Alkali metals, alkaline earth metals
**8**-Acetylene

## Chapter 6
**1**-Fluorine, chlorine, bromine, oxygen
**2**-Halogens, peroxide salts, oxysalts
**3**-D
**4**-Aluminum persulfate $Al_2(SO_5)_3$
  Lithium chlorite
  Sodium fluorate
  Magnesium hypophosphite $Mg_3(PO_2)_2$
  Copper I chlorate
**5**-(App. 22)&(App. 23)

## Chapter 7
**1**-Acute
**2**-Chronic
**3**-Inhalation, ingestion, absorption, injection
**4**-Cancer, teratogenic, mutagenic
**5**-Living organisms
**6**-B, D, E
**7**-Internal or external

**8**-(App. 24)&(App. 25)
**9**-Caution, danger, warning
**10**-STEL, TLV-TWA, PEL, IDLH
**11**-Amount, biological

## Chapter 8
**1**-Electrons, nucleus
**2**-More neutrons
**3**-Alpha, beta, gamma
**4**-Alpha, beta
**5**-Gamma radiation
**6**-B
**7**-Ionizing, non-ionizing
**8**-B, C, A
**9**-Time, distance, shielding
**10**-C

## Chapter 9
**1**-Acids, bases
**2**-Inorganic, organic
**3**-$H^+$ ionization
**4**-Water, percentage
**5**-Neutral
**6**-Acidic
**7**-Basic
**8**-Neutralization, dilution
**9**-(App. 26)

## Chapter 10
**1**-HOT
**2**-Consumer commodities
**3**-Other hazard classes

# REFERENCES

1. Killen, William D., *Fire Command,* Eight Die in Chlorine Tanker Derailment, Quincy, MA: National Fire Protection Association.
2. *NFPA Quarterly,* October 1964, Quincy, MA: National Fire Protection Association.
3. Sullivan, Bill, *Firehouse Magazine,* Burns Illustrate Need For Bunker Gear, July 1994.
4. *Introduction to Chemistry,* Chisholm and Johnson, Usborn Publishing Ltd., London, England.
5. *Milwaukee Fire Department Basic Training Manual,* 1986.
6. *Toxics A–Z,* John Harte, Cheryl Holdren, Richard Schneider, Christine Shirley, University of California Press, 1991.
7. National Transportation Safety Board, Anhydrous hydrogen fluoride release from NATX 9408, Elkhart, IN, February 4, 1985.
8. Dektar, Cliff, *Fire Engineering,* Peroxide Blast Shatters Chemical Plant, January 1979.
9. Klem, Thomas J., *NFPA Journal,* High-Rise Fire Claims Three Philadelphia Fire Fighters, September/October 1991, Quincy, MA: National Fire Protection Association.
10. *JEMS,* August 1982, Proper Use of Highway Flares.
11. Hill, Irma, *Crescent City Remembers,* Scheiwe's Print Shop, Crescent City, IL, 1995.
12. Ryczkowski, John J., *American Fire Journal,* Kingman Revisited, July 1993.
13. *North American Emergency Response Guidebook 1996,* U.S. Department of Transportation, U.S. Government Printing Office (1996).
14. *Code of Federal Regulations, (CFR) 49, Parts 100-177,* American Trucking Association, October 1, 1994.
15. *Hawley's Condensed Chemical Dictionary,* Van Nostrand Reinhold Company, New York, 12th edition, 1993.
16. *National Fire Academy Chemistry of Hazardous Materials Instructor Guide,* U.S. Government Printing Office (1994).
17. *National Fire Academy Chemistry of Hazardous Materials,* Student Manual, U.S. Government Printing Office (1994).
18. *Chemical & Engineering News,* Top 50 Industrial Chemicals for 1995, April 8, 1996.

19. *National Fire Academy Initial Response to Hazardous Materials Incidents: Basic Concepts,* Student Manual, U.S. Government Printing Office (1992).
20. *National Fire Academy Initial Response to Hazardous Materials Incidents: Concept Implementation,* U.S. Government Printing Office (1992).
21. *NFPA Fire Protection Handbook,* Quincy MA: National Fire Protection Association, 17th edition, 1992.
22. *NIOSH Pocket Guide to Chemical Hazards,* U.S. Government Printing Office (1990).
23. *Fire Protection Guide to Hazardous Materials,* Quincy, MA: National Fire Protection Association, 11th edition, 1994.
24. *Toxicology,* Michael A. Kamrin, Lewis Publishers, 1988.
25. *Fundamentals of Chemistry,* James E. Brady and John R. Holm, John Wiley and Sons, 3rd edition, 1988.
26. *Hazardous Materials Emergencies Response and Control,* John R. Cashman, Technomic Publishing Co., Inc., 1st edition, 1983.

# INDEX

## A

Absorption of toxic chemicals, 210
acet-, 26, 136, 147
Acetaldehyde, 137, 148, 211
Acetic acid, 30, 110, 137, 229, 268, 269, 270, 274, 280–281
Acetone, 30, 110, 145–146
   skin absorption of, 211
   top 50 industrial chemicals, 159–160
Acetylene, 76–77, 78, 167
   chemical incompatibilities, 28
   flammable range, 69, 126
   liquefied gases, 66
Acids
   chemical incompatibilities, 28
   diprotic, 268
   heat generation, 195, 196
   inorganic, 193–196, 268–273
   mon\oprotic, 268
   organic, 27, 268, 273–276
   polyprotic, 268
   triprotic, 268
Acrolein, 211, 228
Acrylaldehyde, 228
Acrylic acid, 268, 275
Acrylonitrile, 30, 158–159
Acryl radical, 136, 137
Activated carbon, 176
Acute exposure, 209
Adhesives, 34
Adipic acid, 30, 32, 281–282
Aerial flares, 60
Air
   chemical incompatibilities, 28
   spontaneous ignition on exposure to, 167
Air burst, nuclear explosion, 46
-al, 147
Alcohols, 27
   ester formation, 148–149
   flammable liquids, 141–145
   flammable range, 126

   polarity, 115, 116
   toxicity, 227–228
Aldehydes, 27, 78, 137
   flammable liquids, 147–148
   flammable range, 126
   structure and polarity, 115, 116
   toxicity, 228
Alkali metals, 8
   chemical incompatibilities, 28
   as flammable solids, 168
Alkaline earth metals, 8
   chemical incompatibilities, 28
   as flammable solids, 168
Alkanes
   flammable gases, 71–74
   flammable liquids, 130–131
Alkenes
   flammable gases, 74–76
   flammable liquids, 132–133
Alkyl halides, 27, 93
   chemical release statistics, 245
   polarity, 116
   toxicity, 224–225
Alkynes, flammable gases, 76–78
Alloys, 13
Allyl alcohol, 208
Alpha particle emission, 251
Aluminum, 168
   molten, 289
Aluminum alkyls, 178
Aluminum nitrate, 186, 191
Aluminum phosphide, 178
Aluminum sulfate, 30
Amines, 27
   flammable gases, 78
   flammable liquids, 138–139
   flammable range, 126
   polarity, 115, 116
   toxicity, 226–227
Ammonia, 269, see also Anhydrous ammonia
   chemical incompatibilities, 28
   combustion products, 98

## Y

## Z